Participatory Forestry

Participatory Forestry: Involvement, Information and Science

Special Issue Editor
Alessandro Paletto

MDPI • Basel • Beijing • Wuhan • Barcelona • Belgrade

MDPI

Special Issue Editor
Alessandro Paletto
Research Centre for Forestry and Wood
(CREA), Italy

Editorial Office
MDPI
St. Alban-Anlage 66
4052 Basel, Switzerland

This is a reprint of articles from the Special Issue published online in the open access journal *Forests* (ISSN 1999-4907) from 2017 to 2018 (available at: https://www.mdpi.com/journal/forests/special_issues/participatory_forestry)

For citation purposes, cite each article independently as indicated on the article page online and as indicated below:

LastName, A.A.; LastName, B.B.; LastName, C.C. Article Title. *Journal Name* **Year**, *Article Number*, Page Range.

ISBN 978-3-03921-331-3 (Pbk)
ISBN 978-3-03921-332-0 (PDF)

Cover image courtesy of Consorcio Associações com Moçambique (CAM).

Contents

About the Special Issue Editor vii

Preface to "Participatory Forestry: Involvement, Information and Science" ix

Seongjun Kim, Guanlin Li and Yowhan Son
The Contribution of Traditional Ecological Knowledge and Practices to Forest Management:
The Case of Northeast Asia
Reprinted from: *Forests* **2017**, *8*, 496, doi:10.3390/f8120496 1

Claudia García-Ventura, Álvaro Sánchez-Medina, M. Ángeles Grande-Ortíz, Concepción
González-García and Esperanza Ayuga-Téllez
Comparison of the Economic Value of Urban Trees through Surveys with Photographs in
Two Seasons
Reprinted from: *Forests* **2018**, *9*, 132, doi:10.3390/f9030132 15

Guillaume Peterson St-Laurent, George Hoberg and Stephen R. J. Sheppard
A Participatory Approach to Evaluating Strategies for Forest Carbon Mitigation in
British Columbia
Reprinted from: *Forests* **2018**, *9*, 225, doi:10.3390/f9040225 28

Xabier Bruña-García and Manuel F. Marey-Pérez
The Challenge of Diffusion in Forest Plans: A Methodological Proposal and Case Study
Reprinted from: *Forests* **2018**, *9*, 240, doi:10.3390/f9050240 50

Eugenio Martínez-Falero, Concepción González-García, Antonio García-Abril and
Esperanza Ayuga-Téllez
Validation of a Methodology for Confidence-Based Participatory Forest Management
Reprinted from: *Forests* **2018**, *9*, 399, doi:10.3390/f9070399 65

Simone Blanc, Federico Lingua, Livio Bioglio, Ruggero G. Pensa, Filippo Brun and
Angela Mosso
Implementing Participatory Processes in Forestry Training Using Social Network
Analysis Techniques
Reprinted from: *Forests* **2018**, *9*, 463, doi:10.3390/f9080463 93

Isabella De Meo, Maria Giulia Cantiani, Fabrizio Ferretti and Alessandro Paletto
Qualitative Assessment of Forest Ecosystem Services: The Stakeholders' Point of View in
Support of Landscape Planning
Reprinted from: *Forests* **2018**, *9*, 465, doi:10.3390/f9080465 108

Laura Secco, Alessandro Paletto, Raoul Romano, Mauro Masiero, Davide Pettenella,
Francesco Carbone and Isabella De Meo
Orchestrating Forest Policy in Italy: Mission Impossible?
Reprinted from: *Forests* **2018**, *9*, 468, doi:10.3390/f9080468 124

Tomislav Laktić and Špela Pezdevšek Malovrh
Stakeholder Participation in Natura 2000 Management Program: Case Study of Slovenia
Reprinted from: *Forests* **2018**, *9*, 599, doi:10.3390/f9100599 143

Nathalie Bodonirina, Lena M. Reibelt, Natasha Stoudmann, Juliette Chamagne,
Trevor G. Jones, Annick Ravaka, Hoby V. F. Ranjaharivelo, Tantelinirina Ravonimanantsoa,
Gabrielle Moser, Arnaud De Grave, Claude Garcia, Bruno S. Ramamonjisoa,
Lucienne Wilmé and Patrick O. Waeber
Approaching Local Perceptions of Forest Governance and Livelihood Challenges with
Companion Modeling from a Case Study around Zahamena National Park, Madagascar
Reprinted from: *Forests* **2018**, *9*, 624, doi:10.3390/f9100624 . . . . **164**

Hanna Fors, Märit Jansson and Anders Busse Nielsen
The Impact of Resident Participation on Urban Woodland Quality—A Case Study
of Sletten, Denmark
Reprinted from: *Forests* **2018**, *9*, 670, doi:10.3390/f9110670 . . . . **196**

Lord K. Ameyaw, Gregory J. Ettl, Kristy Leissle and Gilbert J. Anim-Kwapong
Cocoa and Climate Change: Insights from Smallholder Cocoa Producers in Ghana Regarding
Challenges in Implementing Climate Change Mitigation Strategies
Reprinted from: *Forests* **2018**, *9*, 742, doi:10.3390/f9120742 . . . . **216**

About the Special Issue Editor

Alessandro Paletto (Ph.D. in Mountain and Forest Economics) is senior researcher at the Council for Agriculture Research and Economics (CREA), Research Centre for Forestry and Wood in Trento (Italy). His main research topics include stakeholder analysis, stakeholder involvement in decision-making processes, public participation in forest planning and management, and biophysical assessment and economic evaluation of ecosystem services. He is a member of the management committee of COST Action CA15206 "Payments of Ecosystem Services (Forests for Water)" and a substitute member of the management committee of COST Action "Capacity Building in Forest Policy and Governance in Western Balkan Region" (CAPABAL). He is a member of the editorial board of the following scientific journals: *Forests, Annals of Forest Research, Heliyon, Forest@, and Dendronatura*. He is the author of 78 Scopus-indexed publications (774 total citations, h-index = 17) and of more than 100 book chapters, conference papers, and other non-indexed publications.

Preface to "Participatory Forestry: Involvement, Information and Science"

The concept of "participatory forestry" has developed in order to bypass the limits of the top-down approach in forest resource management and to facilitate decision implementation by resolving conflicts during the decision-making process. From a theoretical point of view, the participatory process in forestry must be inclusive, voluntary, fair, and transparent to all participants; based on participants acting in good faith; and complementary to legal requirements. From a practical point of view, in many cases, the participatory process is based on formal but not substantial involvement of stakeholders and/or citizens (i.e., passive participation, participation in information giving). Therefore, the scientific community has a duty to improve techniques and methods for efficient public involvement in decision-making processes and to disseminate research results to targeted audiences and the wider public.

In the post-positivist approach, public participation is considered to be an instrument to ensure the democratization of the policy-making process and to improve the quality of information provided to make appropriate policy decisions. Conversely, in the traditional policy-making process, decisions are made through a top-down approach in which public involvement is absent or limited to a consultation of organized groups. Citizens involvement in policy formulation and the implementation process is considered to be the basis of participatory democracy theory. The main advantages of public participation in the policy-making process are related to the fact that (i) stakeholder involvement encourages the collection of inputs for the development of shared policies, programs, and actions; (ii) stakeholders' skills, knowledge, and awareness increase, gaining citizenship capacity (decision effectiveness); (iii) participation constitutes a prerequisite for public acceptance of decisions (social acceptance); and (iv) participation reduces the potential conflicts between stakeholders with different interests.

In the last few decades, public participation in forestry has become increasingly important in order to include traditional and local knowledge, social needs, and opinions in the decision-making process, but, in many contexts, the involvement of the local community and stakeholders is still not a very common practice. The key points of the debate are as follows: the level of inclusiveness and the modality of involvement. Regarding the level of inclusiveness, it is essential to distinguish between public involvement and stakeholder participation. In the first, the social actors involved in the process are the public in general, while, in the second, the social actors are the stakeholders, who can be defined as any group of organized people who share a common interest or stake in an issue (i.e., associations, organizations, and institutions). This distinction is based on the substantial difference between the interest group participation approach and the direct citizen participation approach. In the interest group participation approach, the equal participation of different interest groups, in order to incorporate all relevant interests, is the key concept. Conversely, in the direct citizen participation approach, the inclusion and representation of citizens in the decision-making process is a prerequisite, while organized groups are considered to be a potential obstacle to democracy.

Regarding the modality of citizen and/or stakeholder involvement, several participatory techniques and methods—e.g., focus groups, public meetings, scenario workshops, world café, fishbowl, open space—which are more or less suitable based on the project objectives and the level of inclusiveness desired, are described in the international literature . The main aim of this

Special Issue titled *Participatory Forestry: Involvement, Information and Science* is to address the topic of public participation and stakeholder involvement in forestry from different scientific perspectives and points of view. The twelve published papers analyze the participatory process in several scientific fields—e.g., forest policy and governance, forest planning and management, and nature conservation—in order to increase scientific knowledge and stimulate the debate on the bottom-up approach in forest resource management.

Alessandro Paletto
Special Issue Editor

forests

MDPI

Review

The Contribution of Traditional Ecological Knowledge and Practices to Forest Management: The Case of Northeast Asia

Seongjun Kim, Guanlin Li and Yowhan Son *

Department of Environmental Science and Ecological Engineering, Korea University, Seoul 02841, Korea; dao1129@hanmail.net (S.K.); jlmeys@naver.com (G.L.)
* Correspondence: yson@korea.ac.kr; Tel.: +82-10-3290-3015

Received: 9 November 2017; Accepted: 8 December 2017; Published: 12 December 2017

Abstract: This study aims to introduce the potential applicability of traditional ecological knowledge and community forestry in Northeast Asia, including China, Japan, and South Korea. In ancient Northeast Asia, forest policies and practices were based on Fengshui (an old Chinese concept regarding the flow of vital forces), with which forests were managed under community forestry. However, these traditional systems diminished in the twentieth century owing to the decline of traditional livelihood systems and extreme deforestation. Recently, legacies from traditional ecological knowledge and community forestry have been revisited and incorporated into forest policies, laws, and management practices because of growing needs for sustainable forest use in China, Japan, and Korea. This reevaluation of traditional ecological knowledge and community forestry has provided empirical data to help improve forestry systems. Although traditional ecological knowledge and community forestry in Northeast Asia have been scarcely theorized, they play a significant role in modifying forest management practices in the face of socioeconomic changes.

Keywords: community forestry; forest history; forest management practice; traditional knowledge

---

1. Introduction

The presence of a relationship between traditional ecological knowledge and forest management practices is undeniable [1]. Traditional ecological knowledge implies legacies inherited across thousands of years of interaction between humans and their surroundings [2], which were used to develop forest management practices in ancient societies [3,4]. Although much of the traditional knowledge and many practices have faded owing to the broad use of modern forestry systems, legacies remain in current societies, with which the applicability of traditional ecological knowledge can be piloted [5]. This contributes to enhancing the understanding of socioeconomic systems and forest ecosystems [6,7], and encourages research that focuses on traditional ecological knowledge in terms of forest management [8,9]. Therefore, many countries have acknowledged the importance of traditional ecological knowledge for forest management [10,11].

China, Japan, and South Korea in Northeast Asia are characterized by a long history of forest management. Ancient Northeast Asian societies relied on forests for timber, fuel, and resources for agriculture, and considered forests to be vital and holy places [12]. This importance of forests forced ancient societies to develop diverse forest policies and practices to foster forest resources. For example, documents have been found reporting afforestation activities, forest resource monitoring, pest control activities, and pruning works in Korea in the third, eighth, fourteenth, and eighteenth centuries, respectively (Table 1). Other records indicating long histories of forest management have also been found in China and Japan [3,4].

Table 1. Major forest management policies and practices of the ancient Korean kingdoms [13–16].

Century	Policies and Practices
Third	*Pinus densiflora* Sieb. and Zucc. afforestation near royal tombs and palaces
Eighth	Afforestation and replanting of *Pinus koraiensis* Sieb. and Zucc., *Juglans mandshurica* Maxim., and *Morus alba* L.
Tenth	*Castanea* spp. and *Pyrus* spp. orchard establishment
Fifteenth	Law defining punishiment aginst ilegal logging and forest fire setting in *Pinus densiflora* forests
Fifteenth	Law establishing locally specialized offices for selecting plantation species according to economic values and local environments
Seventeenth	Promotion of community forestry groups for patrolling local *Pinus densiflora* forests
Eighteenth	Encouragement of pruning activities of *Pinus densiflora* seedlings
Nineteenth	Promotion of land clearing and weeding activities in young *Pinus densiflora* plantations

In the twentieth century, Northeast Asia faced rapid socioeconomic changes accompanied by forest transitions. The economies of China and Japan grew rapidly, and the demand for timber and charcoal grew as well. Accordingly, forests in China were frequently overexploited beyond the allowable quotas as a result of economic growth-oriented development [17,18]. Although forests in Japan were relatively well preserved, many naturally regenerated forests were replaced by monoculture plantations for timber or other land use types [9,19]. South Korea experienced severe deforestation during the same period, owing to Japanese colonization and the Korean War [20]. However, the national forest restoration programs in South Korea successfully restored forest ecosystems with liberalization of timber imports and substitution of energy sources [21].

Because China, Japan, and South Korea have successfully fostered the nations' forest resources in contrast to the other Northeast Asian countries [22], the case of the three countries could be helpful to improve the sustainability of the forestry sector [19]. Nonetheless, current forestry systems rarely integrate knowledge from these countries because of the poor understanding of the legacies from traditional ecological knowledge and practices in ancient Northeast Asian societies. Limitations in the historical, quantitative data in this region further hinder assessment of the applicability of traditional ecological knowledge and practices for the forestry sector and the actual use of the legacies accordingly [7].

The current study aims to provide an overview of the potential applicability of traditional ecological knowledge and practices, especially community forestry, related to forest management in Northeast Asia. Specifically, we explore traditional knowledge and practices in ancient Northeast Asia, and the legacies from the traditions that remain in modern China, Japan, and South Korea. Due to the limited availability of historical data, quantitative approaches were not used for this objective.

2. Traditional Ecological Knowledge and Practices in Ancient Northeast Asia

2.1. The Fengshui Concept within Forestry

Ancient societies in Northeast Asia dealt with local socioeconomic conditions and natural environments managed on the basis of vast experience and old beliefs. Fengshui (an old Chinese concept concerned with the flow of vital forces and substances across a landscape) was one of them. Whether Fengshui entails an understanding of natural phenomena remains a topic of debate because its rationale and reliability have been scientifically untested [23,24]. Nonetheless, it is important to note that this concept, as part of traditional ecological knowledge, affected forest management practices in ancient Northeast Asia [3,25,26]. The issues regarding Fengshui itself were not addressed, in order to focus on its relationship to ancient Northeast Asian forestry.

According to original Chinese traditions, the main focus of Fengshui was land planning and management through harmonizing human settlement with all of its natural surroundings [12]. As one way to achieve harmonization, the Fengshui concept specified the optimal location for afforestation. The basic afforestation strategy was to locate human settlements on sites surrounded by mountains and hills, and to fill up the bare "entrances" and "backyards" of the settlement through afforesting them [23]. Such forests were believed to improve the landscape by adjusting the flow of vital forces and substances [4,23]. Afforested lands were regarded as holy, religious, and even scary places according to Fengshui. Consequently, Fengshui acted as a cultural background to foster and protect local forests for the villagers, and helped in the acquisition of forest products in ancient Chinese society [27]. This attitude to forests based on Fengshui migrated to other countries, such as Japan and Korea, and influenced forest management practices in those countries as well [7].

Fengshui-based forestry in ancient Northeast Asia was adjusted to local experience and knowledge on topography, climate, and tree species. Ancient villagers living on riversides changed the basic afforestation strategy to prevent flooding and soil erosion by planting water-tolerant tree species between the village and river [26]. In coastal areas, they altered the strategy to establish windbreak forests, consisting of salt-tolerant species such as *Pinus thunbergii* Parl., alongside the sea coast [28]. It was also applied to more practical forest management purposes than the original land planning and management concepts. One significant example of the practical Fengshui application was the community forest (afforested lands near villages shared for various purposes by the villagers). Community forests, such as Fengshui forests in China, Satoyama (village mountains) in Japan, and Maeulsoop (village groves) in Korea, were widespread in ancient Northeast Asia because of the common cultural afforestation practiced on the basis of Fengshui [7]. These forests allowed water, organic matter, and nutrients to circulate between the forests, croplands, and village, as they connected these natural and anthropogenic components according to Fengshui [4,29] (Figure 1). Therefore, community forests in Northeast Asia were important for the livelihoods of ancient society, such as the supply of timber, fuel, fodder, and compost for agriculture and residence [4,30].

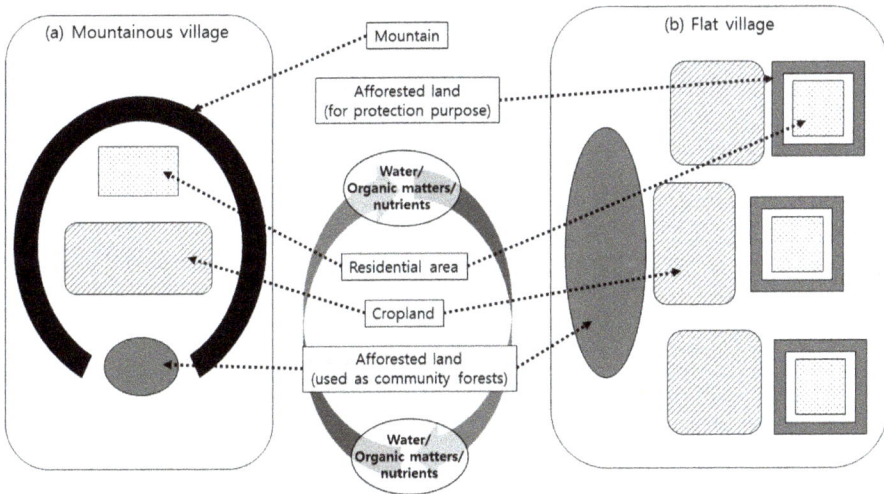

Figure 1. Types of the traditional community forests based on Fengshui in Northeast Asia (modification of the conceptual figures in [4,30]).

Community forests based on Fengshui were also implemented to protect both natural environments and human settlement (Figure 2). Some community forests on the mountain behind villages fulfilled this role by reducing soil erosion and landslide from steep slopes [3,19]. Like Satoyama—implemented for protection purposes in Japan—and "Bibo" (help and improvement) forests in Korea, the forests for such protective purposes enclosed houses, villages, or coastlines [25,31,32]. These forests were important for livelihoods in ancient Northeast Asia because they controlled village microclimates by preventing flooding [3,26], acting as windbreaks [25,28,31], and connecting forest patches [32,33].

Figure 2. A community forest in Gwangyang, South Korea. The forest is located near croplands and residential area to prevent strong wind and flooding.

2.2. The Historical Forest Policies and Practices

Many old records show that ancient kingdoms in Northeast Asia may have established diverse forestry policies and laws based on their experience and knowledge. For example, ancient policies and laws specified the way to manage forests in accordance with the characteristics of tree species and the site conditions. In ancient China, fast-growing water-tolerant species such as *Salix* spp. were afforested on hill slopes near rivers to prevent flooding and landslide in the tenth century [3]. Although naturally regenerated oak forests were widespread in Japanese Satoyama, *Pinus densiflora* Sieb. and Zucc., trees were instead selected in dry and infertile conditions [9,10]. In the seventeenth century, *Cryptomeria japonica* D. Don, *Chamaecyparis obtusa* Sieb. and Zucc., and *Pinus densiflora* trees were recommended in Japan for plantations in moist valleys, on mountainsides, and on dry mountain ridges, respectively, according to these species' tolerance to dry conditions [14,34]. In addition, an ancient Korean kingdom in the fifteenth century had a law establishing locally specialized offices, which selected the plantation species in accordance with the local environmental conditions and the value of each species [15]. This ancient Korean kingdom planted *Pinus densiflora* trees in dry and infertile areas in the fifteenth century as this species is more tolerant to water and nutrient shortages than competitive broadleaf trees [35]. This kingdom also applied intensive forest management practices, such as land clearing, weeding, and pruning practices, to young *Pinus densiflora* plantations in the eighteenth and nineteenth centuries according to the growth characteristics of *Pinus densiflora* [28].

Ancient kingdoms also concentrated on forest protection, by considering the productive and protective functions of forests [13,24]. One ancient Chinese kingdom had a law defining allowable periods for fire setting for slash and burn farming to reduce hazardous forest fire in the eleventh century [3]. Between the fifteenth and twentieth centuries, an ancient Korean kingdom politically protected *Pinus densiflora*, and did not permit private logging in selected lands, especially those containing mature *Pinus densiflora* forests such as "Geumsan" or "Bongsan" (forbidden mountains) [14]. Because government did not have enough labor to protect all forbidden forests from illegal logging and poaching, the ancient kingdom needed to cooperate with local groups. This situation forced the kingdom to strengthen the authority of local groups for self-organization and punishment [16], and resulted in the reinforcement of community forestry systems by villagers (see below).

2.3. Community Forestry

Forestry in ancient Northeast Asia featured community forestry systems for managing and utilizing the shared forests afforested according to the concept of Fengshui. Because community forests were important in terms of religion and livelihood, local groups attempted to sustain them by sheltering them from natural and anthropogenic disturbances [36]. Local groups established self-regulatory rules for community forests defining the permitted ways of wood harvesting, duties of forest patrols, and even stricter punishment for illegal forest use than government laws [4,13]. As local villagers mostly provided the labor for implementing government policies and laws (see Table 1 and Section 2.2), traditional community forestry was important for silviculture as well [35].

In China, there were local groups, such as Cuiguimingyue (village regulation and customary law) in Guizhou Province, that were oriented specifically to forestry [7]. Local villagers organized Cuiguimingyue by themselves, and developed rules to sustain local forest resources and conserve forests for symbolic and religious purposes [27]. This local group attempted to protect community forests through prohibiting logging of old trees, fuelwood collection from other villager's private forests, and setting forest fires [37]. Cuiguimingyue immediately punished villagers who violated the regulations protecting the forests, as government officers had no capacity to monitor these issues arising in the individual villages [27].

Rural villagers in Japan also shared resources from local forests. Villagers cooperated to collect wood for fuel, charcoal, and construction, and to protect local forest resources from overexploitation [19,30]. This form of local groups also dealt with post harvesting forest regeneration, mainly by coppicing of hardwood species [10]. The local groups in Japan focused on the collaboration of forestry and agriculture. Villagers collected fallen leaves and twigs as compost to fertilize cropland soils, and strengthened the community agreement by regulating the allowable period and amount of compost harvesting [30,38].

In Korea, Songgye (private *Pinus densiflora* cooperative) operated as a local pine forestry cooperative from at least the seventeenth century to the early twentieth century because of the ancient Korean kingdom's *Pinus densiflora* conservation policy (Table 1) [13]. Although Songgye originated from wood protection policies of the ancient Korean kingdom for government uses, it also acted as a local group for community forestry [13,39]. The primary objective of this group was to patrol local *Pinus densiflora* forests and issue fines for illegal logging of *Pinus densiflora*, thus protecting the forests from natural and anthropogenic disturbances [36]. This local group contributed to constructing an emergency contact network for watching forest fires especially in dry seasons [13]. Members of Songgye often cooperated to construct and repair trails to *Pinus densiflora* forests near their village [40]. It also led the villagers to cooperate in wood collection and distribute the collected wood at low prices [13]. Although this form of local group is not currently active, such activities contributed to sustaining local pine forests and their associated livelihoods in ancient Korea [39].

3. Implementation of Traditional Ecological Knowledge and Community Forestry in Modern China, Japan, and South Korea

3.1. China

In the mid-twentieth century, the political priority of the Chinese government was to reconstruct the nation's economy, and overcome poverty and backwardness [17]. This growth-oriented economic policy elevated timber and charcoal use for construction and steel making, and promoted large-scale logging operations [41]. Such logging operations frequently resulted in timber overexploitation beyond the allowable quotas [18], and the destruction of the traditional Fengshui forests [27]. In addition to contributing to global climate change, this overexploitation of the forests has led to serious deforestation and desertification in China [41].

Since the mid-1980s, the Chinese government has been concentrating on forest protection through rehabilitating natural forests, afforesting marginal croplands, decelerating desertification, and establishing timber forests with low environmental impacts [41]. To reduce private forest overexploitation, the Chinese government attempted to revisit the applicability of Chinese community forestry traditions. In particular, the community forestry and forest protection cultures in the southern part of China were reconsidered as management alternatives for forestry and biodiversity conservation [7,27,42].

Anhui, Sichuan, and Yunnan Provinces have received model projects incorporating traditional community forestry [43]. These projects used local traditional ecological knowledge and practices to adjust modern knowledge and techniques to local environments. For example, such an approach has been piloted from the early twenty-first century in Yunnan Province, where villagers relied on the traditional shifting cultivation system, consisting of a short cropping phase and long forest fallowing phase [44–46]. In this region, traditional shifting cultivation was combined with the rotational tree planting and harvesting part of the modern forestry concept [46]. Project staff and workers asked farmers to share indigenous knowledge of native, economically valuable crops and tree species, and actively participate in selecting tree rotation age, monitoring pilot sites, and evaluating the piloted approach [44]. They have educated local farmers on how to establish local societies for community forestry, and provided alternative energy sources and energy saving techniques to prevent forest overexploitation [43]. Through these activities encouraging villagers' participation, modern knowledge and techniques (e.g., rotation age and species selection) have been further adjusted in the context of local traditional ecological knowledge and thus contributed to forest resource protection as well as local economy [44,46]. Accordingly, the model projects in Yunnan Province have provided information to reinforce the applicability of the traditions to the forestry sector.

China recently established and modified policies and laws to support governmental projects aiming at the protection and sustainable use of forest resources [43]. The Chinese government issued the modified version of the Forest Law of the People's Republic of China in 1998, which declares the rights of autonomy and economic benefit for ethnic minorities in the forestry sector in addition to the general principle for the nation's forestry [47]. It helps facilitate local villagers' participation in the protection of forests on the basis of their traditional ecological knowledge and practices, such as Cuiguimingyue [48]. Other policies and laws, such as the Law of the People's Republic of China on Land Contract in Rural Areas and the Collective Forest Tenure Reform, define rights and ways for villagers' participation in resolving conflicts and making decisions in relation to the management and use of forests [43,45,48].

3.2. Japan

The post-war economic growth in Japan has revolutionized energy sources, chemical fertilizers, and building materials, and diminished the usefulness of the traditional Satoyama system for obtaining timber, fuelwood, charcoal, and compost [29]. Commercial monoculture plantations or urban areas replaced many natural forests, although forests in Japan were relatively well established [9,49]. However, inexpensive foreign timber imports and an aging rural population have reduced domestic timber production since the 1960s [34]; consequently, both traditional community forests and monoculture plantations were abandoned without taking into account public interests and appropriate management operations [19]. This forest underutilization has become a serious issue in Japan, as the structure of unmanaged forest can shade out understory vegetation and cause soil erosion and biodiversity loss [7,30,50].

To tackle afforestation, traditional ecological knowledge and practices were used in terms of the selection of plantation species. Similar to the ancient kingdoms, plantation species selection in Japan was originally based on topography, climate, and soil properties as well as the characteristics of the tree species [15]. This species selection system was adjusted in the twentieth century with consideration given to profitability and manageability (such as pest tolerance) to relieve forest underutilization and abandonment [15].

Since the twentieth century, Japan has actively revisited traditional ecological knowledge and practices in terms of the sustainability of the forestry sector. The Japanese government has initiated projects targeting the harmonization of human society with nature and, since 1994, has identified the restoration of the traditional Satoyama system as an important strategy [9]. The applicability of the traditional community forestry system has been piloted at several prefectures including Okayama, Kanagawa, Kyoto, and Ishikawa to activate the management and use of abandoned community forests [9,38]. These projects have been combined with activities of local cooperatives to pilot traditional concepts and community forestry for diverse rural societies in mountainous, grassland, and suburban areas. In Ishikawa prefecture, the local government introduced Ishikawa Forest Environment Tax in 2006 and Satoyama Creation Fund in 2011, both of which aim to assist the restoration of the traditional satoyama landscape and management of abandoned plantations [51].

The Japanese government and the United Nations University jointly started the Satoyama Initiative at the Conference of the Parties (COP 10) in 2010. This initiative has promoted the reassessment of community forest traditions around the world, and attempted to integrate such traditions into modern forestry systems [10,29]. The concept of the Satoyama Initiative has migrated to other countries that share the community forestry tradition, and stimulated a reevaluation of traditional ecological knowledge and practices related to forests [10,52].

Monitoring studies have shown that the remaining Satoyama landscapes contribute to ecosystem functioning. Traditional coppicing in Satoyama has higher potential for carbon sequestration than modern forestry systems, consisting of intensive thinning activities [53]. Furthermore, high landscape connectivity among forests, croplands, and water sources within conserved Satoyama significantly served as sites for nesting and foraging of birds, such as *Butastur indicus* Gmelin [54]. Because of the high landscape connectivity, Satoyama also provides habitats for the entire lifecycle of insects, including *Luciola cruciata* Motschulsky, *Ranatra chinensis* Mayrt., and *Apis cerana japonica* Radoszkowski [30,54,55]. This reflects the potential negative impacts of the decline of Satoyama as a result of forest underutilization and urbanization on biodiversity and proper ecosystem functioning [54,56].

3.3. South Korea

Japanese colonization and the Korean War extensively destroyed forests in the Korean peninsula until the mid-twentieth century [20,21] (Figure 3). The traditional Maeulsoop have declined because of colonization and warfare [13], and subsequent construction-oriented village development [28]. Accordingly, the traditional community forestry groups, such as Songgye, became obsolete [39]. The South Korean government has developed diverse forest restoration programs since the 1960s, including large-scale reforestation activities and nursery management combined with a substitution of energy sources from fuelwood to fossil fuel [21,57].

Figure 3. (a) Deforested lands in the mid-twentieth century (Photo from the Korea Forest Service) and (b) revegetated forests after the national forest restoration programs in South Korea.

The forest restoration programs utilized much traditional ecological knowledge and practices. Like the rules of the ancient kingdoms, the South Korean government listed the locally specialized plantation species taking into account climate, topography, and soil fertility after the national soil survey conducted during the 1970s [15]. The fast-growing or nitrogen-fixing species were recommended for degraded and infertile lands but more economically valuable species were selected for relatively fertile lands (Table 2). The government also adopted intensive weeding, pruning, and pest control practices from the ancient policies (Table 1), and widely applied such practices to the young plantations from 1973 to 1978 to increase the survival rate of planted seedlings [58]. The current South Korean forestry system has inherited these forest management practices and tree species selection concept [15].

Table 2. Plantation species for the national forest restoration programs of South Korea in the 1970s [15].

Site Index *	Cool Temperate Central Region	Cool Temperate Southern Region	Warm Temperate Region
(1)	*Larix kaempferi* (Lamb.) Carr., *Castanea crenata* Siebold & Zucc., *Populus* spp.	*Cryptomeria japonica* D. Don, *Paulownia coreana* Uyeki, *Castanea crenata*	*Cryptomeria japonica*, *Paulownia coreana*, *Castanea crenata*
(2)	*Larix kaempferi*, *Castanea crenata*, *Populus* spp., *Pinus koraiensis*	*Chamaecyparis obtusa* Sieb. and Zucc., *Populus* spp., *Castanea crenata*	*Chamaecyparis obtusa*, *Cryptomeria japonica*, *Paulownia coreana*
(3)	*Populus* spp., *Pinus koraiensis*, *Pinus thunbergii* Parl., *Robinia pseudoacacia* L.	*Populus* spp., *Pinus thunbergii*, *Chamaecyparis obtusa*, *Robinia pseudoacacia*	*Pinus thunbergii*, *Chamaecyparis obtusa*
(4)	*Pinus rigida* Mill., *Pinus thunbergii*, *Robinia pseudoacacia*	*Pinus rigida*, *Pinus thunbergii*, *Robinia pseudoacacia*	*Pinus rigida*, *Pinus thunbergii*, *Robinia pseudoacacia*
(5)	*Alnus japonica* (Thunb.) Steud., *Pinus rigida*	*Alnus japonica*, *Pinus rigida*	*Alnus japonica*, *Pinus rigida*

Site Index * was based on topography (e.g., altitude, slope, and aspect) and physicochemical properties of soils (e.g., parent material, depth, texture, moisture, organic matter, and hardness).

The other important strategy originating from traditional ecological knowledge and practices of the ancient Korean kingdoms was the participation of the public in forest restoration programs. The South Korean government established a national tree planting period between 21 March and 20 April, to encourage public participation [20]. Since 1951, a local participatory group, Sallimgye (private forest cooperative), has been established under the government's policies and laws [59]. The structure and activity of Sallimgye were similar to those of the traditional Songgye in ancient Korea [60]. The group consisted of local villagers living near a target reforestation site, who primarily provided labor to carry out government reforestation activities. This local group led villagers' participation in seed collection, nursery operation, and forest patrol, while the allocated benefits from local forests, including fuelwood, timber, and fruit, went to the villagers [57].

The forest restoration program with traditional ecological knowledge and practices benefited livelihoods. Areas reforested with fast-growing trees (e.g., *Robinia pseudoacacia* L.) provided more than 29 million Mg of fuelwood for heating and cooking from 1973 to 1988, and those with *Castanea crenata* Sieb. and Zucc. produced 30 thousand Mg of chestnuts for consumption from 1976 to 1987 [57]. These forest products afforded additional energy and food sources to local people during that period [57]. Reforestation also contributed to the protection of residential and agricultural areas by reducing the risk of natural disasters, such as landslide and flooding [61].

After intensive economic growth and urbanization, raising awareness of sustainability and ecosystem services resulted in a reevaluation of traditional community forests. To confirm the value and applicability of traditional practices, empirical data from Songgye, including old documents and oral histories, have been collected and analyzed from the early twenty-first century [40]. The structure of Songgye in the twentieth century was analyzed to find factors leading to the decline of Songgye and assess the potential applicability of traditional community forestry in current societies [39]. In addition, ecological monitoring found that conservation of Maulsoop could be applicable to reduce wind speed and evaporation; contribute to water conservation during spring, a dry season of the

Korean peninsula [31]; and protect plant diversity—one of the strategies for improving landscape connectivity [33].

Due to the reevaluation of traditional ecological knowledge and practices, the Korea Forest Service has restored more than 70 Maeulsoop since 2003, which are principally used for recreation, education, and tourism [62]. The restoration consists of not only reforesting degraded Maeulsoop, but also collecting diverse information including oral or documented history, area, ownership, and species composition of each Maeulsoop to modify restoration strategies according to natural and socioeconomic conditions [62]. A nongovernmental organization, Forest for Life, has collaborated with the Korea Forest Service and local committees to preserve the traditional Maeulsoop and surrounding landscape [63]. This organization has provided subsidies for the restoration of several Maeulsoop, such as Yondangsoop in Gyeonggi Province in 2005 [62], and has attended town meetings and workshops to exchange perspectives among participants concerning the restoration project [63].

4. Implications

Socioeconomic changes stimulated the implementation of traditional ecological knowledge and community forestry to improve forest management systems in China, Japan, and South Korea. Post-war economic growth in China and Japan accompanied forest overexploitation in China and forest underutilization in Japan (Figure 4). In South Korea, the national level of deforestation after colonization and warfare forced the country to develop strategies for restoring the nation's forest ecosystems (Figure 4). In the twentieth century, these changes acted as critical issues, to which previous forest management practices had to adapt.

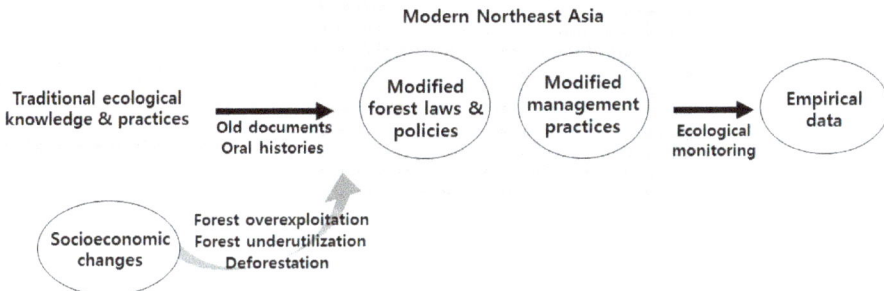

Figure 4. Role of traditional ecological knowledge within the forestry sector in modern Northeast Asia.

Given these changes, traditional ecological knowledge and practices were applied to adjust forest management practices (Figure 4). The Chinese government encouraged the community forestry tradition for the national cropland afforestation program [44,46]. Japan and South Korea implemented the old plantation species selection concept using environmental conditions to relieve forest underutilization and improve the forest restoration strategy [15]. Especially in South Korea, the national forest restoration programs used diverse traditions related to plantation species selection, intensive weeding and pruning practices, and community forestry in the mid-twentieth century [16,58]. For such applications, traditional ecological knowledge and practices tended to combine with modern knowledge of profitability and manageability or large-scale soil monitoring activities.

One important trend was that China, Japan, and South Korea reevaluated the applicability of the community forestry tradition for restoring and sustaining the nations' forests. The reevaluation included not only restoring community forests and traditional community forestry, but also installing political alternatives and paying subsidies to support the restoration. All three countries recently established laws and policies to promote public participation in forestry and the restoration of community forests [38,48,60]. With this political support, community forestry traditions have been implemented and have helped address forest overexploitation in China, forest underutilization in Japan, and deforestation in South Korea [9,54].

Monitoring of the pilot sites has provided empirical data that supports the importance of the contribution of traditional ecological knowledge and practices to the forestry sector (Figure 4). Monitoring studies have detected higher biodiversity and habitat connectivity in traditionally protected forests than in disturbed forests [33,42,54]. Such traditionally protected forests also appear to enhance village microclimates by controlling wind speed and air humidity [25,31]. The restored community forests have been used for recreation and environmental education concerning socioecological system resilience as well as biodiversity conservation [7,30]. The pilot projects have also contributed to resolving the loss of social connections and community forestry activities, left behind by the degradation of rural society [9,37,62]. These recent experiences demonstrate the importance of traditional ecological knowledge in the forestry sector [1].

Several factors aided in the implementation of traditional ecological knowledge and practices. A decentralized system played a role because it enabled villagers to directly distribute and benefit from the revisited traditional ecological knowledge and practices [9,57]. The partnership between the government and private sectors stimulated the establishment of the decentralized system, and technical and financial support for the projects [51,62,64]. Localizing modern techniques using traditional and indigenous knowledge also assisted the reevaluation and implementation of the traditional systems [44,46]. These implications may be helpful for the other Northeast Asian countries, which share a similar culture but suffer from severe loss of forest resources [22]. Particularly, the North Korean government has been promoting community forestry to overcome extreme deforestation, but it has not been scaled up to a national level owing to political, institutional, and financial constraints [65]. We suggest that the experiences of utilizing traditional ecological knowledge and practices in China, Japan, and South Korea are applicable to deal with the constraints hindering forest restoration projects in North Korea.

It is noteworthy that the recent implementation of traditional ecological knowledge in China, Japan, and South Korea is not a strict revival of traditions. The primary objective of projects applying traditional ecological knowledge and practices was the improvement of the local economy and forest resources [64,66]. Many restored community forests have been used for more practical purposes (e.g., education, recreation, and cultivation) than traditional ceremonial purposes [46,62]. This situation changed people's perspectives of traditional ecological knowledge and practices from old religious aspects to well-being, and led to the decline of several traditions that did not fit into such alteration [64]. Thus, the consequences of the reevaluation of traditional ecological knowledge and practices should be further assessed in the context of long-term socioeconomic changes.

Applying traditional ecological knowledge and practices to current forest management practices still remains challenging despite its benefits. Less intensive forestry systems based on the traditional concept might not be enough to substitute the modern forestry system, which has satisfied the demands for industrial and commercial forest products [67]. The traditional concept is limited to the pilot level due to the lack of scientific understanding on the link between traditional ecological knowledge and the socioecological system resilience [7]. These issues should be thoroughly considered before applications of the traditional concept at broader scales. However, the previous trends in China, Japan, and South Korea reflect the importance of traditional ecological knowledge and practices—incorporated into modern forestry systems—in the face of socioeconomic changes to date.

5. Conclusions

Traditional ecological knowledge and practices in Northeast Asia, consisting of the Fengshui concept, historical experiences, and community forestry systems, have established a way of managing and using forest resources in ancient Northeast Asia. These traditional systems had declined in the twentieth century owing to a decreasing demand for the traditional livelihood systems. Key concepts of traditional ecological knowledge and practices were recently revived and combined with forest management practices in the face of current socioeconomic changes in this region. The empirical data from monitoring such activities supported the development of forest management alternatives with traditional ecological knowledge and practices. Before implementing traditional ecological knowledge and practices beyond the pilot level, several challenges should be overcome: the results of revisiting traditional ecological knowledge and practices should be clarified along with the long-term socioeconomic changes; the interaction between traditional ecological knowledge and socioecological system resilience should be scientifically understood. However, the reevaluation of traditional ecological knowledge in China, Japan, and South Korea will be useful to overcome constraints hindering forest restoration in the other Northeast Asian countries.

Acknowledgments: The present study was supported by the Korea Forest Service (S211216L030120), the Ministry of Environment of Korea (2014001310008), and a Korea University Grant in 2017. We thank Hiroyuki Muraoka of the Gifu University in Japan for his helpful comments during the manuscript preparation.

Author Contributions: S.K. and Y.S. conceived the key concept of the paper. S.K. wrote the paper and led the data collection and review. G.L. and Y.S. contributed to the data collection and review.

Conflicts of Interest: The authors declare no conflict of interest. The funder played no role in the process of the current study.

References

1. Berkes, F.; Colding, J.; Folke, C. Rediscovery of traditional ecological knowledge as adaptive management. *Ecol. Appl.* **2000**, *10*, 1251–1262. [CrossRef]
2. Berkes, F. Traditional ecological knowledge in perspective. In *Traditional Ecological Knowledge: Concepts and Cases*; Inglis, J.T., Ed.; International Program on Traditional Ecological Knowledge, International Development Research Centre: Ottawa, ON, Canada, 1993; pp. 1–9, ISBN 1-895926-00-9.
3. Guan, C. Study on knowledge and practice on function of conserving water and soil of forests in ancient China. *Sci. Soil Water Conserv.* **2004**, *2*, 105–110. (In Chinese)
4. Youn, Y.-C.; Liu, J.; Sakuma, D.; Kim, K.; Masahiro, I.; Shin, J.-H.; Yuan, J. Northeast Asia. In *Traditional Forest-Related Knowledge*; Parrotta, J.A., Trosper, R.L., Eds.; Springer: Dordrecht, The Netherlands, 2012; pp. 281–313, ISBN 978-94-007-2143-2.
5. Park, H.; Lee, J.Y.; Song, M. Scientific activities responsible for successful forest greening in Korea. *For. Sci. Technol.* **2017**, *13*, 1–8. [CrossRef]
6. Bürgi, F.; Gimmi, U.; Stuber, M. Assessing traditional knowledge on forest uses to understand forest ecosystem dynamics. *For. Ecol. Manag.* **2013**, *289*, 115–122. [CrossRef]
7. Lee, E.; Krasny, M.E. Adaptive capacity in community forest management: A systematic review of studies in East Asia. *Environ. Manag.* **2017**, *59*, 34–49. [CrossRef] [PubMed]
8. Ramakrishnan, P.S. Traditional forest knowledge and sustainable forestry: A north-east India perspective. *For. Ecol. Manag.* **2007**, *249*, 91–99. [CrossRef]
9. Watanabe, T.; Okuyama, M.; Fukamachi, K. A review of Japan's environmental policies for Satoyama and Satoumi landscape restoration. *Glob. Environ. Res.* **2012**, *16*, 125–135. [CrossRef]
10. Berglund, B.E.; Kitagawa, J.; Lagerås, P.; Nakamura, K.; Sasaki, N.; Yasuda, Y. Traditional farming landscapes for sustaining living in Scandinavia and Japan: Global revival through the Satoyama Initiative. *Ambio* **2014**, *43*, 559–578. [CrossRef] [PubMed]
11. Herrmann, T.M.; Torri, M.-C. Changing forest conservation and management paradigms: Traditional ecological knowledge systems and sustainable forestry: Perspectives from Chile and India. *Int. J. Sustain. Dev. World Ecol.* **2009**, *16*, 392–403. [CrossRef]
12. Coggins, C.; Chevrier, J.; Dwyer, M.; Longway, L.; Xu, L.; Tiso, P.; Li, Z. Village Fengshui forests of southern China: Culture, history, and conservation status. *Asia Netw. Exch.* **2012**, *19*, 52–67. [CrossRef]
13. Chun, Y.W.; Tak, K.-I. Songgye, a traditional knowledge system for sustainable forest management in Choson dynasty of Korea. *For. Ecol. Manag.* **2009**, *257*, 2022–2026. [CrossRef]
14. Kong, W.-S.; Lee, S.H.; Bang, J.-Y.; Hong, S.-C.; Choi, S.-K. *The History of Korean Forests*; The Society for Forests and Culture: Seoul, Korea, 2014; pp. 134–185, ISBN 978-89-93453-21-8. (In Korean)

15. Korea Forest Research Institute. *History of the Tree Selection System and Its Application*; Korea Forest Research Institute: Seoul, Korea, 2009; pp. 7–73, ISBN 978-89-8176-543-9. (In Korean)

16. Kwon, S.-K. An analysis of Bongsan policy in the late era of Chosun dynasty. *Korean Policy Sci. Rev.* **2007**, *11*, 81–104. (In Korean)

17. Dai, L.; Zhao, W.; Shao, G.; Lewis, B.J.; Yu, D.; Zhou, L.; Zhou, W. The progress and challenges in sustainable forestry development in China. *Int. J. Sustain. Dev. World Ecol.* **2013**, *20*, 394–403. [CrossRef]

18. Xu, J.; Tao, R.; Amacher, G.S. An empirical analysis of China's state-owned forests. *For. Policy Econ.* **2004**, *6*, 379–390. [CrossRef]

19. Takeuchi, K.; Ichikawa, K.; Elmqvist, T. Satoyama landscape as social-ecological system: Historical changes and future perspective. *Curr. Opin. Environ. Sustain.* **2016**, *19*, 30–39. [CrossRef]

20. Tak, K.; Chun, Y.; Wood, P.M. The South Korean forest dilemma. *Int. For. Rev.* **2007**, *9*, 548–557. [CrossRef]

21. Bae, J.S.; Joo, R.W.; Kim, Y.-S. Forest transition in South Korea: Reality, path and drivers. *Land Use Policy* **2012**, *29*, 198–207. [CrossRef]

22. Fang, J.; Guo, Z.; Hu, H.; Kato, T.; Muraoka, H.; Son, Y. Forest biomass carbon sinks in East Asia, with special reference to the relative contributions of forest expansion and forest growth. *Glob. Chang. Biol.* **2014**, *20*, 2019–2030. [CrossRef] [PubMed]

23. Han, K.-T. Traditional Chinese site selection—Fang Shui: An evolutionary/ecological perspective. *J. Cult. Geogr.* **2001**, *19*, 75–96. [CrossRef]

24. Wu, Y.-B. Origin of Fengshui woods and its enlightenment to modern society. *J. Landsc. Res.* **2011**, *3*, 96–99.

25. Chen, B.; Nakama, Y.; Kurima, G. Layout and composition of house-embracing trees in an island Feng shui village in Okinawa, Japan. *Urban For. Urban Green.* **2008**, *7*, 53–61. [CrossRef]

26. Whang, B.-C.; Lee, M.W. Landscape ecology planning principles in Korean Feng-shui, Bi-bo woodlands and ponds. *Landsc. Ecol. Eng.* **2006**, *2*, 147–162. [CrossRef]

27. Yuan, J.; Liu, J. Fengshui forest management by the Buyi ethnic minority in China. *For. Ecol. Manag.* **2009**, *257*, 2002–2009. [CrossRef]

28. Park, J.-C. Concept and case of village grove. *Keimyung Korean Stud. J.* **2006**, *33*, 233–262. (In Korean)

29. Takeuchi, K. Rebuilding the relationship between people and nature: The Satoyama Initiative. *Ecol. Res.* **2010**, *25*, 891–897. [CrossRef]

30. Miyaura, T. Satoyama—A place for preservation of biodiversity and environmental education. *Die Bodenkultur* **2009**, *60*, 23–29.

31. Koh, I.; Kim, S.; Lee, D. Effects of Bibosoop plantation on wind speed, humidity, and evaporation in a traditional agricultural landscape of Korea: Field measurements and modeling. *Agric. Ecosyst. Environ.* **2010**, *135*, 294–303. [CrossRef]

32. Lee, K.-S. Bee-bo forest: Traditional landscape ecological forest in Korea. In *Landscape Ecological Applications in Man-Influenced Areas: Linking Man and Nature Systems*; Hong, S.-K., Nakagoshi, N., Morimoto, Y., Eds.; Springer: Dordecht, The Netherlands, 2008; pp. 389–394, ISBN 978-1-4020-5487-7.

33. Koh, I.; Reineking, B.; Park, C.-R.; Lee, D. Dispersal potential mediates effects of local and landscape factors on plant species richness in Maeulsoop forests of Korea. *J. Veg. Sci.* **2015**, *26*, 631–642. [CrossRef]

34. Iwamoto, J. The development of Japanese forestry. In *Forestry and the Forest Industry in Japan*; Iwai, Y., Ed.; University of British Columbia Press: Vancouver, BC, Canada, 2002; pp. 3–9, ISBN 0-7748-0882-9.

35. Han, J.-S. The development of forests and the formation of policy for cultivating pine trees in the period from Taejo to Sejong of Joseon. *Sahak Yonku* **2013**, *111*, 42–81. (In Korean)

36. Gillett, P.L. The village gilds of old Korea. *Trans. Korea Branch R. Asiat. Soc.* **1913**, *4*, 13–44.

37. Yuan, J.; Wu, Q.; Liu, J. Understanding indigenous knowledge in sustainable management of natural resources in China: Taking two villages from Guizhou Proince as a case. *For. Policy Econ.* **2012**, *22*, 47–52. [CrossRef]

38. Hasegawa, M.; Pulhin, J.M.; Inoue, M. Facing the challenge of social forestry in Japan: The case of reviving harmonious coexistence between forest and people in Okayama prefecture. *Small-Scale For.* **2013**, *12*, 257–275. [CrossRef]

39. Yu, D.J.; Anderies, J.M.; Lee, D.; Perez, I. Transformation of resource management institutions under globalization: The case of Songgye community forests in South Korea. *Ecol. Soc.* **2014**, *19*, 2. [CrossRef]

40. Kang, S.-B. The practice of Songgye on the villages around Guksa-bong, Gyeryongsan—Focus on the Songgye of Hyanghan-ri, Gyeryong-shi. *Korean J. Folk Stud.* **2009**, *24*, 97–121. (In Korean)

41. Yu, D.; Zhou, L.; Zhou, W.; Ding, H.; Wang, Q.; Wang, Y.; Wu, X.; Dai, L. Forest management in Northeast China: History, problems, and challenges. *Environ. Manag.* **2011**, *48*, 1122–1135. [CrossRef] [PubMed]

42. Gao, H.; Ouyang, Z.; Chen, S.; Koppen, C.S.A. Role of culturally protected forests in biodiversity conservation in Southeast China. *Biodivers. Conserv.* **2013**, *22*, 531–544. [CrossRef]

43. State Forestry Administration. *National Report on Sustainable Forest Management*; China Forestry Publishing House: Beijing, China, 2013; pp. 175–210.

44. He, J.; Zhou, Z.; Weyerhaeuser, H.; Xu, J. Participatory technology development for incorporating non-timber forest products into forest restoration in Yunnan, Southwest China. *For. Ecol. Manag.* **2009**, *257*, 2010–2016. [CrossRef]

45. He, J.; Sikor, T. Looking beyond tenure in China's collective forest tenure reform: Insights from Yunnan Province, Southwest China. *Int. For. Rev.* **2017**, *19*, 29–41. [CrossRef]

46. Liang, L.; Shen, L.; Yang, W.; Yang, X.; Zhang, Y. Building on traditional shifting cultivation for rotational agroforestry experiences from Yunnan, China. *For. Ecol. Manag.* **2009**, *257*, 1989–1994. [CrossRef]

47. *The Forest Law of the People's Republic of China*; Democracy and Law of China Press: Beijing, China, 2008; ISBN 9787802192225. (In Chinese)

48. Liu, J.; Innes, J.L. Participatory forest management in China: Key challenges and ways forward. *Int. For. Rev.* **2015**, *17*, 477–484. [CrossRef]

49. Ichikawa, K.; Okubo, N.; Okubo, S.; Takeuchi, K. Transition of the Satoyama landscape in the urban fringe of the Tokyo metropolitan area from 1880 to 2001. *Landsc. Urban Plan.* **2006**, *78*, 398–410. [CrossRef]

50. Igarashi, T.; Masaki, T.; Nagaike, T.; Tanaka, H. Species richness of the understory woody vegetation in Japanese cedar plantations declines with increasing number of rotations. *J. For. Res.* **2016**, *21*, 291–299. [CrossRef]

51. Yiu, E. Noto peninsula after GIAHS designation: Conservation and revitalization efforts of Noto's satoyama and satoumi. *J. Resour. Ecol.* **2014**, *5*, 364–369. [CrossRef]

52. Yao, Z.; Xin, Z.-J.; Wu, Y.-M.; You, H.L. Environmental management mode of satoyama in Japan and its enlightenment to construction of new country-side in China. *J. Ecol. Rural Environ.* **2017**, *33*, 769–774. (In Chinese)

53. Terada, T.; Yokohari, M.; Bolthouse, J.; Tanaka, N. "Refueling" satoyama woodland restoration in Japan: Enhancing restoration practice and experiences through woodfuel utilization. *Nat. Cult.* **2010**, *5*, 251–276. [CrossRef]

54. Katoh, K.; Sakai, S.; Takahashi, T. Factors maintaining species diversity in Satoyama, a traditional agricultural landscape of Japan. *Biol. Conserv.* **2009**, *142*, 1930–1936. [CrossRef]

55. Fujiwara, A.; Washitani, I. Dependence of Asian honeybee on deciduous woody plants for pollen resource during spring to mid-summer in northern Japan. *Entomol. Sci.* **2017**, *20*, 96–99. [CrossRef]

56. Kobayashi, S.; Abe, S.; Matsuki, R. Genetic structure of a Japanese brown frog (*Rana japonica*) population implies severe restriction of gene flow caused by recent urbanization in a Satoyama landscape. *Mitochondrial DNA* **2013**, *24*, 697–704. [CrossRef] [PubMed]

57. Lee, K.-J. *Saemaul Undong and Foreset Rehabilitation in Korea: Saemaul Income Boosting Project and the Role of the Village Forestry Cooperative*; Ministry of Strategy and Finance: Sejong, Korea, 2013; pp. 315–328. (In Korean)

58. Korea Forest Service. *Lessons Learned from the Republic of Korea's National Reforestation Programme*; Korea Forest Service: Daegu, Korea, 2014; pp. 1–42, ISBN 92-9225-579-7.

59. Choe, B.-T. The developments and character of setting up the forestry mutual-aid society in 1945~1960s. *Sahak Yonku* **2008**, *90*, 291–336. (In Korean)

60. Park, M.S.; Lee, H. Legal opportunities for public participation in forest management in the Republic of Korea. *Sustainability* **2016**, *8*, 369. [CrossRef]

61. Lee, J.; Lim, C.-H.; Kim, G.S.; Markandya, A.; Chowdhury, S.; Kim, S.J.; Lee, W.-K.; Son, Y. Economic viability of the national-scale forestation program: The case of success in the Republic of Korea. *Ecosyst. Serv.* **2018**, *29*, 40–46. [CrossRef]

62. Korea Forest Service. *Restoration Cases of the Traditional Maeulsoop*; Korea Forest Service: Daegu, Korea, 2015; pp. 1–171. (In Korean)

63. Lee, E.; Krasny, M.E. The role of social learning for social-ecological systems in Korean village groves restoration. *Ecol. Soc.* **2015**, *20*, 42. [CrossRef]

64. Liu, J.; Liu, X. Transforming nomadic traditions to biodiversity-friendly livelihoods from the perspective of traditional forest-related knowledge: The successful story of H village of Yunnan Province in China. In *Satoyama Initiative Thematic Review Volume 2*; Ichikawa, K., Subramanian, S.M., Chakraborty, S., Eds.; United Nations University Institute for the Advanced Study of Sustainability: Tokyo, Japan, 2016; pp. 13–25, ISBN 978-92-808-4573-0.

65. He, J.; Xu, J. Is there decentralization in North Korea? Evidence and lessons from the sloping land management program 2004–2014. *Land Use Policy* **2017**, *61*, 113–125. [CrossRef]

66. Koike, O. Rehabilitation, conservation, and utilization of satoyama ecosystems and human well-being: A case of Kanagawa Prefecture in Japan. *Yokohama J. Soc. Sci.* **2017**, *22*, 41–54.

67. Fuchigami, Y.; Hara, K.; Uwasu, M.; Kurimoto, S. Analysis of the mechanism hindering sustainable forestry operations: A case study of Japanese forest management. *Forests* **2016**, *7*, 182. [CrossRef]

forests

MDPI

Article

Comparison of the Economic Value of Urban Trees through Surveys with Photographs in Two Seasons

Claudia García-Ventura [1], Álvaro Sánchez-Medina [2], M. Ángeles Grande-Ortíz [3], Concepción González-García [3] and Esperanza Ayuga-Téllez [2,*]

[1] Escuela Técnica Superior de Ingenieros de Montes, Forestal y del Medio Natural, Universidad Politécnica de Madrid, Avda. de las Moreras s/n., 28040 Madrid, Spain; claudia.gventura@upm.es
[2] Buildings, Infrastructures and Projects for Rural and Environmental Engineering (BIPREE), Universidad Politécnica de Madrid, Ciudad Universitaria s/n., 28040 Madrid, Spain; alvaro.sanchezdemedina@upm.es
[3] SILVANET. Research Group for Sustainable Environmental Management, Universidad Politécnica de Madrid, Ciudad Universitaria s/n., 28040 Madrid, Spain; m.angeles.grande@upm.es (M.A.G.-O.); concepcion.gonzalez@upm.es (C.G.-G.)
* Correspondence: esperanza.ayuga@upm.es; Tel.: +34-910-671-582

Received: 22 January 2018; Accepted: 8 March 2018; Published: 10 March 2018

Abstract: Urban trees are generally considered to be a public asset and are an important part of a city's heritage. The aim of this work is to analyse the influence of season on the economic appraisal of various trees in Madrid. Photographs were taken of 43 individual tree specimens in summer and winter. The survey was designed to compare differences of opinion in the economic assessment of trees. The trees were assessed by eight valuation methods used worldwide. A total of 78 agroforestry engineering students answered a written survey, and the variables considered were: percentage of students who always evaluated the tree equally (%0), percentage of students who assigned more value to the summer photograph (%S), and percentage of students who assigned more value to the winter photograph (%W). The results were analysed by the statistical test of equal proportions and ANOVA to detect differences according to tree type (evergreen or deciduous), species, and other groupings made by the authors in previous works. W and S percentages are similar. The ANOVA analysis rejects the equality of percentages of S and W between groups. The Welch test rejects the equality of the percentage of S, W, and O between species.

Keywords: appraisal; urban trees; public opinion; photography; summer-winter

1. Introduction

The economic value of urban trees is a monetary reference for the benefits they offer the public. It reflects a variety of factors such as the value of the land where the trees are located, their historic importance, quality and state of health, the social and environmental benefits they afford, and the costs associated with their maintenance.

These economic values are obtained from assessment equations or formulas that consider—with greater or lesser weight—a combination of the above mentioned factors. The appraisal of urban trees is not an exact science, as it depends on the purposes of the assessment and the assessor's experience [1,2]. Each method or formulation takes different variables into account and provides different results. Numerous methods are used to evaluate urban trees in different regions of the world, and most define the value in monetary terms based on an expert's perception of the tree. The assessment involves establishing a measurable and objective criterion considering aspects or variables such as whether the tree stands alone or in a group, its physical deterioration, species and variety, size, age, and state of health and location, among others [3], along with environmental, social, and psychological variables [4].

The assessment can therefore be used as one of the bases for making decisions on the management of these trees and as an instrument for public administration and for society itself. In fact, policymakers need to have a clear idea of the socioeconomic value of urban asset [5], which necessarily includes urban trees. This would supplement the maximum probable (or potential) cost estimates of actions designed to mitigate disasters in cities (including climate change) and to prepare effective risk management strategies. So, what is the most appropriate method for assessing urban trees in a particular city?

To answer this question, several authors have conducted studies comparing methods for appraising urban trees. The valuation methods used show important differences [6–11], and the conclusion is that no method can be used under all conditions [1], making it advisable to combine capitalization and parametric methods [12]. There is therefore no simple solution to the question, and each method must be applied after an exhaustive study of multiple factors including the availability of reliable databases, the proposal of objects for assessment, and possible social repercussions.

This is where the aim of the assessment can be directed towards seeking a method that is closely aligned with social perception. What percentage of the population would choose one method as being the most closely aligned with reality? The answer necessarily involves conducting a survey of the public.

The survey technique is frequently used in issues related to the urban environment for the purposes of territorial management and planning, and the results serve to outline the managers' future lines of action. Various works highlight the concern with evaluating the opinion of the public. Examples include questioning the effects of biodiversity on the perception of urban green spaces [13] by evaluating the risk to urban trees at the municipal level through surveys of residents [14], while other studies evaluate the urban forest factors that citizens consider to be the most beneficial [15], or their recreational preferences [16]. Citizens are also consulted at the level of prevention to assess their reactions to the negative effects of urban trees in the specific case of a storm [17]. In all cases, the knowledge of citizens' opinions allows future actions to be planned for managing the environment and can be used to select indicators for establishing new urban forests [18]. However, there are no studies that compare different appraisals using evaluation methods that consider citizens' opinions or perceptions on this subject.

The aim of this study is precisely to assess this aspect. The tool chosen was the photographic survey. Most studies show that photographs can be used as a valid substitute for aesthetic judgements [19–22]. Other authors have verified the validity of photographs to assess not only aesthetic, but also biological, aspects; in the study indicating the validity of photo-based scenic beauty assessments [23], group-averaged on-site and photo-based assessments were very similar. Repeat photography has been applied in different works and found to be an efficient, effective, and useful method to identify region-wide trends in land-use change [24], and provides a reliable and consistent measurement of phenophase to monitor plant phenology [25]. Similarly, digital photography has been proved useful for observing the seasonal change in aboveground green biomass and foliage phenology [26]. It has been demonstrated that the use of photography to evaluate the perception of forest vegetation and management in urban woodlands can serve as a useful quantitative method and a complement to conventional methods [27]. Last year, similar results were reported using new technologies in a comparison of landscapes in Devon (UK) and Asturias (Spain) through on-site visits and images taken by UAV (unmanned aerial vehicles), revealing a high degree of consensus between both assessments [28]. Applying the same technology, UAV was used to measure within-season tree height growth in a mixed forest stand, and the results closely agreed with published field observations for four tree species [29]. However, the photographic survey with images of trees may be conditioned by the season of the year in which the photographs are taken.

The aim of the present study is to conduct surveys of a population group using photographs of different trees to determine whether there are statistically significant differences in the assessment of the specimens depending on the season of the year in which the photographs are taken.

2. Materials and Methods

2.1. Study Area

The study took place in the Forestry and Environmental Engineering School at the Madrid Polytechnic University in Ciudad Universitaria (Madrid). It has an area of 8.57 ha, of which 7.62 ha are forested. This green space aims to meet two requirements: to contribute to the regeneration of the forest in the Ciudad Universitaria and to show students the forest species of most interest. One characteristic feature of this space is that it has a high diversity. There are 2978 individuals in the arboretum inventory, corresponding to 129 different tree species [8].

2.2. Survey Design

The survey was designed to compare the differences in opinion expressed by the public in regard to the economic value of the specimens, and the influence of the season on this appraisal. The respondents were shown photographs of 42 tree specimens belonging to 12 species. The photographs were taken with a digital camera (Canon EOS 450D, 18–55 mm, Canon Inc., Tokyo, Japan) from the same point of view and by the same person in two different seasons: one in June, which we call the summer photo (S); and another in December, which we call the winter photo (W).

Of the 42 specimens selected, 27 are evergreen species and 15 are deciduous. The 42 trees were divided into four groups based on other characteristics (see [8]), as follows: Freq group (most abundant species in the city of Madrid, 21 specimens); Max group (species with the greatest economic value in all the methods analysed, seven specimens); Min group (species with the lowest economic value in all the methods analysed, seven specimens) and Sin group (species considered as singular trees, seven specimens).

Table 1 shows the selected species and the number of specimens, in addition to the leaf type and the group to which they belong.

Table 1. Specimens selected in the study grouped by their factors.

Species (SP)	Leaf Type (Type)	Group	Number of Specimens
Cupressus arizonica Greene	perennial	Freq	7
Pinus pinea L.	perennial	Freq	7
Platanus × *hybrida* Brot.	deciduous	Freq	7
Quercus suber L.	perennial	Max	7
Ailanthus altissima (Mill.) Swingle	deciduous	Min	7
Cedrus deodara (Roxb.) G.Don	perennial	Sin	1
Pinus halepensis Mill.	perennial	Sin	1
Chamaerops humilis L.	perennial	Sin	1
Eucalyptus globulus Labill.	perennial	Sin	1
Populus alba var. *bolleana* (Lauche) Otto	deciduous	Sin	1
[1]*Quercus canariensis* Willd.	perennial	Sin	1
Sequoiadendron giganteum (Lindl.) J. Buchholz	deciduous	Sin	1

[1] The species Quercus canariensis has marcescent leaves but has been classified for the statistical analysis as evergreen, as there are no observable differences between the summer (S) and winter photographs (W).

2.3. Appraisal Methods Used

The tree specimens selected were appraised by eight valuation methods used in different parts of the world. These methods are classified in three types according to their formulation: parametric, mixed, and capitalisation (see [30]).

2.3.1. Parametric Methods

- North American method (CTLA). It defines the "base value" as the expression of the nursery's unit price according to the cross-section of the trunk, and uses corrective indexes to maintain or reduce its value, but never to increase it.

- Burnley method (Burnley). Its main variable is the tree size measured as the volume of an inverted cone, considering the height and crown area. It also includes a monetary value, designated the "base value". The final figure may be modified by factors that can reduce the base value. This is used mainly in Australia [31].
- Formulaic Expert Method (FEM). This method selects six main criteria (dimension, species, individual, state, location, and outstanding consideration), and the monetary value of the tree is the result of multiplying the total score of the main criteria of a tree by a monetary assignment factor (MAF) derived from the three-year average sales price per square metre of a mid-sized residential home. Used for singular trees in Hong-Kong.
- Copima method (Copima). This is based on the price of the species in the local market, corrected by multiplicative indexes that increase the value of the specimen. It is used in the municipalities of Concepción, La Pintana, and Maipú (Chile) [2].

2.3.2. Mixed Methods

- Granada standard (NG). In the first versions of this method, the base value was obtained for each species with a regression model based on the tree age. Since the last review, the method uses the trunk circumference measurement (measured 1 m from the ground) modified by intrinsic and extrinsic factors. It is used in Spain.
- Contato method (Contato). This classifies the tree based on its diameter, height, and crown area. The base value of each tree is calculated according to its age using the capitalisation formula, and is modified with corrective multiplicative factors that can increase or decrease the end value. Used in Argentina.
- New Zealand method (Standard Tree Evaluation Method) (STEM). This is one of the most widely used methods, and applies a system of points to assess 20 tree attributes in three general categories: state, functions, and outstanding qualities (special merit). The total score (P, with a maximum of 540 points) is multiplied by the wholesale cost of a five-year-old tree (with no specific indication of the tree species). To this is added the wholesale cost of planting and maintaining the tree until it reaches the same age as the replaced specimen.

2.3.3. Capitalisation Methods

- Capitalisation method (Capitalis.). This is based on the capitalisation of the replacement and maintenance costs throughout the life of the tree. Two methodologies can be distinguished. The first uses the replacement costs, and involves finding specimens of the same species, age, and physical and ornamental characteristics on the market, and whose transplantation is technically feasible. This tree must also have a high possibility of rooting without compromising its normal development.
- The second is based on maintenance costs, assuming that the tree chosen to replace the tree being appraised is younger. That is, it is estimated that the internal yield rate (r) for transforming the future tree into the current tree represents an intermediate situation where the substitute tree has a somewhat lower age and dimensions than those of the current tree.

STEM has the highest values in all the examples, while FEM comes second, with the highest values in 95% of cases with the eight methods used. The highest valuations for these two methods are justified by the calculation criteria. FEM has land price as a multiplicative factor, which considerably raises the valuations. With STEM, the final value depends on a parameter resulting from multiplying the difference in age between the tree and its substitute by the annual maintenance cost. For example, in the sequoia specimen (Figure 1), the highest value corresponds to the STEM method (€2,424,146) and the next highest to the FEM (€175,116), and differs significantly from the lowest value (€5832, CTLA). The difference between the rest of the methods becomes more pronounced as the age of the tree increases.

The Burnley, Contato, Capitalisation, Copima, NG, and CTLA methods have lower values. This is because Burnley and Contato calculate the basic value from the tree height and diameter variables that include crown measurements. Burnley considers crown and trunk size; Contato uses crown area and age. These variables are the most suitable for explaining the variability of urban trees without concurring in problems of collinearity [32]. Capitalistion and Copima base their valuation on market prices such as the cost of the plant in the greenhouse, whereas NG and the CTLA calculate the basic value with the measurement of the trunk section, and thus present values of a similar order.

2.4. Conducting the Survey

The survey consisted of five pages of DIN A4 (210 × 297 mm) paper. The first page contained questions on personal details such as sex and age range (18–30, 30–45, 45–60, or over 60), and the following pages included a photograph of each specimen with an identification number, and, on the right, the valuation options in euros obtained with the different methods ordered from lowest to highest. The first survey was made in May (with the photos we have called summer), and the second survey in October, with the winter photos. Figure 1 contains an example of the three specimens showing the photographs of both seasons together.

ID: 10	Valor (€)	ID: 321	Valor (€)	ID: 358	Valor (€)
	47		2.825		5.832
	112		3.570		7.258
	119		8.757		14.845
	304		12.766		53.917
	499		15.407		83.762
	809		18.900		104.848
	84.264		79.071		175.116
	177.601		989.904		2.424.146

(a) Summer photographs

ID: 10	Valor (€)	ID: 321	Valor (€)	ID: 358	Valor (€)
	47		2.825		5.832
	112		3.570		7.258
	119		8.757		14.845
	304		12.766		53.917
	499		15.407		83.762
	809		18.900		104.848
	84.264		79.071		175.116
	177.601		989.904		2.424.146

(b) Winter photographs

Figure 1. View of the survey with three of the specimens presented for evaluation through photographs taken in summer (S) (**a**) and winter (W) (**b**).

The respondents were first-year and Master's students in Forestry Engineering at the School of Forestry and Environmental Engineering at the Madrid Polytechnic University. The opinion of a total of 78 students was collected.

2.5. Statistical Methods

The data from the completed surveys were entered in a MS Excel database (2010, Microsoft, Redmond, Washington, D.C., USA) for their subsequent statistical processing with the STATGRAPHICS Centurion XVII software (2014, Statpoint Technologies, Inc., The Plains, VA, USA).

The variables used were: proportion of students who assigned the same economic value to the same specimen in both seasons (%0), proportion of students who assigned more value to the winter photograph (%W), and proportion of students who assigned more value to the summer photograph (%S).

Two statistical analyses were performed: the *t* test for paired samples, comparing the three above mentioned variables for each specimen. The null hypothesis was that the proportions were equal. The hypothesis of normality was also verified with the Shapiro-Wilk test.

The valuations for the three variables were compared with a simple ANOVA, and box plots were obtained. The factors were species, leaf type, and group. The null hypothesis was the equality of proportions for the different factor levels (SP, Type, Group). The equality of variances between factor levels was verified with Levene's test. In the case of the SP factor, in which only one specimen was valued for some species, the robust Welch test was used with a null hypothesis that was equal to equal mean proportions. This test assumes the inequality of variances; it is adequate when the number of specimens differs widely between factor levels, and the sample sizes are small [33]. The levels with statistically significant differences were obtained using the multiple range test with limits by Fisher's LSD (Least Square Differences).

A significance level of 5% was used in all the statistical tests.

3. Results

The survey was given to 40 undergraduate and 38 Master's students, 27 of whom were women and 51 men.

Figure 2 shows the percentage difference in the selection of methods according to the season. The last selected methods are FEM and STEM. Burnley is the most selected method in both seasons and in similar percentages, although it is greater in summer. The same occurs with Copima, Contato, and NG, while the opposite occurs with CTLA and Capitalis.

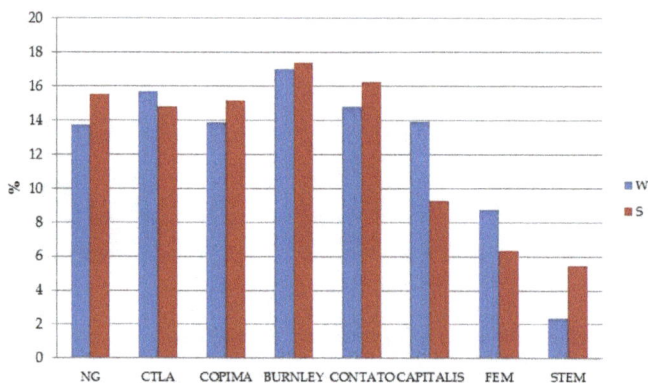

Figure 2. Percentage of methods selected according to season.

The answers were grouped as follows: number of students who left some valuation unanswered (Unanswered); number of students who chose a method with more value in the summer photos (S); number of students who chose the same value (0) in both seasons (S and W); and number of students who chose a method with more value in the winter photos (W).

The pie chart in Figure 3 shows that the number of students who assigned a greater value to the specimens in the S photos (36%) is very similar to the number of students who assigned more value to the specimens in the W photos (34%). A total of 28% of the students assigned the same value, and only 2% of cases had photos with no valuation.

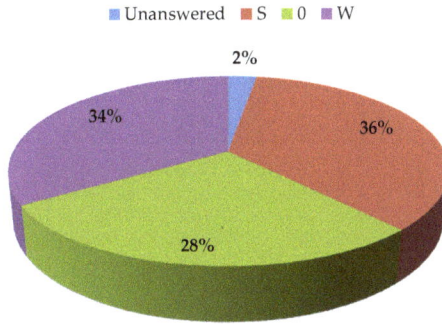

Figure 3. Percentage of students according to the differences shown in the valuation.

The hypothesis that the three variables considered (%S, %0 and %W) have a normal distribution cannot be rejected (Shapiro-Wilk test with *p*-value > 0.05). Table 2 contains a descriptive summary of these variables.

Table 2. Descriptive statistics of the variables.

Statistical	%S	%0	%W
Minimum	20.5	16.7	20.5
Maximum	50	39.7	52.6
Mean	35.6	28.1	33.8
SD	7.37	5.65	8.08
CV	20.70%	20.06%	23.93%

SD standard deviation, CV coefficient of variation.

Similar values can be seen in regard to the measures of variability (SD and CV). The mean values show the lowest value for respondents who rate the photos equally and the highest for those who rate the summer photos higher.

Student's *t* test was conducted for the paired samples of percentages calculated in the same tree as shown in Table 3.

Table 3. *t*-test results.

Null Hypothesis	Alternative Hypothesis	*p*-Value	Estimated Mean Difference
%0 = %S	%0 < %S	0.000021	7.5%
%0 = %W	%0 < %W	0.001457	5.5%
%S = %W	%S > %W	0.198641	1.8%

The results of Table 3 show:

- Significant differences between the percentage of students who rate the specimen more highly in summer (% 0); the difference in percentage is estimated at 7.5% in favour of %S.
- Significant differences between the percentage of students who rate the specimen more highly in winter (%W) and the percentage of students who rate the specimens the same in both seasons (%0); the difference in percentage is estimated at 5.6% in favour of %W.

- Finally, no significant differences are observed between %S and %W.

A simple ANOVA (Table 4) is done to analyse the possible influence of species (SP), leaf type (Type), and group factors (Group) defined in Table 1 in each proportion variable (%S, %0, and %W).

Table 4. ANOVA results (p-value).

Factor	%S	%0	%W
SP	0.0881	0.7097	0.1572
Type	0.2151	0.9849	0.2102
Group	0.0248	0.9726	0.0166

The results of Table 4 show that for the survey data:

- there is no influence of the Type factor in any of the percentage variables;
- there is no influence of the SP factor in the classes %0 and %W, or in %S for a level of 0.05;
- there is no influence of the Group factor in %S;
- there is a possible influence of the SP factor in %S at a 10% significance level;
- there is a possible influence of the Group factor in %S and %W at a 5% significance level.

The graphs in Figures 4 and 5 show the distribution of the values of the variables %S and %W according to the Group factor levels with statistically significant differences in the ANOVA test (Table 4).

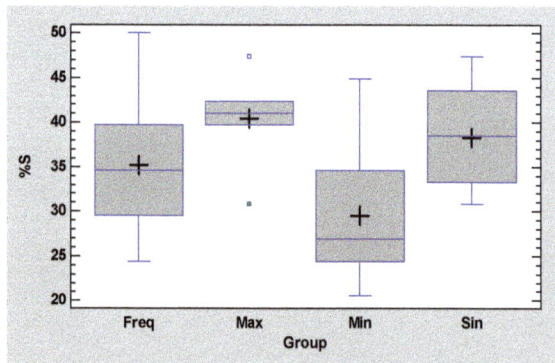

Figure 4. Box chart for %S according to Group.

Figure 4 shows that the specimens that scored highest in all the methods were also rated more highly in the S photos by more students.

Figure 5 show that the specimens with the lowest scores in all the methods (min) had a higher average %W than the rest.

Tables 5 and 6 and show the comparisons by pairs of levels for the Group factor for the variables %S and %W.

Table 5. Significant results to 95% of the multiple range test (LSD) for %S according to Group.

Comparison of Groups	Differences
Max-Min	10.9714
S-Min	8.78571

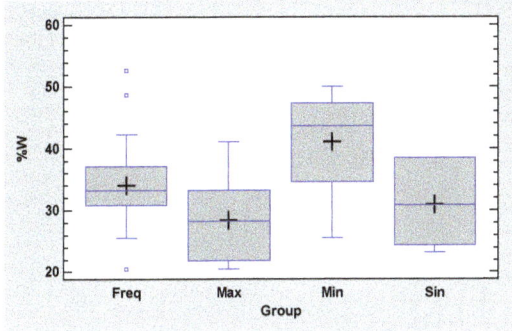

Figure 5. Box chart for %W according to Group.

Table 5 shows that the average of %S is significantly lower for specimens with lower economic valuations in all the methods than for the group of maximum valuations and the group of singular trees.

Table 6. Significant results to 95% of the multiple range test (LSD) for %W according to.

Comparison of Groups	Differences
Max-Min	−12.6286
S-Min	−10.0286
Freq-Min	−6.98247

Table 6 shows that the average of %W is significantly higher for the specimens with lower economic valuations in all the methods than for the rest of the Groups.

The equalities of variance were verified for all the ANOVA analyses with the Levene test to a confidence level of 95%. This equality was not rejected for Type and for Group, but could not be accepted for SP in any of the % variables. The box chart in Figure 6 illustrates these differences in the dispersion of the groups of %S, and shows that due to the different sample size of %S according to the SP factor levels, there is homogeneity of variances, which invalidates the ANOVA for this factor.

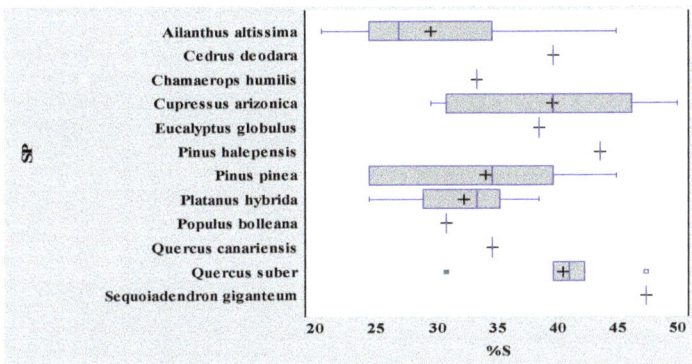

Figure 6. Box chart for %S according to species.

The equality between the medians of each variable of % between levels of the SP factor can be rejected with the Welch test (Table 7).

Table 7. Results of the Welch test for the SP factor (*p*-value).

%S	%0	%W
0.0001	0.0001	0.0001

Multiple comparisons are made between %S values by SP pairs resulting in significant differences, as shown in Table 8. The results with the same tests were done for the variable %0 (Table 9) and %W (Table 10).

Table 8. Significant results of the multiple range test (LSD) for %S according to SP.

Comparison of SP	Difference
Ailanthus altissima—Cupressus arizonica	−10.0857
Ailanthus altissima—Quercus suber	−10.9714
Ailanthus altissima—Sequoiadendron giganteum	−17.9143
Cupressus arizonica—Platanus × hybrida	7.34643
Platanus× hybrida—Quercus suber	−8.23214
Platanus × hybrida—Sequoiadendron giganteum	−15.1750

Comparison of SP shows the SP pair in which the differences in %S are valued.

Difference is the value of the difference in percentage, with negative values when the percentage variable is greater in the second species in the pair and positive otherwise.

Table 8 shows that the median of %S for *Ailanthus altissima* is lower than for the species *Cupressus arizonica* (negative difference), *Quercus suber*, and *Sequoiadendron giganteum*. The median of %S is significantly higher for *Cupressus arizonica* than for *Platanus × hybrida*. The median of %S is significantly lower for *Platanus × hybrida* than for *Quercus suber* and *Sequoiadendron giganteum*.

Table 9. Significant results to 95% of the multiple range test (LSD) for %0 according to SP.

Difference in SP	Difference
Cedrus deodara—Chamaerops humilis	−17.9
Chamaerops humilis—Cupressus arizonica	12.9
Chamaerops humilis—Sequoiadendron giganteum	17.9

Table 9 shows that the median of %0 for *Chamaerops humilis* is lower than for the species *Cedrus deodara*, *Cupressus arizonica*, and *Sequoiadendron giganteum*.

Table 10. Significant results to 95% of the multiple range test (LSD) for %W according to SP.

Difference in SP	Difference
Ailanthus altissima—Chamaerops humilis	16.6143
Ailanthus altissima—Cupressus arizonica	9.88571
Ailanthus altissima—Pinus halepensis	17.9143
Ailanthus altissima—Quercus suber	12.6286
Platanus × hybrida—Quercus suber	8.63929

The results of Table 10 show that the median of %W for *Ailanthus altissima* is higher than for the species *Chamaerops humilis*, *Cupressus arizonica*, *Pinus halepensis*, and *Quercus suber*. The mean of %W is significantly higher for *Platanus × hybrida* than for *Quercus suber*.

4. Discussion and Conclusions

The survey results reveal differences between the seasons for the Capitalisation, FEM, and STEM methods. There are barely any differences in the selection percentages for the rest of the methods.

The calculation formulas in the mixed and parametric methods in this study take location into account. The photographs of the specimens show very little of the area around the tree, so the environment does not influence the interviewees' responses. Similarly, the tree age cannot be seen in the photos, so no relation can be established between this variable and the results of the survey.

The results show that a similar proportion of students award the highest value to the specimen photographed in summer (%S = 36) and in winter (%W = 34), without any significant differences. The %0 percentage is not very different from the previous ones, but in this case, there are significant differences. The CV is similar in all three cases.

The ANOVA analysis reveals significant differences between the Groups of trees for the percentage of students who value the summer trees more highly, and for the percentage who value the winter trees more highly. However, there are no differences between Groups for the respondents who value the specimens equally.

The group of trees with the lowest scores in all the methods are the specimens of *Ailanthus altissima*. The ailanthus has the highest percentage of students who value the trees more highly in the winter photos. These are in turn the specimens with the lowest percentage of students who assess the summer trees more highly. This may be due to the fact that they are mostly young specimens that shed their leaves in winter, which may improve their valuation when comparing them with other specimens with deciduous leaves at this time of year. In any case, this supposition is still unverified.

The opposite occurs with *Quercus suber*. These trees achieve high scores with all methods. Cork oaks have the highest percentage of students who rate the summer trees more highly. These are also the specimens with the lowest percentage of students who rate the winter trees more highly.

There are no differences for the factors among the students who rated the specimens in the winter and summer photos the same. The data in this work offer no explanation for this result, which was also obtained in a previous work by the authors [34].

It is also worth highlighting the lack of any significant differences between the specimens with perennial and deciduous leaves for any of the groups of students. This result contradicts the authors' results in a previous study [32], possibly due to the student sample, which had a greater proportion of undergraduate than Master's students, and where there was an imbalance in their knowledge of trees. On this occasion, the sample is balanced between both. Differences in educational level affect the respondents' preferences, as can be concluded from other works [35,36], and can even be seen between university students with different educational levels [37].

Differences between species were detected with the Welch test for the three groups of students. In the first group (%S), a lower percentage of students value the ailanthus trees more highly in the summer photos than the cypress, cork oak, and sequoia trees. Plane trees are rated more highly in the summer photos by a lower percentage of students than cypress, cork oak, and sequoia trees. The rest of the species do not show any significant differences.

Fan palms are rated equally (%0) by a higher percentage of students than cedar, cypress, and sequoia trees. That is, fan palms have more similar scores in both seasons of the year.

In the last group of students (%W), the ailanthus has the highest percentage of students who value the trees more in the winter photos than cypress, cork oak, fan palm, and Aleppo pine. Plane trees have a higher percentage of students who value the trees more in winter photos than the cork oak.

These results do not point to any clear conclusions in regard to the differences in valuation between the seasons. The percentages of students are similarly distributed between the three options. It would therefore be advisable to increase the number of specimens per species and to make comparisons between groups that are internally more homogeneous. To verify the results between evergreen and deciduous trees, a study should be designed based solely on that factor, and with fewer photos. It is also necessary to increase the number of answers and maintain a balance between different educational levels. Clearer results can be obtained if the number of valuation methods is reduced.

Acknowledgments: The writers would like to thank Prudence Turner for the linguistic revision of the manuscript.

Author Contributions: M. Ángeles Grande-Ortíz and Esperanza Ayuga-Téllez conceived and designed the experiments; Claudia García-Ventura and Álvaro Sánchez-Medina performed the experiments; Concepción González-García and Esperanza Ayuga-Téllez analyzed the data; Claudia García-Ventura wrote the paper.

Conflicts of Interest: The authors declare no conflict of interest.

References

1. Hegedüs, A.; Gaál, M.; Bérces, R. Tree appraisal methods and their application. First results in one of Budapest's districts. *Appl. Ecol. Environ. Res.* **2011**, *9*, 411–423. [CrossRef]
2. Ponce-Donoso, M.; Vallejos-Barra, O.; Daniluk-Mosquera, G. Comparación de fórmulas chilenas e internacionales para valorar el arbolado urbano. *Bosque* **2012**, *33*, 69–81. [CrossRef]
3. Caballer, V. *Valoración de Árboles Frutales, Forestales Medioambientales y Ornamentals*; Ediciones Mundi-Prensa: Madrid, Spain, 1999; 247p, ISBN 84-7114-783-1.
4. Duinker, P.N.; Ordóñez, C.; Steenberg, J.W.; Miller, K.H.; Toni, S.A.; Nitoslawski, S.A. Trees in Canadian cities: Indispensable life form for urban sustainability. *Sustainability* **2015**, *7*, 7379–7396. [CrossRef]
5. Marin, G.; Modica, M. Socio-economic exposure to natural disasters. *Environ. Impact Assess.* **2017**, *64*, 57–66. [CrossRef]
6. Watson, G. Comparing formula methods of tree appraisal. *J. Arboric.* **2002**, *28*, 11–18.
7. Ponce-Donoso, M.; Moya, L.; Bustos-Letelier, O. Evaluation of formula for the appraisal of urban trees in municipalities of Chile. *Sci. For.* **2009**, *37*, 321–329.
8. García-Ventura, C. Comparación de métodos de valoración de arbolado urbano y su aplicación al arboreto de la ETSI de Montes. In *Trabajo Fin de Grado Ingeniero de Montes*; ETSI de Montes, Universidad Politécnica de Madrid: Madrid, Spain, 2013; 256p.
9. López-Aguillón, R.; López-García, M. Evaluación y comportamiento paisajístico de especies nativas en Linares, NL, 16 años de evaluación. *Rev. Mex. Cienc. For.* **2013**, *4*, 164–173.
10. Ponce-Donoso, M.; Vallejos-Barra, O. Valoración de árboles urbanos, comparación de fórmulas. *Rev. Fac. Cienc. Agrar.* **2016**, *48*, 195–208.
11. Ponce-Donoso, M.; Vallejos-Barra, O.; Escobedo, F.J. Appraisal of Urban Trees Using Twelve Valuation Formulas and Two Appraiser Groups. *Arboric. Urban For.* **2017**, *43*, 72–82.
12. Contato-Carol, M.L.; Ayuga-Téllez, E.; Grande-Ortiz, M.A. A comparative analysis of methods for the valuation of urban trees in Santiago del Estero, Argentina. *Span. J. Agric. Res.* **2008**, *6*, 341–352. [CrossRef]
13. Gunnarson, B.; Knez, I.; Hedblom, M.; Sang, O. Effects of biodiversity and environment-related attitude of urban green space. *Urban Ecosyst.* **2017**, *20*, 37–49. [CrossRef]
14. Koeser, A.K.; Hauer, R.J.; Miesbauer, J.W.; Peterson, W. Municipal tree risk assessment in the United States: Findings from a comprehensive survey of urban forest management. *Arboric. J.* **2016**, *38*, 218–229. [CrossRef]
15. Wang, Y.C.; Lin, J.C.; Liu, W.Y.; Lin, C.C.; Ko, S.H. Investigation of visitors' motivation, satisfaction and cognition on urban forest parks in Taiwan. *J. For. Res. Jpn.* **2016**, *21*, 261–270. [CrossRef]
16. Japelj, A.; Mavsar, R.; Hodges, D.; Kovac, M.; Juvancic, L. Latent preferences of residents regarding an urban forest recreation setting in Ljubljana, Slovenia. *For. Policy Econ.* **2016**, *71*, 71–79. [CrossRef]
17. Conway, T.M.; Yip, V. Assessing residents' reactions to urban forest disservices: A case study of a major storm event. *Landsc. Urban Plan.* **2016**, *153*, 1–10. [CrossRef]
18. Barron, S.; Sheppard, S.R.J.; Condon, P.M. Urban forest indicators for planning and designing future forests. *Forests* **2016**, *7*, 1–17. [CrossRef]
19. Shafer, E.; Richards, T. A Comparison of Viewer Reactions to Outdoor Scenes and Photographs of Those Scenes, Upper Darby, Pennsylvania. USDA, Northeastern Forest Experiment Station, Research Paper NE-302. 1974. Available online: https://www.fs.fed.us/ne/newtown_square/publications/research_papers/pdfs/scanned/OCR/ne_rp302.pdf (accessed on 9 January 2018).
20. Shuttleworth, S. The use of photographs as an environment presentation medium in landscape studies. *J. Environ. Manag.* **1980**, *11*, 61–76.

21. Kane, P. Assessing landscape attractiveness: A comparative test of two new methods. *Appl. Geogr.* **1981**, *1*, 77–96. [CrossRef]
22. Palmer, F.J.; Hoffman, R.E. Rating reliability and representation validity in scenic landscape assessment. *Landsc. Urban Plan.* **2001**, *54*, 149–161. [CrossRef]
23. Hull, R.B.; Stewart, W.P. Validity of photo-based scenic beauty judgments. *J. Environ. Psychol.* **1992**, *12*, 101–114. [CrossRef]
24. Kull, C.A. Historical landscape repeat photography as a tool for land use change research. *Norsk Geogr. Tidsskrift* **2005**, *59*, 253–268. [CrossRef]
25. Crimmins, M.A.; Crimmins, T.M. Monitoring plant phenology using digital repeat photography. *Environ. Manag.* **2008**, *41*, 949–958. [CrossRef] [PubMed]
26. Inoue, T.; Nagai, S.; Kobayashi, H.; Koizumi, H. Utilization of ground-based digital photography for the evaluation of seasonal changes in the aboveground green biomass and foliage phenology in a grassland ecosystem. *Environ. Manag.* **2015**, *25*, 1–9. [CrossRef]
27. Heyman, E. Analysing recreational values and management effects in an urban forest with the visitor-employed photography method. *Urban For. Urban Green.* **2012**, *11*, 267–277. [CrossRef]
28. Harding, S.P.; Burch, S.E.; Wemelsfelder, F. The Assessment of Landscape Expressivity: A Free Choice Profiling Approach. *PLoS ONE* **2017**, *12*, e0169507. [CrossRef] [PubMed]
29. Dempewolf, J.; Nagol, J.; Hein, S.; Thiel, C.; Zimmermann, R. Measurement of Within-Season Tree Height Growth in a Mixed Forest Stand Using UAV Imagery. *Forests* **2017**, *8*, 231. [CrossRef]
30. Grande-Ortiz, M.A.; Ayuga-Tellez, E.; Contato-Carol, M.L. Methods of Tree Appraisal: A Review of Their Features and Application Possibilities. *Arboric. Urban For.* **2012**, *38*, 130–140.
31. Moore, G.M. Amenity tree evaluation: A revised method. In *The Scientific Management of Plants in the Urban Environment: Proceedings of the Burnley Centenary Conference*; Centre for Urban Horticulture: Melbourne, Australia, 1991; pp. 166–171.
32. Sánchez-Medina, A.; Ayuga-Téllez, E.; Contato-Carol, M.L.; Grande-Ortiz, M.A.; González-García, C. Selection of dasometric variables used in appraisal methods of urban trees according to its collinearity. *Arboric. Urban For.* **2017**, *43*, 122–131.
33. Brown, M.B.; Forsythe, A.B. The small sample behavior of some statistics which test the equality of several means. *Technometrics* **1974**, *16*, 129–132. [CrossRef]
34. Sánchez-Medina, A.; García-Ventura, C.; Ayuga-Téllez, E. Comparación del valor económico del arbolado urbano mediante encuestas con fotografías en dos estaciones del año. In *VIII Congreso Ibérico de Agroingeniería: "Retos de la Nueva Agricultura Mediterránea", Orihuela-Algorfa, España, 1–3 Junio 2015*; Abadía Sánchez, R., Rocamora Osorio, C., Puerto Molina, H., Eds.; Universitas Miguel Hernández de Elche: Elche, Spain, 2015; pp. 560–569.
35. Koniak, G.; Sheffer, E.; Noy-Meir, I. Recreation as an ecosystem service in open landscapes in the Mediterranean region in Israel: Public preferences. *Isr. J. Ecol. Evol.* **2011**, *57*, 151–171. [CrossRef]
36. Chen, Z.; Xu, B.; Deveraux, B. Assessing public aesthetic preferences towards some urban landscape patterns: The case study of two different geographic groups. *Environ. Monit. Assess.* **2016**, *188*, 4. [CrossRef] [PubMed]
37. Costa, C.G.F.; Figueiredo Bezerra, R.F.; Santander Sa Freire, G.S.S. Evaluation of students' perception of urban Green areas in Fortaleza City, Ceará State, Brazil. *J. Braz. Soc. Urban For.* **2013**, *8*, 68–84.

forests

MDPI

Article

A Participatory Approach to Evaluating Strategies for Forest Carbon Mitigation in British Columbia

Guillaume Peterson St-Laurent [1,*], George Hoberg [2] and Stephen R. J. Sheppard [3]

[1] Institute for Resources, Environment, and Sustainability, University of British Columbia, 2202 Main Mall, Vancouver, BC V6T 1Z4, Canada

[2] School of Public Policy and Global Affairs, University of British Columbia, 6476 NW Marine Drive, Vancouver, BC V6T 1Z2, Canada; george.hoberg@ubc.ca

[3] Faculty of Forestry, University of British Columbia, 2900–2424 Main Mall, Vancouver, BC V6T 1Z4, Canada; stephen.sheppard@ubc.ca

* Correspondence: peterson.guil@gmail.com; Tel.: +1-778-828-7968

Received: 26 March 2018; Accepted: 19 April 2018; Published: 23 April 2018

Abstract: To be successful, actions for mitigating climate change in the forest and forest sector will not only need to be informed by the best available science, but will also require strong public and/or political acceptability. This paper presents the results of a novel analytical-deliberative engagement process that brings together stakeholders and Indigenous Peoples in participatory workshops in the interior and coastal regions of British Columbia (BC) to evaluate a set of potential forest carbon mitigation alternatives. In particular, this study examines what objectives are prioritized by stakeholders and Indigenous Peoples when discussing forest carbon mitigation in BC's forests, as well as the perceived effectiveness of, and levels of support for, six forest-based carbon mitigation strategies. We start by describing the methodological framework involving two series of workshops. We then describe the results from the first round of workshops where participants identified 11 objectives that can be classified into four categories: biophysical, economic, social, and procedural. Afterwards, we discuss the second series of workshops, which allowed participants to evaluate six climate change mitigation strategies against the objectives previously identified, and highlight geographical differences, if any, between BC's coastal and interior regions. Our results effectively illustrate the potential and efficacy of our novel methodology in informing a variety of stakeholders in different regions, and generating consistent results with a surprising degree of consensus on both key objectives and preference for mitigation alternatives. We conclude with policy recommendations on how to consider various management objectives during the design and implementation of forest carbon mitigation strategies.

Keywords: climate change mitigation; forest management; forest carbon; preferences; deliberative-analytical process; British Columbia

1. Introduction

The management of forests has great potential to reduce greenhouse gas emissions and/or increase carbon removals from the atmosphere [1,2]. The way we manage our forests, the types of wood products we produce, and how we use and ultimately dispose of those products, can significantly influence the carbon balance of our forest sector. As with many other jurisdictions, the government of British Columbia (BC), Canada, envisions an important role played by forest management in reaching their climate change mitigation targets [3,4]. BC's 55 million hectares of forests (more than any European country except Russia), 95% of which are publicly owned, have an important role to play in the global carbon cycle [5]. Nevertheless, while the province has ambitious climate policies (e.g., the first revenue neutral carbon tax in North America), a recent study indicates that the province has few effective policies targeting forest carbon management [6]. In addition, BC's forest carbon offset program, which is arguably the most significant policy dealing with climate change mitigation in the forests to date, faces important barriers and limitations [7,8]. While the Climate Leadership Plan announced by the province in 2016 promised a Forest Carbon Initiative, very little concrete information has been provided as to how this will translate into on-the-ground actions other than one general funding announcement [4,9].

To be successful, actions for mitigating climate change will not only need to be informed by the best available science, but will also require political and social acceptability, particularly since forest management entails a diversity of multifaceted, interconnected, and competing values [10]. In effect, sustainable forest management necessitates an integrated approach that balances economic (e.g., timber harvesting, forage), environmental (e.g., biodiversity, erosion, carbon sequestration), social (e.g., recreational, employment), and cultural (e.g., well-being, spiritual) uses, values, and interests that all need to be considered [10–12]. Because of this complexity, decision-making on forest management benefits from involving interested or affected actors "to ensure that all relevant information is included, that it is synthesized in a way that addresses the parties' concerns, and that those who may be affected by a risk decision are sufficiently well informed and involved to participate meaningfully in the decision" [13] (p. 30). Such a participatory approach ensures not only that all values are considered, but also that the resulting strategies benefit from public acceptability, legitimacy, and credibility.

This paper presents the results of a novel analytical-deliberative engagement process that brings together stakeholders and Indigenous Peoples in participatory workshops across the province to evaluate a set of potential forest carbon mitigation alternatives. By doing so, we answer two main research questions:

1. What main objectives do stakeholders and Indigenous Peoples prioritize when discussing carbon mitigation in BC's forest sector?
2. What are BC's stakeholders and Indigenous Peoples' preferences for, and perceived acceptability, credibility, and effectiveness of, climate change mitigation options in BC's forest sector?

Indigenous Peoples in BC and in Canada have distinct legal rights and title to the land and natural resources and do not view themselves as normal stakeholders (for more details, see [14,15]). We will therefore make a distinction between Indigenous Peoples and non-indigenous stakeholders throughout the paper. Importantly, this distinction is only used in the manuscript—Indigenous People participants were included and treated as all other participants during the engagement process.

This engagement is part of a larger research project, the Forest Carbon Management Project (hereafter "larger research project"; this project is funded by the Pacific Institute for Climate Solutions), which aims to generate recommendations for regionally specific forest carbon mitigation activities for BC's forests, while maximising the environmental, economic, and social benefits. We begin by providing an overview of forest carbon management in BC, followed by a presentation of our methodological framework with detailed summaries of analytical-deliberative processes in general, and our analytical-deliberative province-wide engagement process specifically. Subsequently, we discuss the results of the engagement process, revealing the values that should be considered when developing and evaluating forest carbon management strategies, as well as the outcomes of the multi-criteria evaluation of six mitigation strategies, highlighting key trade-offs and geographical differences between BC's coastal and interior regions. Finally, we present individual participants' levels of support for the different strategies, including regional and sectoral discrepancies, and discuss implications for forest carbon management.

2. Forest Carbon Management in British Columbia

The forest carbon cycle comprises various pools amassing or releasing carbon, including components of forest ecosystems, such as biomass (e.g., plants and trees), soil, dead organic matter and litter, and pools from the forest product sector (e.g., wood in construction) [1,2]. A forest's net

» (GHG) emissions result from the difference in the transfer of carbon between these pools. While a forest can represent a carbon sink when it removes more carbon than it emits into the atmosphere, it can also act as a carbon source when it emits more carbon than it removes. Taking into account all carbon pools, BC's forests acted as a carbon sink between 1990 and 2002, but became a net carbon source in 2003 and have emitted more than they sequestered ever since [5]. This shift from sink to source is mainly due to an increase in wildfires and the large number of trees killed by the mountain pine beetle outbreak [16,17].

Incremental activities that reduce GHG emissions from forests or increase removals of carbon from the atmosphere compared to business-as-usual activities are considered forest carbon mitigation actions. The evaluation of an action's net effect on GHG emissions or removals requires a systems perspective that accounts for its impact on: (1) forest ecosystem emissions and removals; (2) storage of carbon in wood products; and (3) substitution effects, which are the impact of substituting wood products for other products (e.g., cement, steel) and fossil fuels (e.g., natural gas, coal) that are more emissions-intensive on a life cycle basis [18–21]. From a climate perspective, there are therefore important alignments and trade-offs between increasing carbon storage in forest ecosystems and seeking to obtain mitigation benefits through the use of wood.

For the purpose of simplicity, possible mitigation strategies in BC's forests can be classified into three main categories (inspired by [19,20]). The first category refers to activities targeting the preservation of existing forest areas through reduced deforestation or the increased expansion of forest areas through afforestation. The second category involves maintaining or increasing forest carbon density (i.e., the quantity of carbon sequestered in a given area of forest). To do so, various sustainable forest management strategies can be implemented, including conservation approaches, silvicultural activities that increase growth rates and carbon uptake of trees, harvesting practices that reduce carbon losses (e.g., avoided slash burning), and efforts to reduce the impacts of natural disturbances. The third set of strategies relates to the use of wood products to generate off-site carbon stocks and replace energy-intensive products and fossil fuels.

3. Methods

3.1. Analytic-Deliberative Processes

Public and stakeholder engagement and participatory processes in environmental and natural resource management have become increasingly prominent over the past several decades [22,23]. Accordingly, a variety of tools and methods have been developed to incorporate the perspectives of various groups in the decision-making process [24]. Of particular interest here are methods combining quantitative, scientific, and analytical methods with deliberative and participatory components, known as analytic-deliberative processes [13,25]. By "reconcil[ing] 'technocratic' and 'citizen-centric' approaches" [26] (p. 300), analytic-deliberative methods offer attractive alternatives to traditional participatory processes. The analytic component implies the use of precise, reproducible, reliable, and agreed upon procedures, enabling the structuring and resolution of factual problems. In contrast, deliberation involves a process of discussion, debate, and rational argumentation where participants share their positions, try to convince each other, and improve their understanding. Within analytic-deliberative, the two components complement each other: the analysis provides information, knowledge, and structure to deliberation, which in turn informs what type of analysis is needed.

One such analytic-deliberative process, multi-criteria decision analysis (MCDA), represents an appealing approach that facilitates and structures the decision-making process. Primarily conceived as technical approaches to be used by a single decision maker [24], MCDA methods have evolved to integrate stakeholder inputs and concerns in the field of forest management [11,27,28]. Insights for developing structuring processes can also be acquired from Structured Decision Making (SDM) [29–31]. One of the main differences between SDM and conventional participatory MCDA is the emphasis put by the former on the problem structuring process, whereas the latter often focuses on finding a prescriptive (and often quantitatively supported) solution. By focusing on the structure rather than the outcome, SDM proposes a transparent, inclusive, and methodical approach that prioritizes participants' identification and in-depth understanding (and sometimes transformation) of their values and how they relate to the problem at stake. While both methods have already been used in Canada and globally to tackle diverse forest management problems [11,27,28], they are not currently in wide usage for decision-making on complex trade-offs for forest carbon solutions.

In general, participatory MCDAs and SDM share at least five steps, although their names and order of occurrence vary:

1. Clarification of the decision context: The main objective is to identify the problem or question being addressed and who should be involved in the process.
2. Structuring of the decision problem: This step involves the identification of evaluation criteria (also termed values, objectives, or points of view) with which the alternatives will be assessed.
3. Identification and outline of alternatives: The objective of this phase is to produce and describe a set of alternatives that can be evaluated and compared by the participants.
4. Elicitation of preferences: The goal of this step is to elicit participants' preferences, that is the evaluation of the importance of the criteria and alternatives.
5. Trade-offs among alternatives: In this final step, the preferences or weightings elicited both for criteria and alternatives are pooled together to rank the alternatives and highlight trade-offs.

3.2. Data Collection

The engagement process presented in this study is separated into three phases based on the five steps of MCDA and SDM (Figure 1), including two cycles of workshops. All the workshops were led and moderated by the first and second authors of this paper.

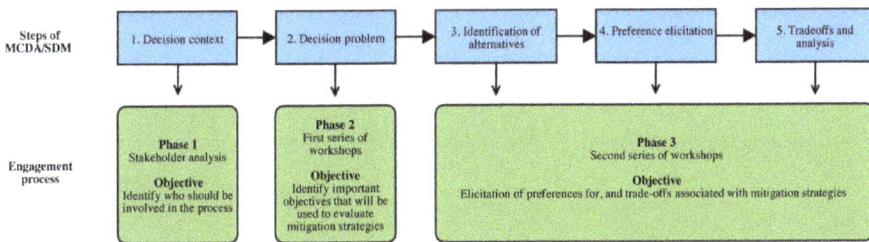

Figure 1. Overview of three phases of engagement process and how they integrate the five steps of multi-criteria decision analysis/Structured Decision Making.

In order to identify regional divergence, the engagement process focuses on four locations in two main forestry regions of the province: the coast (Vancouver and Nanaimo, BC, Canada) and the interior (Kamloops and Prince George, BC, Canada, Figure 2). While variations exist within these two regions, the coastal region is principally composed of older mild temperate rainforests characterized by infrequent disturbance patterns and a high carbon density; the drier and cooler interior region is typically the home of younger forests that face a higher frequency of fires and pest outbreaks that often prevent landscapes from reaching and maintaining the maximum carbon storage characterizing old-growth forests [32,33]. For the purpose of this paper, we will compare the combined results of the coastal (Vancouver and Nanaimo) and interior (Kamloops and Prince George) forestry regions.

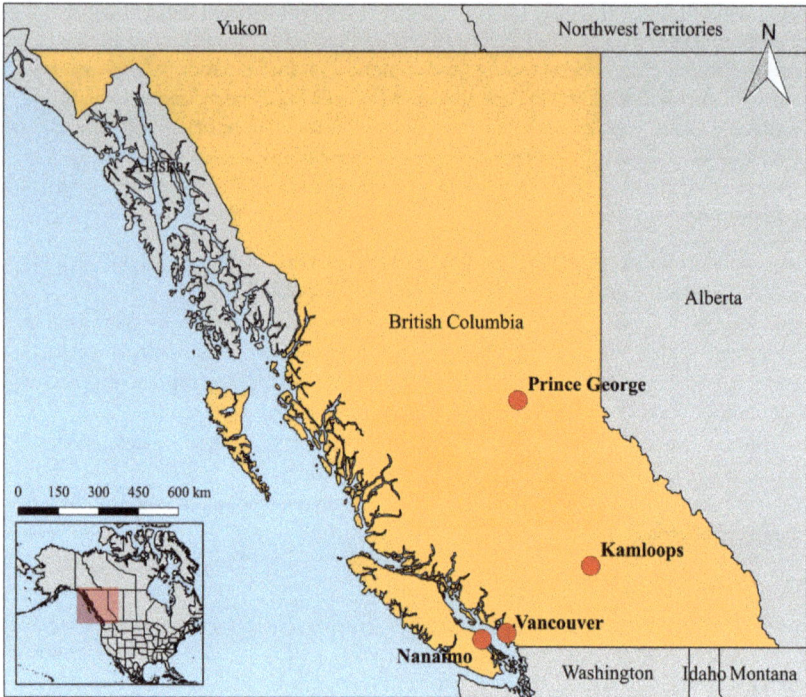

Figure 2. Map showing the locations where the engagement process was carried out.

3.2.1. Stakeholder Analysis

Carrying out a stakeholder analysis [34,35] is a key step in ensuring that all concerned parties are considered during the consultation process [28]. We first created a list of categories of potential forest users and/or individual/groups that may be interested in or affected by mitigation activities in BC's forests: (1) forest industry; (2) other forestry professionals (community forests, consultants); (3) regional/local government; (4) Indigenous People; (5) non-government organizations (NGOs); (6) academia; (7) carbon offset companies; and (8) bioenergy companies. We then identified and classified potential organizations and individual participants in each category for the four areas previously identified areas. To do so, we used a variety of sources, including demographic information, previous government and other agency consultation processes, forest companies, and NGOs. We selected additional participants with the help of iterative identification—when previously unknown individuals contacted for the purpose of this study point out other potential participants [36]. While efforts have been made to ensure representativity (i.e., we made sure that we had at least five invitees for each groups of actors in each region), we did not exclude any potential participants, meaning that invitations were unevenly distributed between categories of actors (i.e., there are more NGOs and forest professionals than carbon offset and bioenergy representatives). Furthermore, final participation strongly depended on participants' availability. Consequently, it is important to acknowledge that unequal sectoral representation may have affected the dynamics of the workshops and the results.

3.2.2. First Series of Participatory Workshops: Objective Identification

The second phase of the engagement process involved a first round of participatory workshops with the goal of identifying a list of important objectives to be considered when generating and evaluating climate change mitigation strategies for BC's forests. In this first round, we held five 3.5-h workshops between February and March 2016 in the two pre-identified regions: coast (two workshops in Vancouver, one in Nanaimo (because of the larger number of interested participants, two workshops were held in Vancouver) and interior (one workshop each in Kamloops and Prince George). A total of 76 participants from different groups of actors with interest in, and knowledge of, BC's forests participated in these workshops (Figure 3). While the different groups of actors were not equally represented in the workshops, the format of the workshops (i.e., deliberation rather than voting model) ensured that all groups were able to state their views.

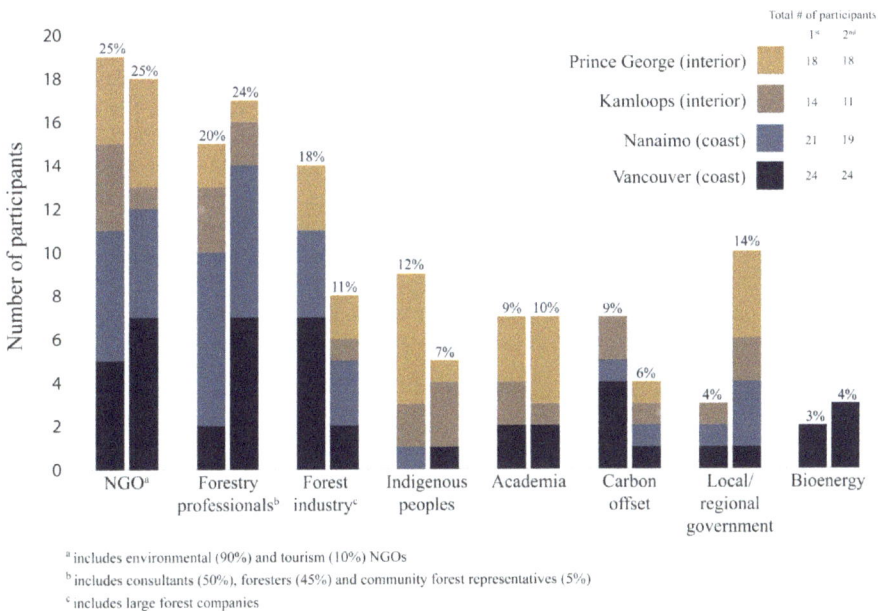

ª includes environmental (90%) and tourism (10%) NGOs
ᵇ includes consultants (50%), foresters (45%) and community forest representatives (5%)
ᶜ includes large forest companies

Figure 3. Representation (number of participants) by region of different groups of actors during the two series of participatory workshops: 1st round February–March 2016, 2nd round May–June 2017. Percent of total participants for each group and total number of participants at each series of workshops is also shown.

Since workshop participants had very different levels of knowledge of forest carbon management (i.e., some were experts working in the field while others had no previous experience with the subject), we started the workshops with a short presentation providing an overview of the different climate change mitigation options available in BC's forests to ensure that all participants had a basic understanding. An illustrated primer on forest carbon mitigation in BC, created for the purpose of this engagement process, was distributed to the participants one week prior to the workshop (supplementary material titled "Primer first series of workshops"). The primer provides a general overview of the carbon cycle, the role of BC's forests in mitigating climate change, and potential mitigation strategies. This primer does not specifically introduce the mitigation strategies evaluated in the second series of workshops.

We then asked participants to identify through deliberation the main objectives that should be considered when evaluating climate change mitigation strategies in BC's forests. Building on the methodology of SDM, objectives were defined as "what really matters" when evaluating forest carbon mitigation strategies [29]. Objectives are usually constructed with a verb that indicates the desired direction (e.g., increase, minimize, maximize) and should focus on "ends", that is the results that we want to achieve, rather than on "means", the actions to be undertaken. Participants were clustered into small groups of three to five individuals, making sure that each contained a diversity of groups of actors. We encouraged each group to reach agreement on a set of objectives and sub-objectives through discussions. Group results were then presented and discussed in plenary. An agreement was reached when all workshop participants agreed on a final list of objectives.

After completion of the workshops, we created an aggregated list of objectives. We first coded the objectives based on their workshop of origin. In an attempt to create a consensus list of objectives without outliers (i.e., objectives that only convey the opinion of a small proportion of participants), we eliminated any objectives that were not identified in more than one workshop. We then combined together the objectives that had the same or very similar meaning, making sure to keep, as much as possible, the wording used by the participants. During this process, some objectives were transformed into sub-objectives, defined as "clarification of what is meant by the objective"; sub-objectives are mostly useful for providing more details about what is meant by a more general objective [29].

Following the workshops, we invited participants to complete an online follow-up survey distributed using Fluidsurvey. A total of 52 out of 76 participants responded to the survey (68% response rate). Participants were asked to describe their overall level of agreement with the aggregated list of objectives using a five-point continuous interval scale. A majority of respondents (84%) either "agreed" or "strongly agreed" with the list of objectives identified during the workshops. Because of this level of agreement, no subsequent modification was made to the list of objectives. We also asked participants to weight the relative importance of the objectives by answering the survey question using a nine-point continuous interval scale, from 'not important at all (0)' to 'very important (8)'.

3.2.3. Second Series of Workshops: Evaluation of Mitigation Strategies

The third phase of the engagement process aimed to allow stakeholders and Indigenous Peoples to evaluate six climate change mitigation strategies (see Table 1 for description) against the objectives previously identified. Five of the strategies affect the management of forest ecosystems, whilst the sixth deals with the use of wood and its allocation to short- and long-lived product types. It is important to note that the strategies being examined in this study are not the full suite of mitigation alternatives, and some options, such as those related to preserving existing forest areas (i.e., reduced deforestation) or creating new forest areas (i.e., afforestation), are not included.

Table 1. Climate change mitigation strategies evaluated by stakeholders and Indigenous Peoples during the participatory workshops of phase 3.

Strategy	Description
Bioenergy	A recovery of a portion of harvest residues for local bioenergy production to replace fossil fuels.
	A reduction of on-site burning of harvest residues (pile-burning of slash).
Higher Utilization	Higher utilization of wood from harvest cut blocks so that more wood is extracted per hectare, thereby lowering the area harvested while keeping the harvest volume unchanged.
	Increased proportion of salvage harvesting, referring to the harvesting of trees in forests affected by natural disturbances such as fire and insects, to replace green tree harvesting.
Longer-lived wood products (LLWP)	The production of a commodity mix shifted towards a greater proportion of longer-lived products (sawnwood, other solid wood, and panels), at the expense of pulp and paper products. Both the baseline harvest volume and the proportion exported for each product are assumed to remain unchanged.
Reduced harvest	A reduction in harvest with a corresponding decrease in production of wood products.
Rehabilitation *	The reforestation of underproductive sites where no trees would otherwise be planted.
Old growth conservation	Prevent the harvesting of old growth forests, defined as stands older than 250 years old.

* Not included in modelling [37].

Four one-day participatory workshops (7.5 h each), held during May and June 2017, revisited the same four locations as the first series of workshops (Figure 2). Participation in the first series of workshops was not a prerequisite to be invited. In total, 72 participants from the eight different groups of actors took part in the workshops (Figure 3), out of which 44 (61%) also participated in the previous series of workshops.

The first hour of the workshops was devoted to presenting the six mitigation strategies (Table 1) that would be evaluated and the set of objectives that would be used to evaluate them. The presentation mainly revisited the information found in a consultation document (supplementary material titled "Primer second series of workshops") that had been provided to the participants in preparation for the event and describes in detail the whole process, the objectives, and the strategies [38]. Five of the evaluated strategies originate from a parallel study [37], independent of the engagement process, that was carried out by a team of multidisciplinary experts from the overarching project. At least one member of this team attended each of the workshops as technical experts. The parallel process involved the development of strategies and the modelling of their climate change mitigation potential, financial, and socioeconomic impacts for the years 2017–2050 (more details on the strategies and the modelling exercise can be found in [37]; Table S1 identifies the outputs of the biophysical and socio-economic modelling that were provided to workshop participants). Late in the process, a sixth strategy, rehabilitation, was added because it had subsequently been proposed by the government of BC as a potential mitigation strategy in the Climate Leadership Plan [4] and was included in the BC government's Forest Carbon Initiative announced in February 2017. Time constraints did not allow for modelling of this rehabilitation strategy to be carried out. However, since it may represent an important mitigation strategy for the BC government, we decided to introduce it in the engagement process, acknowledging that the lack of modelling results represents a limitation.

We asked participants to evaluate in small groups (three to six people) each of the six strategies against the objectives. Each group purposefully contained a diversity of actors and was assigned a moderator who took detailed notes of the main issues and trade-offs discussed. The groups in Vancouver and Nanaimo evaluated the strategies for the coastal region, whereas participants in Kamloops and Prince George evaluated the strategies for the interior region. To keep the process straightforward for participants, a simplified value measurement technique inspired by the Simple Multi Attribute Rating Technique (SMART) was used [39–42]. The small mixed groups were asked to evaluate through consensus rating the performance of each strategy against each objective with the use of an 11-point continuous scale (from 'very good' to 'very poor'; modified from [43]).

To conclude the workshops, results of the evaluation exercise were presented back to, and discussed with, the participants in plenary. Participants were also allowed to raise questions and/or comments about the engagement process, the modelling methodology, and results, and to converse about their experience during the workshop (e.g., what they learned/observed, what they liked/disliked). Before leaving, participants filled in a short survey where they were asked to provide their individual levels of support for the strategies using a five-point continuous interval scale.

3.3. Data Analysis

Both descriptive and inferential statistics were performed in R Studio (version 1.0.153, Integrated Development for R. RStudio Inc., Boston, MA, USA, http://rstudio.org/). Because of the ordinal nature of continuous interval scales, we used nonparametric tests (Mann–Whitney U test and Kruskal–Wallis one-way analysis) to compare the perceived importance of the objectives by regions (first series of workshops), and individual levels of support for the various mitigation strategies by regions and groups of actors (second series of workshops). This was to permit analysis of differences between the original stakeholder groups represented by participants.

For analysis purposes, the 11-point continuous linguistic scale used for the group evaluation of the mitigation strategies was transformed into numeric scores. A cumulative derived preference for each strategy was calculated by summing, for each objective, the product of the aggregated individual weighted importance of an objective (score out of eight; first series of workshops) and the mean score of the strategy when evaluated against that objective (second series of workshops). The resulting score was the transformed into a percentage by dividing it with the maximum possible score of a strategy.

4. Results and Discussion

4.1. Final Aggregated List of Objectives and Their Relative Importance

Over 30 objectives were identified by participants during the five workshops, and an aggregated list of 11 objectives and their sub-objectives was later generated by the research team based on the results of the workshops (Figure 4). The objectives can be classified into four categories based on their primary focus: biophysical, economic, social, and procedural. For the purpose of this paper, we kept the objectives and their categorization as close as possible to the participants' definitions and classification during the workshops, even though some sub-objectives could better fit into another category (e.g., increasing sustainable economic opportunities and local government revenues would be better classified as economic instead of social sub-objectives).

The objectives identified by our workshop participants are generally comprehensive and similar to general studies on forest values (for examples see [12,44,45]). In particular, the economic, biophysical, and social categories of objectives consistently appear in most studies focused on general forest values, although the terminology used varies (e.g., ecological, environmental, cultural, socio-cultural). Interestingly, procedural objectives are less frequently highlighted by these studies, even though the category appears in many well-known lists of criteria and indicators, often under a category referring to policy, legal, and/or institutional issues [46–48]. Other studies, by analysing public participation in forest management, specifically focus on procedural objectives [49].

BIOPHYSICAL

INCREASE CLIMATE CHANGE ADAPTATION AND FOREST RESILIENCE.
Increase the natural capacity of BC's forests to:
· adapt to climate change
· respond to climate change perturbations by resisting damage and recovering quickly

MAINTAIN ECOSYSTEM SERVICES.
· maintain water quality and quantity
· maintain air quality
· maintain soil quality
· maintain recreational, cultural and spiritual opportunities

MAINTAIN EXISTING BIODIVERSITY.
· ensure biodiversity conservation
· ensure the protection of natural old growth forests

MAXIMIZE THE CLIMATE CHANGE MITIGATION POTENTIAL.
· maximize carbon sequestration from BC forests and forest sector
· minimize greenhouse gases emissions from BC forests and forest sector

PROCEDURAL

ENSURE EVIDENCE-BASED DECISION MAKING
Ensure that future forest carbon mitigation strategies make use of:
· the best available science
· First Nations traditional knowledge.

RECOGNIZE INDIGENOUS PEOPLES RIGHTS AND CLAIMS TO FOREST LANDS.
· recognize Indigenous Peoples existing titles and claims
· respect Indigenous Peoples rights
· ensure inclusion of Indigenous Peoples in decision-making

SOCIAL

ENSURE SOCIAL LICENSE AND POLITICAL FEASIBILITY.
Ensure that future forest carbon mitigation strategies make use of:
· ensuring participation and public sense of ownership
· promoting public awareness
· maximizing administrative flexibility, adaptability and feasibility

INCREASE RESILIENCE OF LOCAL COMMUNITIES.
· increase sustainable economic opportunities
· increase local government revenues
· increase local participation in decision-making

ECONOMIC

INCREASE ECONOMIC OPPORTUNITIES FOR INDIGENOUS PEOPLES.
· increase generated revenues
· increase employment
· increase professional development

INCREASE PROVINCIAL NET ECONOMIC BENEFITS
· increase industry competitiveness
· maximize efficiency of resource use
· increase the production of value-added products (e.g., long-lived wood products)

INCREASE PROVINCIAL SOCIOECONOMIC BENEFITS.
· maximize direct employment from the forests and forest sector
· maximize indirect employment from the forests and forest sector

Figure 4. List of the 11 key objectives and sub-objectives for evaluating BC's forests carbon mitigation strategies by categories. The sub-objectives reflect what was discussed in the workshops and may only partially describe the scope of the objectives, meaning that specific issue (e.g., aesthetic values) may not be represented in the list.

The perceived importance of the 11 objectives based on a nine-point continuous interval scale aggregated across all participants is illustrated in Figure 5. All of these objectives are rated on the higher end of the importance scale, though the biophysical objective tended to be the highest rated and the economic objective the lowest rated. These results are in line with previous studies evaluating public priorities for forest management in BC, which generally find a preference for environmental values over social and economic values [12,45]. Our study also highlights the high importance of procedural objectives, indicating that stakeholders and Indigenous Peoples are as concerned about the fairness, effectiveness, and inclusiveness of the decision-making process as they are about its outcomes [49,50].

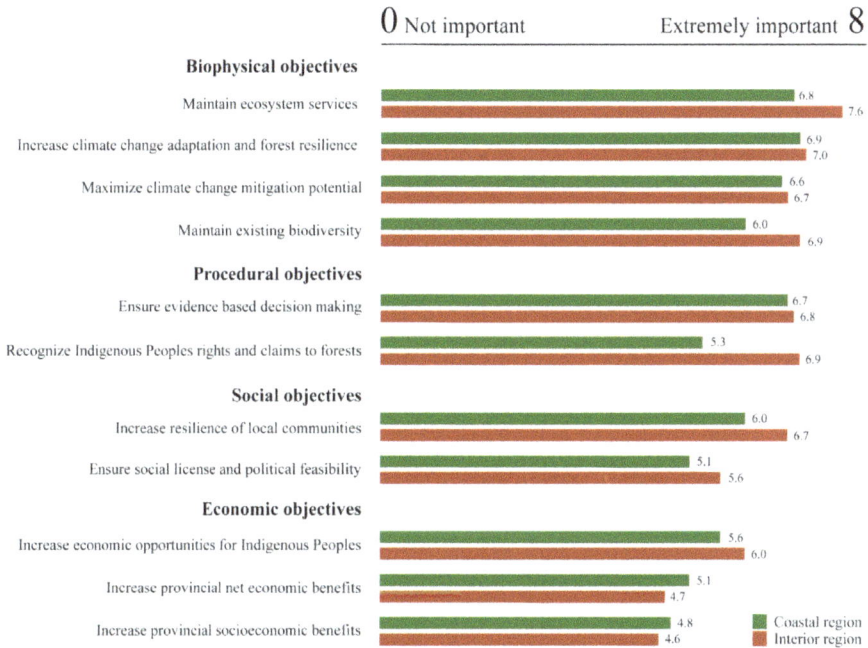

Figure 5. Weight of the objectives for the coast and the interior regions, where 0 = not at all important and 8 = very important.

Perhaps the most important result lies in the finding that mitigating climate change, while perceived as important, is not considered as the highest priority, but as just one of many objectives that should be considered when developing forest carbon management strategies. Three other objectives rank higher in terms of relative importance in the Coastal region, and five other objectives in the Interior. This result is consistent with discussions worldwide about the important of considering co-benefits (e.g., poverty reduction, biodiversity conservation, and increase in soil and water quality) when using forests to mitigate climate change [51,52].

In general, there is little divergence in the weighting of the objectives between the two regions, with the only statistically significant difference found for "recognize Indigenous Peoples rights and claims to forest land" (Mann–Whitney $U = 201.5$, $p = 0.02$), where the average score in the interior (7.92) was significantly higher than on the coast (6.33). This result can be partly explained by the fact that Indigenous People's attendance was strongly weighted toward the interior (all indigenous participants but one). One could also expect that an overall greater representation of indigenous participants, which was relatively low at 12%, would have increased the perceived importance of the two objectives associated with indigenous peoples.

4.2. Evaluation of the Six Mitigation Strategies

4.2.1. Group Evaluation: Performance of the Strategies against the Objectives

This section presents results from the evaluation of the strategies against the objectives carried out during the second series of participatory workshops. As noted above, at each workshop, groups of participants with a diversity of actors were asked to develop consensus evaluations. In this section, the group evaluations are discussed, along with views of participants more generally. Importantly, participants in all workshops agreed that the procedural objectives should be excluded from the evaluation because they are mainly on the process and are not relevant to the assessment of the strategies. Figure 6 highlights the performance of the strategies against each of the nine other objectives (Table 1) for the coastal and interior regions, whereas Figure 7 shows the cumulative derived preference for each strategy. The next sections describe in detail the performances of the strategies against the objectives, the regional differences, if any, and the main points of discussion during group deliberation.

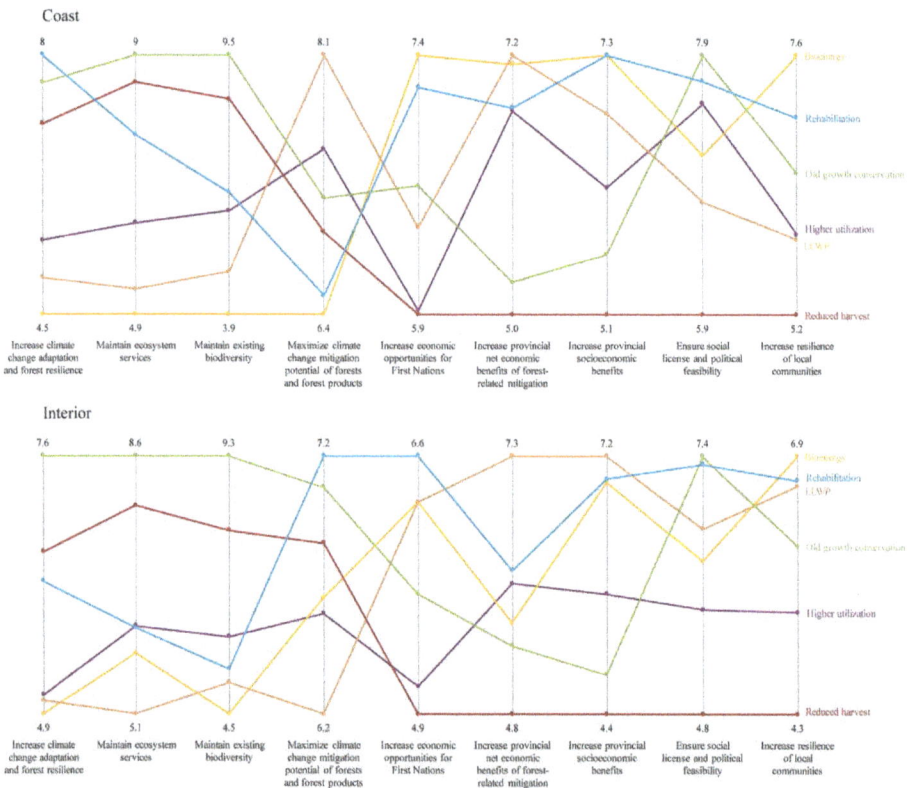

Figure 6. Average score of the six strategies on performance against each of the nine objectives for the coast (Nanaimo and Vancouver) and the interior (Kamloops and Prince George). The values are normalized (from 0 to 1) from the average score (from 0 = very poor to 10 = very good), with each vertical axis representing the scores of the strategies against one objective. The minimum and maximum average scores (out of 10) across scenarios are also shown in numbers at the top and bottom of the graph.

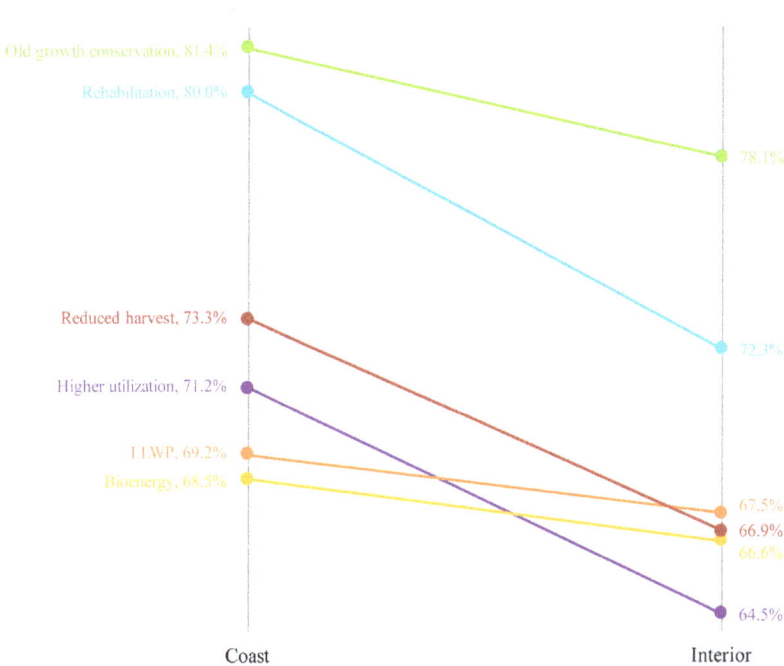

Figure 7. Average cumulative derived preference for the six strategies on the coast and the interior. The derived preference of each strategy was calculated by summing, for each objective, the product of the average performance of a strategy against an objective (Figures 6 and 7) with the weight of this objective (Figure 5). The resulting score was transformed into a percentage by dividing it with the maximum possible score of a strategy.

Bioenergy

The evaluation of the bioenergy strategy was mainly consistent in both regions. The strategy scored relatively low against the biophysical objectives, mainly because of concerns that removing an excessive amount of coarse woody debris, which holds a key role in providing nutrients and habitats to plants, animals, and insects, could lead to soil disturbance and negatively affect ecosystem services and ecological habitats (for instance, see [32]). Consequently, many participants agreed that sustainable harvesting of residues should be ensured by identifying the ecologically valuable proportion of the biomass that should be left on the forest floor (discussed in [53]). Participants did not appear to associate bioenergy use from forest thinning for fire risk reduction as a climate change adaptation strategy, or a potential win-win strategy (see [54]).

Participants generally recognized the mitigation potential of limiting wood waste at harvest and using a portion of harvest residues for energy to replace local fossil fuel use, especially since the carbon would otherwise be emitted to the atmosphere, either progressively through decay or immediately through slash-and-burn, as discussed by Miner, et al. [55] and Smyth, et al. [56]. Participants insisted that bioenergy should only be used to replace emission-intensive fossil fuels such as coal and diesel, but not clean energy (for example, see [57]). Many groups also discussed the fact that the strategy would be most effective in the context of off-the-grid remote communities that are using emission-intensive energy sources like diesel (discussed in [58]). Although it was not analysed or presented to participants, many wished to stress that standing live trees should not be harvested to produce bioenergy. Standing live trees are typically not harvested for bioenergy in BC since they typically provide greater mitigation potential when used in long-lived wood products rather than to produce energy, which offers no or very little emission reductions (discussed in [5,59]). Some participants believed that the GHG emissions generated by the transportation of harvest residues to the processing plant should be considered to ensure the climate effectiveness of the activity.

The strategy scored relatively high against economic and social objectives. Many groups anticipated that this strategy could lead to the creation of a new industry, involving the construction of a "bioenergy grid" and numerous processing plants with the potential to stimulate economic and local development and generate employment in many regions. Some participants also highlighted technical issues associated with transportation (i.e., extensive distance between residues and processing plants) and the high upfront cost of building new processing plants, and questioned the cost-effectiveness of the strategy compared to other options. Finally, while many participants believed that the strategy would receive public and political support, some feared that bioenergy is still perceived as a "dirty industry" because of air pollution.

Higher Utilization

Most groups were supportive of the strategy, mentioning that there is a great need to improve harvesting practice to reduce waste that could otherwise be utilized. However, the strategy's low cumulative derived preference (Figure 7) can be partly attributed to the fact that many participants expressed doubts regarding its technical feasibility. While they agreed that gains at harvest can be made in some areas (i.e., it is possible to extract and make greater use of harvested material), reservations were expressed regarding the assumption that the same quality of wood products could be produced with the extra wood extracted. This feeling was stronger in the interior, where many participants judged that forest companies already maximize the amount of wood they can cost-effectively extract from harvested areas.

Similar to the bioenergy strategy, some participants explained that removing an excessive amount of wood debris could negatively affect climate change adaptation and forest resilience, ecosystem services, and biodiversity. Nonetheless, some participants advocated the strategy for its potential to lower the area harvested each year (because more wood is extracted per hectares, a given harvest volume can be achieved from less area). Higher utilization scored relatively high in terms of its climate change mitigation potential, with many participants highlighting the benefits of reducing both the amount of harvesting waste left to decay or to be burned on site and the area harvested while maintaining carbon transfer into wood products (discussed in [5,56]).

Many groups claimed that "reducing waste" in the forest sector would receive significant social licence due to the negative perception associated with big slash piles. The strategy's relatively low performance against the economic and social objectives is partly due to the technical reservations previously discussed and the high perceived cost associated with implementing the strategy.

Longer-Lived Wood Products (LLWP)

Participants from both regions highlighted the fact that producing a higher proportion of longer-lived wood products to the detriment of pulp and paper products did not really affect forest management on the ground and therefore had no real impact on the first three biophysical objectives (most groups gave a neutral score).

Participants from both regions agreed that there should be a positive climate change mitigation potential. While the strategy scored higher than all other strategies in relation to mitigation potential on the coast, it was perceived as a relatively less effective strategy on average in the interior. This discrepancy can be explained by the concerns brought up by various participants in regards to what percent of the carbon removed through harvesting was considered to remain stored and the extent to which there would be substitution in wood products. In effect, without contesting the benefits associated with the production of wood products, some participants highlighted the uncertainties associated with factors used to calculate carbon storage and displacement effects (for instance, see [53,56,60]), expressing a concern that the current numbers may overestimate the benefit of using harvested wood products. Others also discussed the issues associated with leakage (i.e., emissions avoided in one location simply occur in another location, see [61]), pointing out that the paper products that are not produced in BC could simply be produced somewhere else if consumption does not diminish.

In general, the strategy performed well against economic and social objectives. Various participants hoped that such a strategy could stimulate innovation and the development of new technologies maximizing the use of lower quality biomass (e.g., harvest residues). However, other participants, mostly on the coast, mentioned that such a strategy would be challenging to implement due to the current lack of technical capacity to allow for the transformation of low quality timber into longer-lived wood products. Others also feared that transitional issues in terms of political feasibility and economic impacts would arise in the short term because of a reduction in the production of pulp and paper, which could lead, for instance, to mill closure and the loss of employment in certain regions.

Reduced Harvest

Reduced harvest, which scored similarly in both regions, was deemed effective at increasing climate change adaptation and forest resilience and maintaining ecosystem services and existing biodiversity because it allowed for the conservation of more forest areas. However, it was pointed out that this strategy could also be detrimental in the long term since it would reduce the possibility of implementing reforestation strategies to adapt to climate change in BC's forests, including the use of genomic-assisted seed selection (i.e., selecting seeds from trees that are genetically adapted to anticipated future climatic conditions; see [62]) and assisted migration (i.e., planting species in areas that are anticipated to be climatically suitable in the future; see [63]).

While reducing harvest scored relatively high in terms of climate change mitigation, mainly because of an expected increase in carbon density at the landscape level, some groups argued that the strategy could significantly reduce the carbon storage and substitution benefits associated with wood products, and that leakage could simply shift harvest to another location. Certain groups also discussed the risk of reversal, that is when the mitigation benefits of a strategy are compromised by natural disturbances such as wildfire or insect outbreaks (see [64]). Such natural disturbances are expected to increase both in intensity and frequency in BC and in Canada [17,65,66], indicating the risk that conserved forests may become carbon sources in future years. This issue was particularly emphasized in the interior region where frequent natural disturbances make it difficult to establish forest carbon reserves through conservation strategies. As Stinson and Freedman [33] (p. 12) explain, the obstacles associated with frequent natural disturbances are common to most of Canada's forests, where "it may not be feasible to establish protected forest-carbon reserves for the purposes of sequestering and maintaining carbon on the landscape".

The strategy scored lowest against all of the economic and social objectives in both regions. In general, reducing harvest was expected to have negative economic and employment impacts at the provincial and local levels. Nonetheless, a few groups, mainly on the coast, believed that reducing harvest could stimulate non-traditional forestry activities (e.g., tourism, use of non-timber forest products) and, in the long term, induce innovation in the forest sector.

Old Growth Conservation

Conserving old growth forests obtained the highest aggregated score in both regions (Figure 7). However, some participants questioned the definition of old growth forests used in the modelling of the strategy (i.e., forests of over 250 years old), pointing out that age class of old growth forests varies greatly between the coast and the interior.

In general, the strategy was considered the most effective at increasing forest resilience and maintaining ecosystem services and biodiversity. It was also considered extremely effective at mitigating climate change, even though issues were raised. First, as with the reduced harvest strategy, various groups discussed the risks associated with reversal. That being said, while natural disturbances are frequent in the interior and could prevent conservation initiatives, various participants highlighted the conservation opportunities found in BC's coastal temperate rainforests that are characterized by very infrequent disturbance natural patterns and a high carbon density [32,33,67].

Second, many groups spent a considerable amount of time discussing the rate of carbon uptake in old growth forests. Disagreement arose over whether or not carbon uptake by old growth forests (and therefore their potential role at mitigating climate change) was undervalued by conventional forest carbon science, which treats older forests as offering a lower carbon uptake than younger maturing forests [19,68]. Under such conditions, "a conservation strategy with no or limited harvest is expected to yield landscapes with high carbon density (but lower uptake rates), whereas a strategy that involved intensive management will yield a forest landscape with a lower carbon density but a higher carbon uptake rate" [19] (p. 306). Notwithstanding the conclusions of these discussions, most groups agreed that conserving old growth forests, particularly the temperate rainforests found on BC's coastal region, should be done in spite of uncertainties associated with its climate effectiveness. That being said, some participants, particularly from the forest industry, indicated that large scale old growth conservation could be detrimental to BC's forest sector (e.g., reduction of annual allowable cut, loss of employment) and should thus be implemented with caution.

Conserving old growth forests also scored the highest in terms of social license, with many participants discussing the historical willingness found in BC's population for old growth forest conservation [69,70]. However, trade-offs between social license and political feasibility were also highlighted, suggesting that possible resistance from the forest industry and forest-dependent communities could complicate the implementation of the strategy. The strategy scored somewhat lower against the economic and social objectives for many of the same reasons as the reduced harvest strategy (e.g., negative impact on economy and employment). Nevertheless, in contrast to the reduced harvest strategy that focuses solely on reducing harvesting, various groups believed that targeting the conservation of old growth forests could provide further economic benefits associated with tourism.

Rehabilitation

Rehabilitation received the second highest cumulative score in both regions (Figure 7). Nonetheless, many participants argued that this strategy was mostly applicable and beneficial in the interior, where various opportunities exist to plant trees in areas that have been damaged by recent wildfires or insect infestation [17] and are not successfully regrowing. In contrast, participants on the coast agreed that a rehabilitation strategy should focus mostly on ecosystem and habitat restoration activities. While various groups from the coast evaluated the strategy with such activities in mind, two groups in the coast workshops (one in Vancouver, one in Nanaimo) decided to skip the evaluation of the strategy because they judged it to be inapplicable.

Participants from both regions agreed that the strategy has the potential to increase climate change adaptation and forest resilience, provided that it includes activities focused on adaptation (e.g., use of improved seeds, assisted migration), as discussed above in the context of the reduced harvest strategy. Similarly, participants mentioned that rehabilitating forests could benefit ecosystem services and biodiversity only if it focuses on re-establishing and/or improving ecological and structural characteristics, species diversity and composition, and ecological processes. However, many groups gave a lower score to the strategy because they feared that the reforestation/rehabilitation actions would focus on monocultures of only a few commercial species.

Participants on the coast considered that the strategy had positive, although limited, potential to generate climate change mitigation. In contrast, participants from the interior assessed it as the most effective strategy for mitigation because of the climate benefits associated with increased carbon sequestration due to accelerated regrowth.

Participants all agreed that rehabilitation would receive strong public support, thereby supporting BC government's Forest Carbon Initiative, which will focus on "the reduction of carbon emissions in the forest sector and the capture of carbon through the restoration of forests damaged by disease and wildfire" [9]. The rehabilitation strategy was also positively appraised in term of its economic and social impact, mainly because participants judged that it has the inherent capacity to stimulate the economy by generating employment and helping to more quickly re-establish the province's timber supply that has been reduced by the recent mountain pine beetle epidemics [71]. However, participants also discussed the high cost associated with the strategy, pointing out the large areas of forests that were affected in recent years and would need to be rehabilitated.

4.2.2. Individual Support for Strategies: Regional and Sectoral Variations

Figure 8 shows the results from the post-workshop survey where participants identified their individual level of support for or opposition to the different mitigation strategies in their respective region (i.e., coast or interior). On average, participants supported each of the six strategies, with significant difference in the levels of support between strategies on the coast (Kruskall wallis: H(5) = 17.5, p = 0.004), but not in the interior (Kruskall wallis: H(5) = 2.8, p = 0.73). The coastal participants tended to be more supportive than the interior participants.

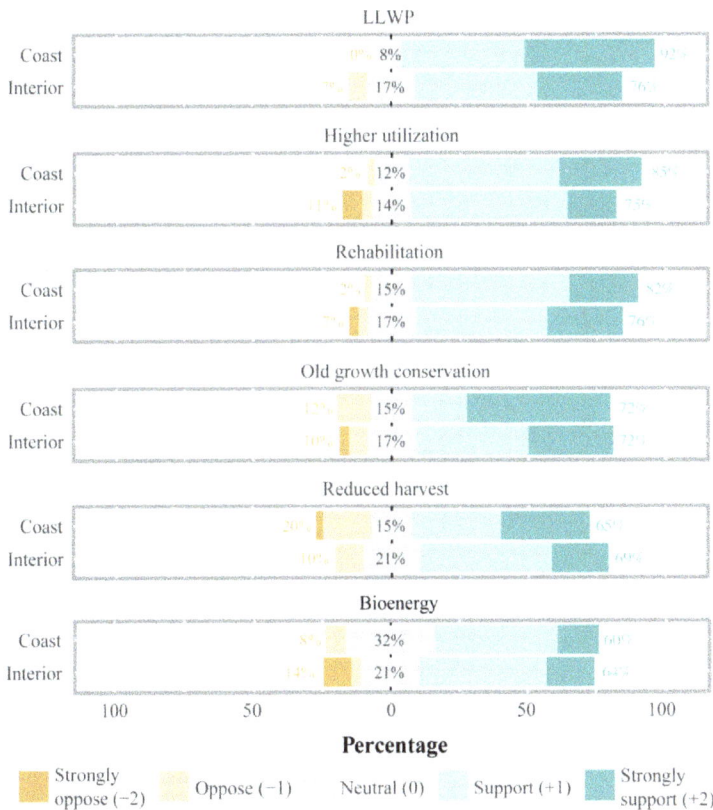

Figure 8. Degree of support for, or opposition to the strategies, as indicated in the individual post-workshop survey.

Long-lived wood products (LLWP), higher utilization, and rehabilitation are the strategies that received the greatest support in both regions, which somewhat differs from the results of the group evaluation where old growth conservation ranked the highest. Two reasons may explain this finding. First, participants may support a strategy even though they do not perceive it as the best performing, especially since participants could support more than one strategy. This is particularly important in the context of forest carbon management in BC, since only an approach that combines various strategies that are adapted to geographical and economic particularities will be able to maximize climate benefit [5,56,68]. Second, as previously explained, the LLWP strategy was judged as having no impact on most biophysical objectives, meaning that its cumulative derived preference averaged lower than other strategies.

Our results suggest that, in general, there is little divergence in the levels of support for each of the strategies between regions and between groups of actors, which is not dissimilar to what was previously found in BC [12]. In effect, no significant difference was found between regions, and only one significant difference between groups arose, for old growth conservation (Kruskall wallis H(7) = 15.4, p = 0.03), where the mean response of NGOs (1.65) was significantly higher than the forest industry (-0.143, p = 0.01). This divergence in support for the old growth strategy also highlights the enduring division between the environmental movement and the forest industry regarding the role and importance of conserving old growth forests in BC [69,70,72]. The fact that NGOs were the best represented group in the evaluation exercise (25% of participants), combined with their advocacy for increasing old growth conservation, might partly explain why the old growth conservation strategy performed so well in the evaluation exercise. A greater participation from the forest industry, which was more likely to oppose the strategy, might have shifted the overall results.

The results in Figure 8 also demonstrate relatively few cases of strong opposition and extreme polarity in response to alternative scenarios. Only with the higher utilization and bioenergy scenarios in the Interior was there a substantive minority (>10% of participants) strongly opposed, and every scenario had substantially more participants who were strongly supportive then strongly opposed. This suggests that the level of controversy or polarity in public preference may not be high with any of these options.

5. Conclusions

This study examined what objectives are prioritized by stakeholders and Indigenous Peoples when discussing forest carbon mitigation in BC's forests, as well as the perceived effectiveness of, and levels of support for, six forest-based carbon mitigation strategies through their participation in an analytical-deliberative engagement process. Our study effectively demonstrates the potential and efficacy of this non-traditional method in informing a variety of stakeholders in different regions, and generating consistent results with a surprising degree of consensus on both key objectives and preference for mitigation alternatives.

Participants identified a total of 11 objectives that can be classified into four categories: biophysical, economic, social, and procedural. On average, participants ascribed a higher level of importance to biophysical and procedural objectives than to economic and social objectives. Conservation-focused strategies (i.e., old growth conservation and reduced harvest) generally performed better against biophysical objectives and were often judged as more adapted for the coastal region (although old growth conservation opportunities also exist and were pointed out in the interior), whereas strategies focused on improving forest management (i.e., higher utilization, bioenergy, rehabilitation, LLWP) scored higher against economic and social objectives and were generally perceived as most relevant in the interior. The greater weights granted to biophysical objectives therefore partly explain why old growth conservation, and reduced harvest to a lesser extent, performed relatively well in terms of their cumulative derived preferences.

On average, all strategies received positive cumulative derived preferences (i.e., all higher than 60%) and individual levels of support, indicating that stakeholders and Indigenous Peoples are inclined to consider alternative practices in managing forests and wood products in the context of climate change mitigation. However, participants also raised various issues that require consideration. Most notably, questions were raised regarding assumptions used in the modelling (e.g., the rate of carbon intake in old forests, displacement factors), the costs associated with implementing certain strategies (e.g., rehabilitation, higher utilization and bioenergy), and technical limitations that might impede some strategies (e.g., engineering capacity to transform low grade harvested material into longer-lived wood products). Additionally, questions were raised about the potential negative impacts of certain strategies on ecosystems and forest health (e.g., bioenergy, higher utilization), issues that were not addressed in the modelling.

In addition, many participants argued that some strategies, including old growth conservation, and to a lesser extent reduced harvest, should be implemented for reasons other than their climate effectiveness (e.g., biodiversity conservation). However, it was also noted that the expected increase in natural disturbances due to climate change could limit the capacity to effectively implement these strategies, particularly in the interior region. This constraint is particularly important in a context where massive fire and insect outbreaks are typically hard to contain [65], and fire suppression can sometimes lead to unwanted consequences, leading to the view that it should be implemented with caution [32]. For instance, one cause of the massive mountain pine beetle outbreak of the 2000s is that the proportion of mature pine, which is more vulnerable to the beetles, tripled between 1910 and 2000 because of historic fire repression [73]. Consequently, participants agreed that forest management strategies should aim as much as possible to rebuild the natural resilience of forests to disturbance and increase their adaptability to climate change.

Very little variation in levels of support for the strategies was found across regions and groups of actors, with the only significant difference found between NGOs and the forest industry in terms of their support for the old growth conservation strategy. That being said, the unbalanced representation of different groups of actors in the workshops, with a marked dominance of NGOs and forestry professionals (accounting together for nearly 50% of participants), might still have influenced and somewhat biased the group evaluation of the strategies.

To conclude, the diversity of mitigation options, combined with their associated uncertainties and complexity, points to a crucial question: how are we to take advantage of alignments and make trade-offs to identify the best portfolio of mitigation strategies? According to recent studies, forest carbon mitigation in Canada [37,56,68], as well as other parts of the world (e.g., Sweden [74]; USA [75]), cannot be maximized by implementing a single strategy, but rather with combinations of options adapted to geographical particularities. Only a systems perspective taking into account all the carbon pools and fluxes will allow identification of the best mitigation scenario(s) and an understanding of the trade-offs between increasing carbon storage in forest ecosystems or increasing it in wood products [2,18,20]. Forest carbon mitigation will also need to be conceived and implemented together with climate change adaptation strategies and with other biophysical, procedural, economic, and social objectives in mind. This paper provides a first overview of these alignments and trade-offs. A sustained consideration of the perceptions of all interested and affected parties during the design and implementation of a provincial forest carbon management strategy, as carried out with this approach to analytic deliberative participation, could help legitimize the process and ensure/sustain its public acceptance and perceived effectiveness. Investment in carbon-focused management actions in public forests will also require increased monitoring and reporting of management outcomes. This will increase the accountability of the managers, particularly in countries where forests are predominantly publicly owned. Transparency and accountability will be a key prerequisite to sustain public support and investments in carbon-focused forest management.

Supplementary Materials: The following are available online at http://www.mdpi.com/1999-4907/9/4/225/s1, Two primers used as consultation documents for the first and second series of workshops respectively; Table S1: Results of the biophysical and socio-economic modelling that were provided to participants during the second series of workshops.

Acknowledgments: This research was financially supported by the Forest Carbon Management Project of the Pacific Institute for Climate Solutions (PICS), while the work of G.P.S.-L. was supported by the Social Sciences and Humanities Research Council of Canada and PICS. Special thanks are due to Michelle Connolly, Caren Dymond, Robin Gregory, Shannon Hagerman, Robert Kozak, Tony C. Lemprière, Sara Muir-Owen, Dennis Paradine, Carolyn Smyth, Werner A. Kurz, and Zach Xu for their support in preparing workshop material, for their different roles during the workshops and for reviewing previous versions of the manuscript. We are grateful to all the workshop participants for the strong interest that they showed in our project.

Author Contributions: G.P.S.-L., G.H., and S.S. conceived and designed the experiments; G.P.S.-L., G.H., and S.S. performed the experiments; G.P.S.-L. analysed the data; G.P.S.-L. wrote the original draft; G.H. and S.S. contributed to reviewing and editing.

Conflicts of Interest: The authors declare no conflict of interest.

References

1. IPCC. *Climate Change 2014: Mitigation of Climate Change. Contribution of Working Group III to the Fifth Assessment Report of the Intergovernmental Panel on Climate Change*; Cambridge University Press: Cambridge, UK; New York, NY, USA, 2014.
2. Food and Agriculture Organization of the United Nations (FAO). *Forestry for a Low-Carbon Future: Integrating Forests and Wood Products in Climate Change Strategies*; Fao Forestry Paper 117; FAO: Rome, Italy, 2016.
3. BC MFLNRO (Ministry of Forests, Lands and Natural Resource Operation). *Our Natural Advantage: Forest Sector Strategy for British Columbia*; Government of British Columbia: Victoria, BC, Canada, 2012.
4. Government of British Columbia. *British Columbia's Climate Leadership Plan*; Government of British Columbia: Victoria, BC, Canada, 2016.
5. BC MFLNRO. *Climate Mitigation Potential of British Columbian Forests: Growing Carbon Sinks*; Government of British Columbia: Victoria, BC, Canada, 2013.
6. Hoberg, G.; Peterson St-Laurent, G.; Schittecatte, G.; Dymond, C.C. Forest carbon mitigation policy: A policy gap analysis for British Columbia. *For. Policy Econ.* **2016**, *69*, 73–82. [CrossRef]
7. Peterson St-Laurent, G.; Hagerman, S.; Hoberg, G. Emergence and influence of a new policy regime: The case of forest carbon offsets in British Columbia. *Land Use Policy* **2017**, *60*, 169–180. [CrossRef]
8. Peterson St-Laurent, G.; Hagerman, S.; Hoberg, G. Barriers to the development of forest carbon offsetting: Lessons from British Columbia, Canada. *J. Environ. Manag.* **2017**, *203*, 208–217. [CrossRef] [PubMed]
9. Forest Enhancement Society of BC. $150-Million Reforestation Investment Will Help Fight Climate Change, Create More Rural Jobs. Available online: http://www.fesbc.ca/reforestation-investment.html (accessed on 23 January 2018).
10. Howlett, M.; Rayner, J.; Tollefson, C. *From Old to New Dynamics in Canadian Forest Policy: Dynamics whithout Change?* 3rd ed.; Boardman, R., VanNijnatten, D.L., Eds.; Oxford University Press: Don Mills, ON, Canada, 2009; pp. 183–196.
11. Diaz-Balteiro, L.; Romero, C. Making forestry decisions with multiple criteria: A review and an assessment. *For. Ecol. Manag.* **2008**, *255*, 3222–3241. [CrossRef]
12. Sheppard, S.R.J.; Meitner, M. Using multi-criteria analysis and visualisation for sustainable forest management planning with stakeholder groups. *For. Ecol. Manag.* **2005**, *207*, 171–187. [CrossRef]
13. Natural Resource Council (NRC). *Understanding Risk: Informing Decisions in a Democratic Society*; National Academy Press: Washington, DC, USA, 1996.
14. Cullen, D.; McGee, G.J.A.; Gunton, T.I.; Day, J.C. Collaborative planning in complex stakeholder environments: An evaluation of a two-tiered collaborative planning model. *Soc. Nat. Resour.* **2010**, *23*, 332–350. [CrossRef]
15. Tindall, D.B.; Trosper, R.; Perreault, P. *Aboriginal Peoples and Forest Lands in Canada*; UBC Press: Vancouver, BC, Canada, 2013.
16. BC MOE. British Columbia Greenhouse Gas Inventory. Available online: http://www2.gov.bc.ca/gov/content/environment/climate-change/reports-data/provincial-ghg-inventory (accessed on 31 August 2017).
17. Kurz, W.A.; Dymond, C.C.; Stinson, G.; Rampley, G.J.; Neilson, E.T.; Carroll, A.L.; Ebata, T.; Safranyik, L. Mountain pine beetle and forest carbon feedback to climate change. *Nature* **2008**, *452*, 987–990. [CrossRef] [PubMed]

18. Kurz, W.A.; Smyth, C.E.; Lemprière, T.C. Climate change mitigation through forest sector activities: Principles, potential and priorities. *Unasylva* **2016**, *246*, 61–67.

19. Lemprière, T.C.; Kurz, W.A.; Hogg, E.H.; Schmoll, C.; Rampley, G.J.; Yemshanov, D.; McKenney, D.W.; Gilsenan, R.; Beatch, A.; Blain, D.; et al. Canadian boreal forests and climate change mitigation. *Environ. Rev.* **2013**, *21*, 293–321. [CrossRef]

20. Nabuurs, G.J.; Masera, O.; Andrasko, K.; Benitez-Ponce, P.; Boer, R.; Dutschke, M.; Elsiddig, E.; Ford-Robertson, J.; Frumhoff, P.; Karjalainen, T.; et al. Forestry. In *Climate Change 2007: Mitigation. Contribution of Working Group III to the Fourth Assessment Report of the IPCC*; Metz, B., Davidson, O.R., Bosch, P.R., Dave, R., Meyer, L.A., Eds.; Cambridge University Press: Cambridge, UK; New York, NY, USA, 2007; pp. 541–584.

21. Smyth, C.; Rampley, G.; Lemprière, T.C.; Schwab, O.; Kurz, W.A. Estimating product and energy substitution benefits in national-scale mitigation analyses for Canada. *GCB Bioenergy* **2016**, *9*, 1071–1084. [CrossRef]

22. Jasanoff, S. Technologies of humility: Citizen participation in governing science. *Minerva* **2003**, *41*, 223–244. [CrossRef]

23. Petts, J.; Brooks, C. Expert conceptualisations of the role of lay knowledge in environmental decisionmaking: Challenges for deliberative democracy. *Environ. Plan. A* **2006**, *38*, 1045–1059. [CrossRef]

24. Garmendia, E.; Gamboa, G. Weighting social preferences in participatory multi-criteria evaluations: A case study on sustainable natural resource management. *Ecol. Econ.* **2012**, *84*, 110–120. [CrossRef]

25. Renn, O. A model for an analytic-deliberative process in risk management. *Environ. Sci. Technol.* **1999**, *33*, 3049–3055. [CrossRef]

26. Burgess, J.; Stirling, A.; Clark, J.; Davies, G.; Eames, M.; Staley, K.; Williamson, S. Deliberative mapping: A novel analytic-deliberative methodology to support contested science-policy decisions. *Public Underst. Sci.* **2007**, *16*, 299–322. [CrossRef]

27. Mendoza, G.A.; Martins, H. Multi-criteria decision analysis in natural resource management: A critical review of methods and new modelling paradigms. *For. Ecol. Manag.* **2006**, *230*, 1–22. [CrossRef]

28. Sheppard, S.R. Participatory decision support for sustainable forest management: A framework for planning with local communities at the landscape level in Canada. *Can. J. For. Res.* **2005**, *35*, 1515–1526. [CrossRef]

29. Gregory, R.; Failing, L.; Harstone, M.; Long, G.; McDaniels, T.; Ohlson, D. *Structured Decision Making*; John Wiley & Sons, Ltd.: Chichester, UK, 2012.

30. Failing, L.; Gregory, R.; Higgins, P. Science, uncertainty, and values in ecological restoration: A case study in structured decision-making and adaptive management. *Restor. Ecol.* **2013**, *21*, 422–430. [CrossRef]

31. Ohlson, D.W.; McKinnon, G.A.; Hirsch, K.G. A structured decision-making approach to climate change adaptation in the forest sector. *For. Chron.* **2005**, *81*, 97–103. [CrossRef]

32. Carlson, M.; Chen, J.; Elgie, S.; Henschel, C. Maintaining the role of Canada's forests and peatlands in climate regulation. *For. Chron.* **2010**, *86*, 434–443. [CrossRef]

33. Stinson, G.; Freedman, B. Potential for carbon sequestration in canadian forests and agroecosystems. *Mitig. Adapt. Strateg. Glob. Chang.* **2001**, *6*, 1–23. [CrossRef]

34. Grimble, R.; Wellard, K. Stakeholder methodologies in natural resource management: A review of principles, contexts, experiences and opportunities. *Agric. Syst.* **1997**, *55*, 173–193. [CrossRef]

35. Harrison, S.R.; Qureshi, M.E. Choice of stakeholder groups and members in multicriteria decision models. *Nat. Resour. Forum* **2000**, *24*, 11–19. [CrossRef]

36. Herath, G. Incorporating community objectives in improved wetland management: The use of the analytic hierarchy process. *J. Environ. Manag.* **2004**, *70*, 263–273. [CrossRef] [PubMed]

37. Xu, Z.; Smyth, C.E.; Lemprière, T.C.; Rampley, G.J.; Kurz, W.A. Climate change mitigation strategies in the forest sector: Biophysical impacts and economic implications in British Columbia, Canada. *Mitig. Adapt. Strateg. Glob. Chang.* **2018**, *23*, 257–290. [CrossRef]

38. Peterson St-Laurent, G.; Hoberg, G.; Kurz, W.A.; Lemprière, T.C.; Smyth, C.E.; Xu, Z. *Evaluating Options for Managing British Columbia's Forest Sector to Mitigate Climate Change*; Pacific Institute for Climate Solutions: Victoria, BC, Canada, 2017.

39. Barfod, M.B.; Leleur, S. *Multi-Criteria Decision Analysis for Use in Transport Decision Making*; Technical University of Denmark: Kgs Lyngby, Denmark, 2014.

40. Kajanus, M.; Kangas, J.; Kurttila, M. The use of value focused thinking and the A'WOT hybrid method in tourism management. *Tour. Manag.* **2004**, *25*, 499–506. [CrossRef]

41. Reynolds, K.M. Prioritizing salmon habitat restoration with the ahp, smart, and uncertain data. In *The Analytic Hierarchy Process in Natural Resource and Environmental Decision Making*; Springer: Dordrecht, The Netherlands, 2001; Volume 3, pp. 199–217.

42. Kangas, J.; Kangas, A. Multiple criteria decision support in forest management—The approach, methods applied, and experiences gained. *For. Ecol. Manag.* **2005**, *207*, 133–143. [CrossRef]

43. Stagl, S. Multicriteria evaluation and public participation: The case of uk energy policy. *Land Use Policy* **2006**, *23*, 53–62. [CrossRef]

44. Waeber, P.O.; Nitschke, C.R.; Le Ferrec, A.; Harshaw, H.W.; Innes, J.L. Evaluating alternative forest management strategies for the champagne and aishihik traditional territory, southwest yukon. *J. Environ. Manag.* **2013**, *120*, 148–156. [CrossRef] [PubMed]

45. Kozak, R.A.; Spetic, W.C.; Harshaw, H.W.; Maness, T.C.; Sheppard, S.R.J. Public priorities for sustainable forest management in six forest dependent communities of British Columbia. *Can. J. For. Res.* **2008**, *38*, 3071–3084. [CrossRef]

46. The Montréal Process. The Montréal Process: Criteria and Indicators for the Conservation and Sustainable Management of Temperate and Boreal Forests. Available online: https://www.montrealprocess.org/Resources/Criteria_and_Indicators/index.shtml (accessed on 28 June 2017).

47. Prabhu, R.; Colfer, C.J.P.; Dudley, R.G. *Guidelines for Developing, Testing and Selecting Criteria and Indicators for Sustainable Forest Management: A C&I Developer's Reference*; CIFOR: Jakarta, Indonesia, 1999; Volume 1.

48. Forest Stewardship Council. FSC Principles and Criteria: International Guidelines Developed through Consensus. Available online: https://ca.fsc.org/en-ca/fsc-certification/forest-management-certification/principles-criteria (accessed on 28 June 2017).

49. Tindall, D.B.; Harshaw, H.W.; Sheppard, S.R.J. Understanding the social bases of satisfaction with public participation in forest management decision-making in British Columbia. *For. Chron.* **2010**, *86*, 709–722. [CrossRef]

50. Harshaw, H.W.; Sheppard, S.R.J.; Jeakins, P. Public attitudes toward sustainable forest management: Opinions from forest-dependent communities in British Columbia. *BC J. Ecosyst. Manag.* **2009**, *10*, 81–103.

51. Freedman, B.; Stinson, G.; Lacoul, P. Carbon credits and the conservation of natural areas. *Environ. Rev.* **2009**, *17*, 1–19. [CrossRef]

52. Simonet, G.; Delacote, P.; Robert, N. On managing co-benefits in REDD+ projects. *Int. J. Agric. Resour. Gov. Ecol.* **2016**, *12*, 170. [CrossRef]

53. Lippke, B.; Oneil, E.; Harrison, R.; Skog, K.; Gustavsson, L.; Sathre, R. Life cycle impacts of forest management and wood utilization on carbon mitigation: Knowns and unknowns. *Carbon Manag.* **2011**, *2*, 303–333. [CrossRef]

54. Blanco, J.A.; Flanders, D.; Littlejohn, D.; Robinson, P.; Dubois, D. *Fire in the Woods or RE in the Boiler? A New Tool to Help Rural Communities Determine If Forest Biomass from Wild RE Abatement Can Sustainably Fuel a District Heating System*; Pacific Institute for Climate Solutions: Victoria, BC, Canada, 2013.

55. Miner, R.A.; Abt, R.C.; Bowyer, J.L.; Buford, M.A.; Malmsheimer, R.W.; O'Laughlin, J.; Oneil, E.E.; Sedjo, R.A.; Skog, K.E. Forest carbon accounting considerations in us bioenergy policy. *J. For.* **2014**, *112*, 591–606.

56. Smyth, C.E.; Stinson, G.; Neilson, E. Quantifying the biophysical climate change mitigation potential of Canada's forest sector. *Biogeosciences* **2014**, *11*, 3515–3529. [CrossRef]

57. McKechnie, J.; Colombo, S.; Chen, J.; Mabee, W.; MacLean, H.L. Forest bioenergy or forest carbon? *Assessing trade-offs in greenhouse gas mitigation with wood-based fuels. Environ. Sci. Technol.* **2011**, *45*, 789–795. [PubMed]

58. Smyth, C.; Kurz, W.A.; Rampley, G.; Lemprière, T.C.; Schwab, O. Climate change mitigation potential of local use of harvest residues for bioenergy in Canada. *GCB Bioenergy* **2016**, *9*, 817–832. [CrossRef]

59. Greig, M.; Bull, G. *Carbon Management in British Columbia's Forests: Opportunities and Challenges*; Forrex Series 24; Forum for Research and Extension in Natural Resources Society (FOREX): Kamloops, BC, Canada, 2008.

60. Lippke, B.; Wilson, J.; Meil, J.; Taylor, A. Characterizing the importance of carbon stored in wood products. *Wood Fiber Sci.* **2010**, *42*, 5–14.

61. Gan, J.; McCarl, B.A. Measuring transnational leakage of forest conservation. *Ecol. Econ.* **2007**, *64*, 423–432. [CrossRef]

62. ONeill, G.; Wang, T.; Ukrainetz, N.; Charleson, L.; McAuley, L.; Yanchuk, A.; Zedel, S. *A Proposed Climate-Based Seed Transfer System for British Columbia*; B.C. Tech. Rep. 099; Government of British Columbia: Victoria, BC, Canada, 2017.

63. Park, A.; Talbot, C. Assisted migration: Uncertainty, risk and opportunity. *For. Chron.* **2012**, *88*, 412–419. [CrossRef]

64. Parker, C.; Merger, E.; Streck, C.; Conway, D.; Tennigkeit, T.; Wilkes, A. *The Land-Use Sector within the Post-2020 Climate Regime*; Norden: Copenhagen, Denmark, 2014.

65. Kurz, W.A.; Stinson, G.; Rampley, G.J.; Dymond, C.C.; Neilson, E.T. Risk of natural disturbances makes future contribution of Canada's forests to the global carbon cycle highly uncertain. *Proc. Natl. Acad. Sci. USA* **2008**, *105*, 1551–1555. [CrossRef] [PubMed]

66. Swift, K.; Ran, S. Successional responses to natural disturbance, forest management, and climate change in British Columbia's forests. *BC J. Ecosyst. Manag.* **2012**, *13*, 1–23.

67. Parfitt, B. *Managing BC's Forests for a Cooler Planet: Carbon Storage, Sustainable Jobs and Conservation*; Canadian Centre for Policy Alternatives: Vancouver, BC, Canada, 2010.

68. Kurz, W.A.; Shaw, C.H.; Boisvenue, C.; Stinson, G.; Metsaranta, J.; Leckie, D.; Dyk, A.; Smyth, C.; Neilson, E.T. Carbon in Canada's boreal forest—A synthesis. *Environ. Rev.* **2013**, *21*, 260–292. [CrossRef]

69. Pralle, S. *Branching out, Digging in: Environmental Advocacy and Agenda-Setting*; Georgetown University Press: Washington, DC, USA, 2006.

70. Cashore, B.; Hoberg, G.; Howlett, M.; Rayner, J.; Wilson, J. *In Search of Sustainability: British Columbia Forest Policy in the 1990s*; UBC Press: Vancouver, BC, Canada, 2001; p. 340.

71. BC MFLNRO. *Discussion Paper: Area-Based Forest Tenures*; Government of British Columbia: Victoria, BC, Canada, 2014.

72. Price, K.; Roburn, A.; MacKinnon, A. Ecosystem-based management in the great bear rainforest. *For. Ecol. Manag.* **2009**, *258*, 495–503. [CrossRef]

73. Taylor, S.W.; Carrol, A.L.; Alfaro, R.I.; Safranyik, L. *Forest, Climate and Moutain Pine Beetle Outbreak Dynamics in Western Canada*; Safranyik, L., Wilson, B., Eds.; Natural Resource Canada: Victoria, BC, Canada, 2006; pp. 67–94.

74. Lundmark, T.; Bergh, J.; Hofer, P.; Lundström, A.; Nordin, A.; Poudel, B.; Sathre, R.; Taverna, R.; Werner, F. Potential roles of swedish forestry in the context of climate change mitigation. *Forests* **2014**, *5*, 557–578. [CrossRef]

75. Malmsheimer, R.W.; Bowyer, J.L.; Fried, J.S.; Gee, E.; Izlar, R.L.; Miner, R.A.; Munn, I.A.; Oneil, E.; Stewart, W.C. Managing forests because carbon matters: Integrating energy, products, and land management policy. *J. For.* **2011**, *109*, S7–S51.

forests

MDPI

Article

The Challenge of Diffusion in Forest Plans: A Methodological Proposal and Case Study

Xabier Bruña-García * and Manuel F. Marey-Pérez

Research group of Projects & Planification/IDEGA, University of Santiago de Compostela, 27002 Lugo, Spain; manuel.marey@usc.es
* Correspondence: ringojbg@yahoo.es; Tel.: +34-982-823-247

Received: 22 March 2018; Accepted: 30 April 2018; Published: 2 May 2018

Abstract: Society's participation in decisions regarding land planning and management is essential for reaching viable and long-lasting solutions. The success of forest plans depends on the involvement of different stakeholders. In turn, stakeholder involvement depends on the representativity achieved in public participation in the development of the plan. The first stage, diffusion, is the key element in the process. This paper describes a methodology for the diffusion stage that obtains six times more participants than a similar process. Its aim is to achieve stakeholder representativity in the forestry sector in forest planning at a subregional level. The methodology is validated and applied in a municipality of Galicia, north-west Spain. It is evaluated in terms of efficiency considering the effort in each stage and the results achieved.

Keywords: participatory process; forest governance; diffusion; social forestry; stakeholder analysis

1. Introduction

Land planning is a complex issue [1,2], an active process that requires careful thought about the future and involves the coordination of all relevant activities for achieving specified goals and objectives [2]. Planning is an integral component of forest management. It is about determining and expressing the goals and objectives of the government, rural communities or companies, and deciding the targets and steps that should be taken to achieve these objectives [2]. Planning is not necessarily a complicated process, but it requires clear objectives. It requires imagination and a willingness to consider all points of view relevant to a given situation. The planning process should lead to a balanced outlook from which proposals for effective management can be written [3].

There is no single universal public participation model, but different ones depending on each particular case [4]. The studies examined in Table 1 propose different solutions to the issue of the number of stakeholders involved. The average is of four categories—generic groups of stakeholders—14 groups and 54 participants. There is a higher variability in the number of participants than in the number of stakeholders. It is interesting to identify the elements used to classify and identify stakeholders in the different cases (Table 1) regardless of the place where they are implemented.

The inclusion of people's opinions about the technical aspects of forest resources and their management has sparked scientific debate [5–11]. In a modern sense, public participation in the devising of forest plans is, preferably, voluntary [12,13]. Within public participation processes, people, individually or in organized groups, can exchange information, express opinions and articulate interests with the aim of influencing the final result of decision-making [12–14]. It is both an opportunity and a need for society today, as it is a useful tool to avoid conflicts, share information and encourage good relations with the planning team [15–19]. Public participation contributes to democratization through the reinforcement of transparency in decision-making: it increases the plurality of aims, encourages mutual learning, and increases the awareness of collective responsibility for natural resource-management questions [20–23].

There are several levels of participation within decision-making processes in forest-resource planning. They range from passive participation, where stakeholders are informed about the decisions made by others, to interactive participation, which involves joint decision-making and shared responsibility [20,24–26].

A thorough analysis and classification of stakeholders is essential in the beginning of a public participation process. If important stakeholders are left out, key issues may be ignored and the assessment of the situation will be incomplete [9,27]. A second risk of leaving out key stakeholders is that those key stakeholders will use their exclusion to attack the legitimacy of the process as a whole [28] Multifunctional and modern management requires coordination and agreement between diverging and contradictory stakeholders. Therefore, mutual communication and participation of all identified stakeholders is necessary [13].

The process of public participation is divided into three consecutive stages: diffusion, debate and proposals. There are many references in the literature regarding the last two stages, debate and proposal selection, especially to the use of multiple criteria decision analysis (MCDA) tools, described by Hwang and Yoon [29], and used by authors such as: Pykäläinen et al. [30], Kangas and Kangas [31], Ananda and Herath [32,33], Maness and Farrell [34], Laukanen et al. [35], Sheppard and Meitner [36], Hiltunen et al. [37], Ananda and Herath [38], and Hiltunen et al. [19].

There is an international agreement that participation is a basic principle in forest-resource plans [39,40], but there are no guidelines regarding who should participate [39,41]. Grimble and Wellard [42], Banville et al. [43] and Nordström et al. [27] define the stakeholder as "someone who is affected by or can affect the situation in some way; that is, the stakeholders have vested interests in the decision problem". Therefore, stakeholders are the people and the institutions that depend on or obtain benefits from forest resources or those that decide, control or regulate access to those resources [36]. According to this, stakeholders may or may not be formally organized and individual or collective. For us, participants are the people taking part in the participation process who are members of at least one stakeholder. Sheppard and Meitner [36] propose the classification of participants according to criteria related to property rights, the history of planning processes, reputation, influence and importance. Gong et al. [44] consider property rights, social capital and contractual rules as key elements. Nordström et al. [27] consider that the aims of stakeholder analysis stem from the identification of the most relevant stakeholders and the determination of the scope of their participation. On the one hand, Tikkanen et al. [45] divide stakeholders in the forestry sector on a regional level into three main groups of cooperation: private forestry-oriented, environment and nature-oriented, and background (the background group is formed by those participants with a mixed orientation between economic and environmental criteria). This clear division may lead the participants to defend their own interests instead of truly collaborating with others. On the other hand, Leskinen [7] observed that each group of actors has similar aims in relation with forestry programs in all the regions studied, regardless of the characteristics of the forests or the socioeconomic structure of the region. This confirms the existence of general categories of stakeholders [18]. Recently, Paletto et al. [46] propose a non-subjective method to identify and classify stakeholders by social network analysis (SNA).

O'Neill [47] and Elsasser [39] determined representativity as the main issue in any public participation process. In this sense, a broad representation of various interests in the planing process is essential (McCool and Gutrie [48]. Sheppard and Meitner [36] present several criteria for an effective participatory process and the first is a broad representation of stakeholders. For Buchy and Hoverman [28], representativity is one of the four principles of good practices in public participation from the perspective of the inclusión of all those who would like to participate and to facilitate their participation. Since there is no clear definition, from the perspective of public participation we understand representativity as a balanced and reliable participation of the people and organization with interests in the territory to be planned. Reliable participation is reached when stakeholders and territories are minimally represented in the process.

Examples of different methodologies applied to public participation processes are those by Sheppard and Meitner [36], Martins and Borges [49], Díaz-Balteiro and Romero [1], Food and Agriculture Organization (FAO) [50], Kangas et al. [21], and Paletto et al. [46]. All these works have in common the fact that the process starts with a diffusion stage and includes an appeal for the participation of the population. Authors such as FAO [50], Beierle and Cayford [51], Higgs et al. [25] pose the key questions of who must participate and when. The process has to guarantee representativity [52], a balanced composition of the participating groups [24,53], and the integration of public participation within the technical process for the development of the plan. "The development of rural landscapes is a major challenge for planning and puts a great degree of responsibility in the hands of decision-makers. But successful planning depends on the acceptance of the public and thus interests of stakeholders play a crucial role" [54]. Diffusion is key in the participative process since it arises from people's interest in taking part in the process from the beginning. The principles for the success of the public participation process stated by Nordström et al. [27] depend on this stage.

Public participation in forest-plan design has been a source of scientific debate [55]. Our contribution in this paper is to answer questions such as who can and must participate in these processes; how many people and how many members of each group can and must participate; how participation should take place; and how much is the cost associated with this process is. To achieve this goal, we propose a methodology of diffusion as a first stage of the public participation process. The methodology is validated in a tactical forest plan in the forest district of Fonsagrada-Os Ancares in the autonomous community of Galicia, north-west Spain.

Table 1. Characteristics of participants in different public participation processes.

Case	Surface	Number of Categories	Stakeholders	Participants	Reference
Ostrobothnia region of the Coastal Forestry Centre	528,000 ha forest	-	5	20	[55]
Arrow Forest District	40,000 km²	-	9	47	[36]
Finnish National Forest Programme	-	-	12	74	[56]
Portuguese Chamusca County	-	4	11	32 out of 22 entities	[57]
Comunità Montana Collina Materana	60,784 ha (19.8% forest)	2	9	63	[58,59]
Västerbotten in northern Sweden	8637 ha	4	33	-	[27]
Distretto Arci-Grighine	55,183 ha (43.20% forest)	2	12	124	[58,59]
NRP in Eastern and Western Lapland, Finland.	1.1 million ha forest	-	8	16	[19]
Regional Forest Programmes in Finland	-	8	40	-	[21]
Average	-	4	14	54	-
Variance		6	150	1419	

2. Material and Methods

2.1. The Case Study Area

In the case of Galicia (north-west Spain) (Figure 1), the forest area of 2,060,453 ha accounts for 69% of the region. It has a high potential for forest productivity [60]. In the year 2010, this reached 6,868,500 m³, which represents 50% of Spanish timber production and about 4.5% of that in the European Union [60]. Galicia is characterized by the presence of small forest owners [61,62], agricultural exploitation [63–65], and a mosaic of cropped and livestock-farmed and forested land [66]. For 40 years there has been an increase in the process of afforestation in different agricultural areas and in the abandonment of rural activities [67], together with a sharp decrease in population in rural areas [67].

Forests have been managed primarily by individual private owners. Over 96.6% of forest lands in Galicia are privately held and about 63.7% of forest land is managed by 672,618 individual private owners, with an average forest land of 1.8 ha per owner [62]. Farmers comprise 29.8% of these owners and 19.5% of them own productive and commercial woodlands [62]. The remaining private forests correspond to communal forests, named Montes Veciñais en Man Común (MVMC), a communal form of private land tenure. There are 2878 MVMC units, with an average area of 230 ha [68]. Woodlands account for 15.1% of the MVMC land and 81.5% of the individual private area [68]. Thus, in contrast with the situation reported for forests owned by individual private owners, Galician communal forests have a huge potential for forest exploitation because of their large average size. Moreover, Galician communal forests have considerable economic power and rely on integrated and professional management.

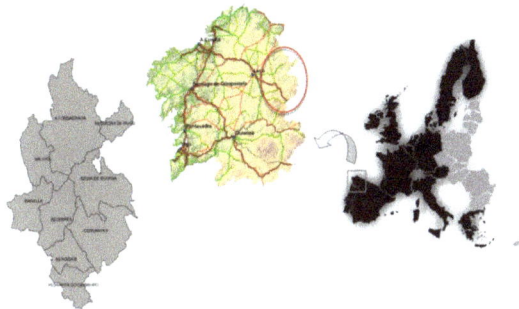

Figure 1. Location of the Fonsagrada-Ancares forest district.

The study area was the Fonsagrada-Ancares forest district, made up of nine municipalities in the eastern part of the province of Lugo (Figure 1) (geographical coordinates: 43°18′, 7°0′ North, 42°36′, 7°3′ South, 42°54′ N, 7°19′ West and 42°51′, 6°44′ East). Forested land covers a total area of 87,828 ha. 63.6%. It is rural area of 1728 km^2 with a population of 19,111 people [69] and a population density of 11.06 people/km^2 that is unevenly spread and far below the European average of 116 people/km^2 [70].

The area is becoming increasingly depopulated due to the lack of services and economic opportunities to replace agriculture and cattle breeding. The population pyramid, in Figure 2, shows the regressive trend characteristic of ageing populations. As a result of the loss of population, the physiographic characteristics and suitable productive capacity, the forest surface has increased to up to 65% of the total surface today [67].

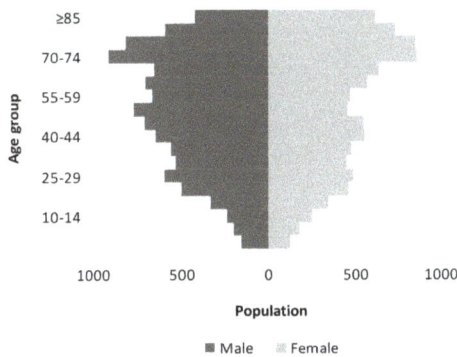

Figure 2. Population pyramid of the Fonsagrada-Os Ancares forest district.

2.2. Methods

The methodology, shown in Figure 3, has been developed considering our experience in several participation processes using different methodologies. We know the importance of incorporating a sufficient number of participants from the first stages of the plan onwards. We also know how difficult it is to incorporate new participants once the process has begun. The proposal consists of four stages that have to be developed in a short period of time and in a coordinated form to solve the problems that may arise. We explain each stage in the following sections.

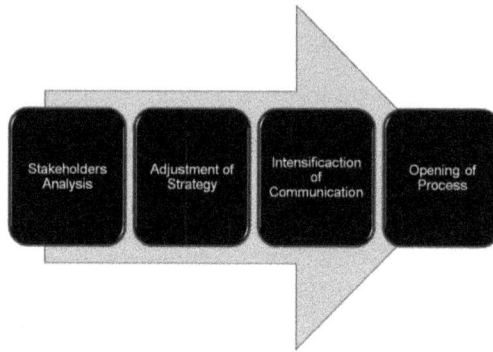

Figure 3. Stages in the diffusion phase.

2.2.1. Stakeholder Analysis

First, the agents involved must be characterized, considering current property rights, planning process history in the area, and previous knowledge about the social framework. Then, the following classification criteria are established: (1) importance of the different stakeholders; (2) experience in participation; (3) necessity of sectorial and territorial representation.

We propose three categories to answer the basic questions about a forest plot: (a) who owns it; (b) who can use it; and (c) who regulates it, as shown in Figure 4. The stakeholders in the area are selected from within these categories. The process of stakeholder selection makes it possible to carry out the diffusion stage directed towards the most representative, who are those directly affected by the decisions of the plan. A minimum number of participants, distributed by sectors and by territory, must be initially determined for a correct representativity in the participative process. This percentage is established as a balance between the statistical and technical rigor of the work and the financial resources and staff availability.

We drew up a census of potential participants in the process distributed by territory and stakeholder. Then, we established a first objective so that a minimum percentage of the population attended the meetings (territory). After that, we established further minimum percentages for the involvement in the process of each stakeholder. This two-fold process aims at territorial and sectorial representativity.

2.2.2. Adjustment of Strategy

Once the stakeholders are identified, diffusion must be carried out with two types of strategy: generic and specific.

(i) Generic: a simple message is broadcast by different media so that it reaches all stakeholders. Its aims are (1) to announce the beginning of the public participation process; and (2) to call for informative meetings on a local level.

(ii) Specific: this strategy is adapted to each stakeholder. It consists of three parts: (1) identification of leaders and representatives; (2) creation of a specific message for each group coordinated with the general message; (3) setting up of a system of communication that guarantees that the leaders receive the message. Figure 5 shows the first two stages of the specific strategies. The letters (A, B, C, D, E ...) represent the different stakeholders. The size of the figures represents the numerical weight of each stakeholder. Their superposition represents the existence of a range of individuals that belong to two or more stakeholders, since excluding criteria are not considered. The letters MG stand for general message and M1, M2, M3 ... represent specific messages.

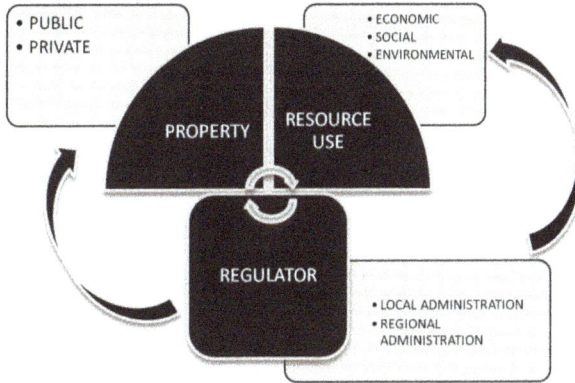

Figure 4. Distribution categories of stakeholders.

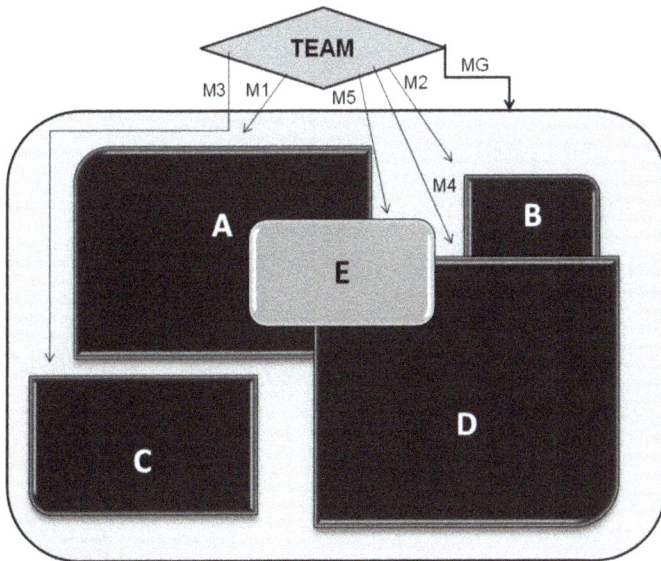

Figure 5. Specific strategy in a public participation process.

2.2.3. Intensification of Communication

Results of the incorporation of participants in the process and their distribution by groups are to be analyzed and compared with the proposed aims. In case the aims are not fulfilled totally or partially, both the message and the method will be revised. This revision is essential to improve the quality of the process and to take corrective measures for its improvement [71].

2.2.4. Opening of Process

Once the presence and the representation of all relevant and representative stakeholders of the area has been guaranteed, the process is open to all those interested to ensure an open and democratic process. There are some proposals for this stage that include the use of the internet in forest-resource planning and teledemocracy [16]. The use of new technologies provides the process with transparency and equity [25].

This can alter its development, but it is necessary to guarantee a real public participation process [72], where all those affected have the chance of participating.

3. Results

3.1. Stakeholder Analysis

Twelve stakeholders have been identified in the process within the three main categories established in the methodology. Both the number of categories (3) and the number of groups (12) are similar to those published by other authors (see Table 1). Within the category of property, we included individual and collective owners, known as comuneros, as they are the types of property present in the area of validation [63]. We included the public administrations that regulate land-use and forest activity within the category of regulators: local administration (municipalities) and forest administration (dependent on the regional administration). We included eight stakeholders that encompass the direct and indirect uses of forest resources within the category of resource users on the territory: farmers, cattle breeders, beekeepers, hunters, companies, owner associations and non-profit organizations (represented by letters A, B, C, D, E in Figure 5). The identification of these stakeholders was carried out with a directed process that considers the social characteristics of the area and the effective use of the forest resources. Maintaining the objective to achieve maximum representativity, we established a guided model based on the social map of the territory, drawn up in the technical part of the district's forestry plan and studies, and previous experiences of the drafting team. Paletto et al. [46], for the Forest Landscape Management Plan (FLMP) in Italy, use social network analysis (SNA) to classify them into three categories: key, primary and secondary stakeholders, a figure equivalent to that of this work.

In our study, the inhabitants are the people who live in the area that is subject to planning and the potential participants are the people or groups that are members of the stakeholders that may or may not live in the area that is subject to planning. The initial population is the whole of the population whom the public participation process is aimed at. It is the sum of all potential participants. We analysed the representativity of the potential participants in each group on this population (Table 2).

Table 2. Initial population by category and stakeholders.

Category	Stakeholder	Subtotal	Total	% Subtotal	% Total
Property	Individual forest owners [1]	9776	18,569	42.64	81
	Collective forest owners [4]	8793		38.35	
Regulator	Local administration [1]	9	241	0.04	1.05
	Forest administration [1]	232		1.01	
Resource user	Farmers and stockbreeders [2]	1983	4116	8.65	17.95
	Hunters [2]	1363		5.95	
	Beekeeper [2]	439		1.91	
	Local forest company [3]	53		0.23	
	Other related companies [3]	110		0.48	
	Forest owners' associations [1]	3		0.01	
	Ecological groups [1] Non-profit organizations [1]	165		0.72	
TOTAL		22,926	22,926	100%	100%

[1] Forest Resource Management Plan; [2] Consellería de Medio Rural, Xunta de Galicia; [3] Database SABI. (Bureau Van Dijk); [4] INE (National Statistics Institute) and IGE (Galician Statistics Institute).

We established a percentage of the census (1% in both cases) for the first objective, who attended the meeting of each municipality, and for the second, a percentage of participants of each stakeholder in the participatory process. Tables 3–5 show the results obtained. We chose this limit after considering the diagnosis of the current state of the population, a lack of participatory culture, demographic and physiographic characteristics, and the distribution of population settings. Once we set the goal of participation, the diffusion process started.

Table 3 shows the results obtained for the real distribution and for the theoretical distribution, which was initially our aim.

Table 3. Initial real and theoretical distribution of representativity in the participation process.

Groups	Representatives	
	Real	Theoretical
Individual forest owners	135	98
Collective forest owners	81	88
Farmers and stockbreeders	44	20
Hunters	8	14
Beekeeper	1	1
Local forest company	10	4
Other related companies	1	1
Local administration	1	0
Forest administration	8	2
Forest owners associations	1	0
Ecological groups	6	0
Non-profit organizations	7	0
Total	303	229

Table 4. Attendance at information meetings.

Information Meeting	Number of People Present	Inhabitants	% Municipality	Initial Population	% Initial Population
Baleira	44	1658	2.65	2230	1.97
Baralla	48	3034	1.58	2546	1.89
Becerreá	72	3264	2.21	3406	2.11
Cervantes	71	1844	3.85	2837	2.5
A Fonsagrada	107	4748	2.25	6317	1.69
Navia de Suarna	57	1552	3.67	1866	3.05
Negueira de Muñiz	27	222	12.16	532	5.08
As Nogais	16	1421	1.13	1552	1.03
Pedrafita do Cebreiro	11	1368	0.80	1240	0.89
TOTAL	453	19,111	2.37	22,526	2.01

Table 5. Attendance at information meetings in the intensification stage.

Information Meeting	Number of People Present	Inhabitants	% Municipality	Initial Population	% Initial Population
As Nogais	24	1421	1.69	1552	1.54
Pedrafita do Cebreiro	2	1368	0.15	1240	0.16

It can be seen that 8 of the 12 groups equal or exceed 1% of participants in the process and 4 do not reach the minimum established. It was necessary to continue with the program and calendar for the elaboration of the plan and take as reference the considerations of Vallejo and Hauselman [73] in which they establish that a participatory process is a compromise between speed and legitimacy, where speed will depend on the number of participants in the process. If the method does not allow progress in the process quickly, the participants will abandon it. In turn, as Figure 5 shows, there is an overlap among groups; some participants are, for example, owners, ranchers, comuneros, beekeepers, etc. and it was considered that it was possible to guarantee the representativity of both groups and territories according to the definition considered.

3.2. Adjustment of the Strategy

The generic strategy was based on meetings held in each municipality; 466 posters were distributed in 425 population nucleuses, 43% of the total, to call for meetings. This informative phase was reinforced with the distribution of 1600 brochures placed at the most accessible points: health centers, administration offices. Attendance data for each meeting can be seen in Table 5 divided by municipality. On the whole, 453 people attended, which amounts to 2.37% of inhabitants in the area and 2.01% of the starting population.

Within a specific strategy, first, we held a meeting with the local administration where we informed them of the start of the forest plan and asked for their collaboration for the process. After that, we sent 329 letters with specific messages about the implications of the forest plan with acknowledgement of receipt to the leaders of stakeholders within the categories of property and users. The message and the means used to reach each stakeholder focused on the interests that each group has with respect to forest planning. Thus, for example, for the forest communities, the importance to the plan of the use of their forest was indicated; for the hunters, the implications of the plan for hunting species was indicated; and for environmental groups the importance of the plan for conservation was indicated.

3.3. Intensification of Communication

Analyzing the results of the previous stage, we noticed the low representation of two municipalities: As Nogais and Piedrafita do Cebreiro. Therefore, we decided to reinforce the diffusion strategy there by means of a new diffusion process and a call for a new informative meeting. The result is shown in Table 6.

With this new process, we substantially increased the total percentage of attendance in the municipality of As Nogais (2.57%). However, in the case of Piedrafita do Cebreiro, attendance was still low (1.10%), just above the established limit as the aim for territorial representation.

Table 6. Registration figures in the participative process by municipality.

Municipality	Number of People Present	Registered	% Registration	Initial Population	% Initial Population
Baleira	44	32	72.73	2230	1.43
Baralla	48	23	47.92	2546	0.90
Becerreá	72	44	61.11	3406	1.29
Cervantes	71	54	76.06	2837	1.90
A Fonsagrada	107	72	67.29	6317	1.14
Navia de Suarna	57	18	31.58	1866	0.96
Negueira de Muñiz	27	16	59.26	532	3.01
As Nogais	40	31	77.5	1552	2.00
Pedrafita do Cebreiro	13	13	100	1240	1.05
TOTAL	479	303	63.26	22,526	1.32

3.4. Opening of the Process

The process started on 22 December 2008 and finished on 28 August 2009. The web page www.planforestaldistritovii.com had 2271 visits, 9.16 visits/day, 1030 different users (5% of the district's population), and an average duration of visit of 5'26''.

After the diffusion stage finished, the registration process started. 303 people registered (1.32% of the initial population). The end of the diffusion process concluded with the possibility of signing up in the participation process in the forest plan. This was for administrative requirements and the convenience of being in contact with people interested in continuing to participate. In Table 7, we can see the results organized by municipality (territorial level) and, in Table 8, organized by stakeholder (sectorial level).

Regarding the age of the participants, this encompassed people aged from 22 to 82, with more participants over 65. Regarding their sex, over 94% were men. The misrepresentation of women can be explained by cultural factors, as happens in other places [14]. In relation to training, 24% of participants had a university degree, especially the younger ones, 23% completed sixth form, and 53% had basic training.

Table 7. Number of people registered in each interest group.

Category	Stakeholders	Registered	%	% Initial Population	Total	% Total	% Initial Population
Property	Individual forest owners	135	44.55	1.38	216	71.29	1.16
	Collective forest owners	81	26.73	0.92			
Regulator	Local administration	1	0.33	11.11	10	3.30	4.15
	Forest administration	9	2.97	3.45			
Resource user	Farmers and stockbreeders	44	14.52	2.22	77	25.41	2.22
	Hunters	8	2.64	0.59			
	Beekeeper	1	0.33	0.23			1.02
	Local forest company	10	3.30	18.87			
	Other related companies	1	0.33	0.91			
	Forest owners associations	6	1.98	3.8			
	Ecological groups	3	0.99	1.9			7.74
	Non-profit organizations	4	1.32	100			
	TOTAL	303	100	-	303	100	-

3.5. Results of the Cost and Effort of Diffusion

The cost of the diffusion process was estimated by measuring the working hours of the technical staff as indicators, with a result of 289 h. In Table 8, we can see the hours classified by stages. The most time-consuming stage is that of the generic strategy, with 135.42 h (47% of resources). The effort in the area of validation is 0.17 h of technical staff per square kilometer and 0.95 h per participant.

Table 8. Summary of diffusion cost by stage measured in technical staff working hours.

Stage	Effort (h Technical Staff)	% Effort
Previous	130	45.00
Outline peparation	40	13.84
Outline meeting	10	3.46
Diffusion preparation	80	27.69
Specific strategy	23.5	8.13
Diffusion meeting	6	2.08
Initial meeting	4	1.38
Meeting with mayors	4.5	1.56
District meeting	3	1.04
Letters	6	2.08
Generic strategic	135.42	46.87
Information meetings	13.07	4.52
Putting up posters	122.35	42.35
Total	288.92	100

4. Discussion

As stated by Aasetre [13], the different interests in the resources to be the subject of planning must be key to defining the structure of participation, so that the sensibilities that influence resource management are included in the planning stage. In the proposed methodology, we identified three universal types of stakeholders: property, forest-resource users and regulators. Other authors establish different numbers: Kangas et al. [21] propose eight; Marques et al. [57] and Nordström et al. [27] propose four; whereas Agnoloni et al. [58] and Ferretti et al. [59] propose two. The structure of categories must be simple enough to be capable of extrapolation to a wide range of forest areas.

They must also be homogeneous and internally coherent enough to be representative of the interests in resources of each group and to represent all the sensibilities and interests in the land together.

The validation of the methodology has classified 12 groups of stakeholders in the study area. These results coincide with those in Sheppard and Meitner [36], Primmer and Kyllönen [56], Hiltunen et al. [19], Agnoloni et al. [58], Marques et al. [57], and Kangas et al. [21]. This may be due to the repetition of a high number of stakeholders in different times and places. This common group is associated with the categories of property and regulator. The higher or lower number of new stakeholders is motivated by the complexity of the category of forest-resource users. This category depends on the historical component of use and the relative importance of forest activity as a socioeconomic activity, together with the level of definition of the classification methodology used by researchers.

Once the structure of participation has been developed, the next question is related to stakeholder representativity. Our methodological proposal takes into account the proposals by Janse and Konijnendijk [18] that considered that the number of participants was conditioned by factors related to participation tools and communication, always adapted to the place and time of the process. As pointed by Côté and Bouthillier [53], it is necessary to consider the dimension of the area and the time available for the process.

There is a high number of participants in the validation, 303, 1.32% of the initial population, which contrasts with an average of 54 in similar works consulted (Table 1). The territorial distribution is considerably higher than the average of the referenced processes at 0.2 participants/km².

The results show that, using the validated methodology, we obtained six times more participants than the average with a similar number of stakeholders. Our effort in the diffusion stage has allowed us to obtain a level of representativity measurable in terms of stakeholders and territories. Gaining participants in the amount and distribution considered as objective was not achieved in the first stage, and it was necessary to resort to the stage of intensification of communication.

It is necessary to assess participant representativity [39,47] and whether participants are evenly distributed [9,27,39]. Herein lies the importance that all interests groups are present and that their representation in the public participation process is proportional to its real representation in the initial population. The statistical validation guarantees the representativity of the process.

The next question answered is whether the diffusion process has been attractive enough. We considered the relationship between the number of people registered and the number of those who attended informative meetings. The global result is that 63.26% of participants in the meeting registered and participated in the process. Purdon [74] stated the need for more transparent participative processes so that they could be more attractive to the population. Leskinen [7] includes environmental aspects which will raise people's interest in participating in planning.

Deming [71] stated that one of the challenges of research in public participation is to achieve efficient methodologies. So far the works revised have not been concerned with this issue. However, the results obtained with the methodology we propose show that the cost or effort is relatively low, making it a highly efficient methodological process.

The methodology proposed for the case study can be replicated in other forest-planning processes taking into account the local characteristics in each area as key elements for the methodology's adaptation: to carry out a correct stakeholder analysis in the area, to establish a representation percentage, a strategy considering the role of each stakeholder, and to increase diffusion in those groups that do not reach the initial representation aim.

Currently, we do not have data about the cost of the diffusion stage of public participation processes in forest planning. Our methodology aims to offer an answer to this question.

A more intensive process could improve performance and innovation [21,51]. In some processes, participants find that too much time is devoted to participation [21], so it is necessary to find a balance between participants' time and the results aimed. Thus, it is necessary to use new technologies to decrease displacements and meeting times.

5. Conclusions

Public participation processes within forest plans are currently being debated. Our work attempts to answer some of the uncertainties that have been formulated. After the implementation of the diffusion stage in the forest plan in the district of Fonsagrada-Os Ancares some answers that help to improve public participation can be provided for application to similar processes.

The diffusion stage has been proved to be an essential phase in the beginning of forest-resource planning processes. It has a high influence in stakeholder representativity that will, in turn, influence the rest of the processes, both regarding the quality of proposals and the actual forest management [75].

The literature reviewed, while highlighting the importance of diffusion in participatory processes, does not clearly establish indicators of appropriate levels. Our contribution focuses on proposing measurable objectives adapted to the characteristics of the area and a method to achieve them.

Diffusion processes are important because they provide the participative process with "raw material" and "fuel". The diffusion stage is a key element in the process because failure at this stage means failure of the overall process. The aim of this is to raise interest and ensure a sufficient number of stable participants for the process to fulfil the representativity criteria. In this case, 63% of those attending the information meetings participated in the whole process.

Once the aim of attracting the attention of a representative percentage of the population has been fulfilled, the challenge is to keep the attention of the highest possible number of participants throughout the process so that they provide information and knowledge right up to the conclusion of the plan.

The minimum number of participants is going to depend on the percentage fixed at the beginning of the process, which must be established considering both the territory and the sector. The validation of participant representativity will be determined by the fulfillment of the minimum percentage (at territorial and sectorial level) and the analysis of theoretical participant distribution in contrast to the actual distribution. In case the minimum percentage is not reached (1% of the initial population in our case), it is necessary to increase diffusion until it is reached. It is also necessary to apply statistical contrasts to verify that results have been achieved, in our case, the T test for dependent samples.

The cost of the process is a limiting factor for its generalization in an area. In this case, it has been considered in every stage and the optimization of financial resources has resulted in an economical process with a total cost of 5 €/km^2 or 25 €/participant.

When a process begins it is necessary to have real public participation and, in this sense, new technologies are very efficient at providing transparency, swiftness and equity in the process. In the case described in this work, the web page of the forest plan ensured the transparency of the process, allowing 5% of the population to follow it.

Author Contributions: Xabier Bruña-García carried out the case study. Manuel F. Marey-Pérez provided expertise in the methodology and analysis of the results. The article was improved by the contributions of all of the co-authors at various stages of the analysis and writing process.

Acknowledgments: We would like to extend our gratitude to all those people who actively contributed their participation in formulating the Forest Plan of the Fonsagrada-Os Ancares District. We want to thank the anonymous reviewers for their contributions.

Conflicts of Interest: The authors declare no conflict of interest.

References

1. Díaz-Balteiro, L.; Romero, C. Making forestry decisions with multiple criteria: A review and an assessment. *For. Ecol. Manag.* **2008**, *174*, 447–457. [CrossRef]
2. Zilberman, D.; Goetz, R.; Garrido, A. *Natural Resource Management and Policy*; Springer: Berlin/Heidelberg, Germany, 2010.
3. Food and Agriculture Organization (FAO). *Principles of Sustainable Tropical Forest Management Where Wood Production Is the Primary Objective*; Forest Resources—Forestry Department: Roma, Italy, 1998.

4. Bruña-García, X.; Marey-Pérez, M.F. Public participation: A need of forest planning. *iForest* **2014**, *7*, 216–226. [CrossRef]

5. Gerber, J. Beyond dualism—The social construction of nature and the natural and social construction of human beings. *Progress Hum. Geogr.* **1997**, *21*, 1–17. [CrossRef]

6. Pelkonen, P.; Pitkänen, A.; Schmidt, P.; Oesten, G.; Piussi, P.; Rojas, E. *Forestry in Changing Societies in Europe*; Silva Network, Part I and II; Silva Network, Part I and II, University of Joensuu:: Joensuu, Finland, 2000.

7. Leskinen, L.A. Purposes and challenges of public participation in regional and local forestry in Finland. *For. Policy Econ.* **2004**, *6*, 605–618. [CrossRef]

8. Grundmann, R. The role of expertise in governance processes. *For. Policy Econ.* **2009**, *11*, 398–403. [CrossRef]

9. Lakicevic, M.; Srdjevic, Z.; Srdejevic, B.; Zlatic, M. Decision making in urban forestry by using approval voting and multicriteria approval method (case study: Zvezdarska forest, Belgrade, Serbia). *Urban For. Urban Green.* **2014**, *13*, 114–120. [CrossRef]

10. Paletto, A.; Giacovelli, G.; Pastorella, F. Stakeholders' opinions and expectations for the forestbased sector: A regional case study in Italy. *Int. For. Rev.* **2017**, *19*, 68–78.

11. De Meo, I.; Ferretti, F.; Paletto, A.; Cantiani, M.G. An approach to public involvement in forest landscape planning in Italy: A case study and its evaluation. *Ann. Silvicult. Res.* **2017**, *41*, 54–66.

12. International Labour Office (ILO). *Public Participation in Forestry in Europe and North America: Report of the Team of Specialists on Participation in Forestry*; Report of the FAO/ECE/ILO Joint Committee Team of Specialists on Participation in Forestry; WP 163; International Labour Office: Geneva, Switzerland, 2000.

13. Aasetre, J. Perceptions of communication in Norwegian forest management. *For. Policy Econ.* **2006**, *8*, 81–92. [CrossRef]

14. Atmiş, E.; Özden, S.; Lise, W. Public participation in forestry in Turkey. *Ecol. Econ.* **2007**, *62*, 352–359. [CrossRef]

15. Hellström, E. *Conflict Cultures—Qualitative Comparative Analysis of Environmental Conflicts in Forestry*; Silva Fennica Monographs; University of Helsinki: Helsinki, Finland, 2001; Volume 2, p. 109.

16. Kangas, J.; Store, R. Internet and teledemocracy in participatory planning of natural resources management. *Landsc. Urban Plan.* **2003**, *62*, 89–101. [CrossRef]

17. Santos, R.; Antunes, P.; Baptista, G.; Mateus, P.; Madruga, L. Stakeholder participation in the design of environmental policy mixes. *Ecol. Econ.* **2006**, *60*, 100–110. [CrossRef]

18. Janse, G.; Konijnendijk, C. Communication between science, policy and citizens in public participation in urban forestry—Experiences from the Neighbourwoods project. *Urban For. Urban Green.* **2007**, *6*, 23–40. [CrossRef]

19. Hiltunen, V.; Kurttila, M.; Leskinen, P.; Pasanen, K.; Pykäläinen, J. Mesta: An internetbased decision-support application for participatory strategic-level natural resources planning. *For. Policy Econ.* **2009**, *11*, 1–9. [CrossRef]

20. Elsasser, P. Rules for participation and negotiation and their possible influence on the content of a national forest program. *For. Policy Econ.* **2002**, *4*, 291–300. [CrossRef]

21. Kangas, A.; Saarinen, N.; Saarikoski, H.; Leskinen, L.A.; Hujala, T.; Tikkanen, J. Stakeholder perspectives about proper participation for Regional Forest Programmes in Finland. *For. Policy Econ.* **2010**, *12*, 213–222. [CrossRef]

22. Lennox, J.; Proctor, W.; Russell, S. Structuring stakeholder participation in New Zealand's water resource governance. *Ecol. Econ.* **2011**, *70*, 1381–1394. [CrossRef]

23. Jansson, M.; Gunnarsson, A.; Martensson, F.; Andersson, S. Children's perspectives on vegetation establishment: Implications for school ground greening. *Urban For. Urban Green.* **2014**, *13*, 166–174. [CrossRef]

24. Germain, R.H.; Floyd, D.W.; Stehman, S.V. Public perceptions of the USDA Forest Service public participation process. *For. Policy Econ.* **2001**, *3*, 113–124. [CrossRef]

25. Higgs, G.; Berry, R.; Kidner, D.; Langford, M. Using IT approaches to promote public participation in renewable energy planning: Prospects and challenges. *Land Use Policy* **2008**, *25*, 596–607. [CrossRef]

26. Garmendia, E.; Stagl, S. Public participation for sustainability and social learning: Concepts and lessons from three case studies in Europe. *Ecol. Econ.* **2010**, *69*, 1712–1722. [CrossRef]

27. Nordström, E.M.; Eriksson, L.O.; Öhman, K. Integrating multiple criteria decision analysis in participatory forest planning: Experience from a case study in northern Sweden. *For. Policy Econ.* **2010**, *12*, 562–574.

28. Buchy, M.; Hoverman, S. Understanding public participation in forest planning: A review. *For. Policy Econ.* **2000**, *1*, 15–25. [CrossRef]
29. Hwang, C.L.; Yoon, K. *Multiple Attribute Decision Making. Methods and Applications: A State of the Art Survey*; Springer: Berlin, Germany, 1981.
30. Pykäläinen, J.; Kangas, J.; Loikkanen, T. Interactive decision analysis in participatory strategic forest planning: Experiences from state owned Boreal forests. *J. For. Econ.* **1999**, *5*, 341–364.
31. Kangas, A.; Kangas, J.; Pykäläinen, J. Outranking methods as tools in strategic natural resources planning. *Silva Fennica* **2001**, *35*, 215–227. [CrossRef]
32. Ananda, J.; Herath, G. Incorporating stakeholder values into regional forest planning: A value function approach. *Ecol. Econ.* **2003**, *45*, 75–90. [CrossRef]
33. Ananda, J.; Herath, G. The use of Analytic Hierarchy Process to incorporate stakeholder preferences into regional forest planning. *For. Policy Econ.* **2003**, *5*, 13–26. [CrossRef]
34. Maness, T.C.; Farrell, R. A multi-objective scenario evaluation model for sustainable forest management using criteria and indicators. *Can. J. For. Res.* **2004**, *34*, 10–14. [CrossRef]
35. Laukkanen, S.; Palander, T.; Kangas, J. Applying voting theory in participatory decision support for sustainable timber harvesting. *Can. J. For. Res.* **2004**, *34*, 1511–1524. [CrossRef]
36. Sheppard, S.R.J.; Meitner, M. Using multi-criteria analysis and visualisation for sustainable forest management planning with stakeholder groups. *For. Ecol. Manag.* **2005**, *207*, 171–187. [CrossRef]
37. Hiltunen, V.; Kangas, J.; Pykäläinen, J. Voting methods in strategic forest planning—Experiences from Metsähallitus. *For. Policy Econ.* **2008**, *10*, 117–127. [CrossRef]
38. Ananda, J.; Herath, G. A critical review of multi-criteria decision making methods with special reference to forest management and planning. *Ecol. Econ.* **2011**, *68*, 2535–2548. [CrossRef]
39. Elsasser, P. Do "stakeholders" represent citizen interests? An empirical inquiry into assessments of policy aims in the National Forest Programme for Germany. *For. Policy Econ.* **2007**, *9*, 1018–1030. [CrossRef]
40. Mayer, P. The MCPFE and COST E19. In *NFP Research: Its Retrospect and Outlook*; Glück, P., Voitleithner, J., Eds.; Institute of Forest Sector Policy and Economics: Wien, Austria, 2004; pp. 183–194.
41. Barstad, J. A Planner's View on National Forest Programmes. In *NFP Research: Its Retrospect and Outlook*; Glück, P., Voitleithner, J., Eds.; Institute of Forest Sector Policy and Economics: Wien, Austria, 2004; pp. 65–82.
42. Grimble, R.; Wellard, K. Stakeholder methodologies in natural resource management: A review of principles, contexts, experiences and opportunities. *Agric. Syst.* **1997**, *55*, 173–193. [CrossRef]
43. Banville, C.; Landry, M.; Martel, J.M.; Boulaire, C. A stakeholder approach to MCDA. *Syst. Res. Behav. Sci.* **1998**, *15*, 15–32. [CrossRef]
44. Gong, Y.; Bull, G.; Baylis, K. Participation in the world's first clean development mechanism forest project: The role of property rights, social capital and contractual rules. *Ecol. Econ.* **2010**, *69*, 1292–1302. [CrossRef]
45. Tikkanen, J.; Leskinen, L.A.; Leskinen, P. Forestry organisation network in Northern Finland. *Scand. J. For. Res.* **2003**, *18*, 547–599. [CrossRef]
46. Paletto, A.; Hamunen, K.; De Meo, I. Social network analysis to support stakeholder analysis in participatory forest planning. *Soc. Nat. Res.* **2015**, *28*, 1108–1125. [CrossRef]
47. O'Neill, J. Representing people, representing nature, representing the world. *Environ. Plan. C Gov. Policy* **2001**, *19*, 483–500. [CrossRef]
48. McColl, S.F.; Guthrie, K. Mapping the dimensións of succeful public participation in messy natural resources management situations. *Soc. Nat. Res.* **2001**, *14*, 309–323.
49. Martins, H.; Borges, J.G. Addressing collaborative planning methods and tools in forest management. *For. Ecol. Manag.* **2007**, *248*, 107–118. [CrossRef]
50. FAO. *Elaboración de una Política Forestal Eficaz—Una Guía*; Development of an Effective Forest Policy—A Guide; FAO: Roma, Italy, 2010.
51. Beierle, T.C.; Cayford, J. *Democracy in Practice: Public Participation in Environmental Decisions*; Resources for the Future: Washington, DC, USA, 2002.
52. Kangas, J.; Kangas, A. Multiple criteria decision support in forest management—The approach, methods applied, and experiences gained. *For. Ecol. Manag.* **2005**, *207*, 133–143. [CrossRef]
53. Côté, M.; Bouthillier, L. Assesssing the effect of public involvement processes in forest management in Quebec. *For. Policy Econ.* **2002**, *4*, 213–225. [CrossRef]

54. Tress, B.; Tress, G. Scenario visualisation for participatory lLandscape planning—A study from Denmark. *Landsc. Urban Plan.* **2003**, *64*, 161–178. [CrossRef]
55. Sipilä, M.; Tyrväinen, L. Evaluation of collaborative urban forest planning in Helsinki, Finland. *Urban For. Urban Green.* **2005**, *4*, 1–12. [CrossRef]
56. Primmer, E.; Kyllönen, S. Goals for public participation implied by sustainable development, and the preparatory process of the Finnish National Forest Programme. *For. Policy Econ.* **2006**, *8*, 838–853. [CrossRef]
57. Marques, A.F.; Borges, J.G.; Lucas, B.; Garcia, J.; Melo, I. A participatory approach to design a toolbox to support forest management planning at regional level. *For. Syst.* **2013**, *22*, 340–358. [CrossRef]
58. Agnoloni, S.; Bianchi, M.; Bianchetto, E.; Cantiani, P.; De Meo, I.; Dibari, C.; Ferretti, F. I piani forestali territoriali di indirizzo: Una proposta metodologica. (The regional forest plans to address: A methodological proposal). *Forest* **2009**, *6*, 140–147. [CrossRef]
59. Ferretti, F.; Dibari, C.; De Meo, I.; Cantiani, P.; Bianchi, M. ProgettoBosco, a Data-Driven Decision Support System for forest planning. *Math. Comput. For. Nat.-Resource Sci.* **2011**, *3*, 27–35.
60. Marey-Pérez, M.F.; Rodríguez-Vicente, V. Forest transition in Northern Spain: Local responses on large-scale programmes of field-afforestation. *Land Use Policy* **2008**, *26*, 139–156. [CrossRef]
61. Rodríguez-Vicente, V.; Marey-Pérez, M. Assessing the role of the family unit in individual private forestry in northern Spain. *Scand. J. For. Res.* **2008**, *23*, 53–77. [CrossRef]
62. Rodríguez-Vicente, V.; Marey-Pérez, M. Analysis of individual private forestry in northern Spain according to economic factors related to management. *J. For. Econ.* **2010**, *16*, 269–295. [CrossRef]
63. Álvarez-López, C.J.; Riveiro-Valiño, J.A.; Marey-Pérez, M.F. Typology, classification and characterization of farms for agricultural production planning. *Span. J. Agric. Res.* **2008**, *6*, 125–136. [CrossRef]
64. Riveiro-Valiño, J.A.; Marey-Pérez, M.F.; Marco-Gutiérrez, J.L.; Álvarez-López, C.J. Procedure for the classification and characterization of farms for agricultural production planning: Application in the Northwest of Spain. *Comput. Electron. Agric.* **2008**, *61*, 169–178. [CrossRef]
65. Riveiro-Valiño, J.A.; Álvarez-López, C.J.; Marey-Pérez, M.F. The use of discriminant analysis to validate a methodology for classifying farms based on a combinatorial algorithm. *Comput. Electron. Agric.* **2009**, *66*, 113–120. [CrossRef]
66. Díaz-Varela, E.R.; Marey-Pérez, M.F.; Rigueiro-Rodríguez, A.; Álvarez-Álvarez, P. Landscape metrics for characterization of forest landscapes in a sustainable management framework: Potential application and prevention of misuse. *Ann. For. Sci.* **2009**, *66*, 301–311. [CrossRef]
67. Marey-Pérez, M.F.; Rodríguez-Vicente, V.; Álvarez-López, C.J. Practical application of multivariant analysis techniques to the forest management of active farmers in the northwest of Spain. *Small Scale For.* **2012**, *11*, 453–476. [CrossRef]
68. Marey-Pérez, M.F.; Gómez-Vázquez, I.; Díaz-Varela, E.R. Different approaches to the social vision of communal land management: The case of Galicia (Spain). *Span. J. Agric. Res.* **2010**, *8*, 848–863. [CrossRef]
69. Xunta de Galicia. Consellería de Medio Ambiente e Desenvolvemento Sostible. *Plan de Ordenación de Recursos Naturais do Parque Natural de Os Ancares*; Xunta de Galicia. Consellería de Medio Ambiente e Desenvolvemento Sostible: Santiago de Compostela, Spain, 2008.
70. OCDE. *Creating Rural Indicators for Shaping Territorial Policy*; OCDE: París, France, 1994.
71. Deming, W.E. *Out of the Crisis*; MIT Press: Cambridge, MA, USA, 1986.
72. Stenseke, M. Local participation in cultural landscape maintenance: Lessons from Sweden. *Land Use Policy* **2009**, *26*, 214–223. [CrossRef]
73. Vallejo, N.; Hauselmann, P. *Gobernance and Multistakeholder Processes*; UNCTAD/IISD: Winnipeg, MB, Canada, 2004.
74. Purdon, M. The Nature of ecosystem management: Postmodernism and plurality in the sustainable management of the boreal forest. *Environ. Sci. Policy* **2003**, *6*, 377–388. [CrossRef]
75. Bruña-García, X.; Marey-Pérez, M.F. Participative Forest Planning: How to obtain knowledge. *For. Syst.* **2018**, *27*. [CrossRef]

forests

MDPI

Article

Validation of a Methodology for Confidence-Based Participatory Forest Management

Eugenio Martínez-Falero [1], Concepción González-García [1], Antonio García-Abril [1] and Esperanza Ayuga-Téllez [2,*]

1 SILVANET, Research Group for Sustainable Environmental Management, Universidad Politécnica de Madrid (UPM), Ciudad Universitaria s/n, 28040 Madrid, Spain; eugenio.mfalero@upm.es (E.M.-F.); concepcion.gonzalez@upm.es (C.G.-G.); antonio.garcia.abril@upm.es (A.G.-A.)
2 Buildings, Infrastructures and Projects for Rural and Environmental Engineering (BIPREE), Universidad Politécnica de Madrid (UPM), Ciudad Universitaria s/n, 28040 Madrid, Spain
* Correspondence: esperanza.ayuga@upm.es; Tel.: +34-910-671-582

Received: 30 April 2018; Accepted: 2 July 2018; Published: 4 July 2018

Abstract: This paper formulates a new strategy for participatory forest management consisting of encouraging public participation as long as it increases empathy among participants. The strategy requires the homogeneous representation of the opinion of a participant (i.e., to determine how they assess a forest plan and identify the best one). Utility assessments are prepared for participants through pair-comparisons between meaningful points in the territory and from value functions based on forest indicators. The best plan is designed by applying combinatorial optimization algorithms to the utility of a participant. The calculating of empathy—of one participant relative to another—is based on the equivalence of their respective utilities when the current forest plan is modified. This involves calculating the opinions that are due to systematic changes in the collective plan for those participants that each participant supposes will affect the utility of the other participants. Calculating empathy also requires knowing the interactions among participants, which have been incorporated through agent-based simulation models. Application of the above methodology has confirmed the association between increases in empathy and convergence of opinions in different scenarios: well and medium-informed participants and with and without interaction among them, which verifies the proposed strategy. In addition, this strategy is easily integrated into available information systems and its outcomes show advantages over current participatory applications.

Keywords: public participation; AI decision-making algorithms; empathetic utility functions; assessment of sustainability

1. Introduction

The paper validates an Artificial Intelligence methodology applicable to forest management. However, as in all AI process, the algorithms must be trained in the available data, never replicated. This requires a type of manager able to understand the performance of algorithms in order to propose modifications when they have to apply to real cases. The paper has been written by foresters for these types of forest managers.

A challenge most of foresters must face is to increase social accountability. Therefore, we start with a question: what would happen if we trust in others to make our own decisions?

Virtually all experiments show that the presence of a large number of decision makers improves individual decisions [1]. The internet certainly allows the processing of information from many participants, which mitigates the negative biases of individual decisions [2].

In addition, opinions often converge when working in trusted environments. In such environments, each person takes the reasons of others into consideration. To do this, all reasons must be known (which requires them to be explicitly described). In addition, if someone considers other reasons be reasonable, they ought to incorporate them into their personal judgments. Consequently, opinions will converge as empathy increases among the people involved in participative processes (participants). Empathy is reinforced using a common language of communication ICTs (Information and Comunication Tecnologies) and when participants have been educated regarding shared ethical and justice criteria.

However, it is more important to consider empathy as a dynamic process than to calculate it, because it is the increase or decrease of empathy that governs the convergence of opinions. This paper shows that it is possible to objectively measure the evolution of empathy throughout participatory processes, and to do it automatically. This inspires a strategy for collective decision-making based on confidence scenarios: to encourage participatory processes while confidence is increasing, and to conclude them as soon as it ceases to increase. Decisions must then be made by applying the available techniques for aggregation of opinions, but first, it is necessary to have an objective representation of the opinions in order to compare and aggregate them.

1.1. Representation of Opinions

This concept has been developed in past research and below we explain it succinctly. Below, we describe the models required for the case of application in detail in the corresponding section.

All opinions have two aspects: the way in which a person assesses a solution and the identification of the best solution for a person. The following notation will be adopted in the rest of this paper: X is a solution to a problem; A is an individual who participates in the collective decision-making; $u_A(X)$ is the assessment of X made by A; and X_A is the best solution for A. Consequently, A's opinion will be represented by the following pair: $o_A = [u_A(X), X_A]$.

Assessment is not easy because multiple variables are usually required to describe solutions (for example, a solution has a cost and a rate of return; generates a specific number of jobs, etc.). Not everyone will have the same answer to questions of the type: Would you prefer a project costing € 2M and returning € 5M or another project costing € 0.5M and returning € 2M? Assessment is usually individual for each person, and it is deduced from personal patterns of comparison. Assessment transforms each person's preferences into a numerical value that determines how a person assesses a situation. When assessment is homogeneous, it allows third parties to make comparisons between individual preferences: all they need to do is compare their respective numerical representations.

From a methodological point of view, there are three procedures for assessment:

- Pair-wise comparison of alternatives—which leads to preference modeling (see, for example, [3]). The evaluator directly compares pairs of alternatives to establish (subjectively) a preference relation between them. The preference is obtained from the global knowledge the evaluator has about alternatives and assessment derives—objectively—from the preference/indifference relations in all the pairs of compared alternatives.
- The aggregation of criteria through the building of a value (or utility) function (first formalized by Debreau [4]) Here, the preference is induced measures of the different consequences of the alternatives being compared (consequences are objectively measured, for example, at a planning level, they will incorporate environmental, social, and economic dimensions), and the meaning that the evaluator attributes to the interrelationship between these consequences—which drives to a subjective aggregation process.
- Analysis of past decisions (developed from the works of Arrow [5]).

The outcomes arising from the application of the three procedures are interchangeable (see [6]), which makes it possible to exchange the concepts of preference, utility, and choice (see, among other authors, Suzumura [7] and Sen [8], for the integration of the available procedures). It is also possible to apply the calculating capabilities of any of these methodologies to determine the outcomes in the others.

Assessment also helps obtain the solution that best fits the preferences of a participant. However, the number of solutions to forest management is normally so high that it is not operative to generate all feasible solutions and then assesses all of them to identify the best one. Instead, combinatorial optimization tools are applied to obtain acceptable solutions. In this context, the best solution is obtained under risk conditions, which means adopting the solution with the highest probability of satisfying the preferences of a specific participant. The building of this subjective likelihood is also based on the assessment procedures.

The design of the "best solution" is solved methodologically (see, for example, [9]) by applying different algorithms: recursive algorithms, heuristic optimization algorithms (mathematical programming, Monte Carlo methods, genetic algorithms, simulations of natural systems, tabu-search, etc.), and non-heuristic algorithms (basically neural network analysis), or a combination of all three.

1.2. Opinion Aggregation Techniques

Opinion aggregation plays a central role in the operation of collective decision making, but can a group choose between different options?

When a large number of people make their judgments to assess a solution, there are as many judgments that overestimate the solution as underestimate it. The result is that the average of assessments tends to approach the real value. However, despite efforts to reach consensus, not all members of a group must accept the aggregated solution. Some advances on these topics are discussed below.

The work by Arrow [10]—inspired by the work of Borda [11] and Condorcet [12]—relates "social choices" to individual preferences through social welfare functions. Unfortunately, results in this area are pessimistic: even some very mild conditions of reasonableness could not be simultaneously satisfied by a social choice. In consequence, aggregation of opinions will usually involve circumventing Arrow's impossibility theorem by relaxing some of its applicability conditions (see, for example, [13,14]).

Social choice is focused on voting, which is the natural way that decisions are made in a democracy. However, voting is not harmless: methods expressing preferences can be manipulated to guarantee a choice of certain alternatives a priori [15]. Also, majority voting systems do not consider the choice of alternatives representing minorities. Currently, there are other social-choice mechanisms based on Web 2.0 tools: document ranking, folksonomy, recommender systems, open software, electronic prediction markets, and wikis. Sadly, these methods also need to relax even the softest reasonableness conditions to achieve collective decisions.

Parallel to social choice, the utilitarian approach [16,17]—inspired by Bentham [18]—used the aggregation of individual utilities to obtain evaluations of social interest. They were concerned with the total utility of the community, but they did not incorporate how utility is distributed or concentrated [19]. However, the arising of the logical positivism [20] allowed the identification of individual gains and losses, which made it possible to progress toward the "interpersonal comparison of utilities". This enables the use of many different types of welfare rules (egalitarianism, envy-freeness, etc.), which differ in that way that fairness and efficiency are handled.

However, is it possible to make comparisons between the utilities of two or more people? There has been progress in this regard (see, for example, [21]) interpersonal utility comparisons have been developed from different postulates (as those proposed by [22]), and very general conditions have been defined for its implementation [23]. At present, the use of computational rules [24] allows aggregation of opinions based on Harsanyi's theory [25,26]. This theory adds the notion of empathetic preferences [27–29] to Von Neuman-Morgenster's hypotheses on utility and configures the paradigm of aggregation described in this paper.

1.3. Available Computer Applications for Collective Decision-Making

The progress in collective decision-making (see, for example, [30]) has led to the implementation of operative computer applications. All of them have important areas that require improvement, but the existing developments are already part of the solution rather than the problem. The applications presented below are grouped based on the decision-making model on which they are based:

1.3.1. Type-A Applications

These models were based on formalizing strategies for games on a network and require the current advances in computer technology and the methodology of information-decision systems. In these models, each participant tries to maximize their personal gain, which leads to a reduction in the earnings of others. There are two fundamental procedures: Group Decision Support Systems (GDSS) [31] which aim to reach an overall negotiated decision among the participants through face-to-face meetings; and the Social Decisions Support Systems (SDS) [32], which visualize the flow of discussion through a network of statements, opinions, arguments, and comments, which helps to gain consensus before voting on an issue.

- A-1: Group Decision Support Systems—GDSS-.

The largest number of applications are in this group: PLEXSYS from the University of Arizona [33]; Colab, designed for Xerox at the Palo Alto Research Center, [34]; Shell GDSS, developed at the University of Minnesota [31]; DECAID (Decision Aids for Groups), designed by [35]; LADN (Local Area Decision Network) [36]; and SMU Decision Room Project [37–39]. However, the most widely used is Loomio, which manages the decision-making in some emerging political parties in Europe [40].

- A-2: Social Decisions Support Systems—SDSS-.

Deliberative Opinion Poll (from Stanford University) [41].

1.3.2. Type-B Applications

In general, conflict resolution requires social collaboration, and this is where the generation of confidence becomes relevant for collective decisions. Applications in this area usually implement models that simulate the propagation of opinions, as is done with infection or contagion processes. In fact, social and environmental agents have always tried to know the propagation of their opinions among the components of the society.

The software available to simulate these models is not strictly for collective decision-making, but also for the analysis and processing of information (structured and unstructured) with multiple receivers. However, some applications have evolved to provide an aggregate solution by focusing on different aspects. We can distinguish between applications [42] that prioritize,

○ organization of information
   -Paramount [43] Decisions
   -Analytica [44]
○ assessment made by multiple agents
   -1000 Minds of IBM [45] and
○ Application of procedures to predict the evolution of systems (analysis of dynamic systems and agent-based simulation models) [46].

1.3.3. Type-C Applications

In the end, although the emphasis may be at one end of the spectrum, systems that consider both personal preferences and social interactions usually fit the population's view better than systems that consider just one of these circumstances. Applications in this field lead to effective solutions, although they are not always efficient. Two programs are the most used:

LIQUIDFEEDBACK. This application, [47] developed at the Massachusetts Institute of Technology (MIT), incorporates social networks of confidence so that the decision can be delegated to others (liquid democracy). It does not ask direct questions about the result, but rather encourages participants to propose alternatives. It has a sophisticated voting system to allow participants to express their opinions without allowing tactical considerations.

ALLOURIDEAS. This application uses a pair-comparison method between alternatives, which each participant must select. The outcome is the sorting of a list of proposals (to which new ones can be added) by the whole group. This application was developed at Princeton University [48].

2. Methodology for Confidence-Based Collective Decision-Making Based

When decisions are reported through digital systems (or the computers capture decisions automatically) the behavior of each person is recorded. The behavioral analysis can be automated and is aimed at representing opinions, which means—as explained on Section 1.1—knowing how a person assesses a solution and which solution is best for such a person.

However, the possibilities of digital management go beyond representing opinions: as Harsany [25,26] proposed, we can also use empathy among opinions.

For our purposes, empathy is the ability of a person to see the world through the eyes of another. If A chooses the best solution for B rather than himself, A reveals a preference for B's opinion over his own opinion. If A is consistent in this behavior, we can say A empathizes with B, but it is a bit more complex than this. Binmore [49] describes properly the additional assumptions of Harsany´s theory, arguing:

Besides our personal preferences, which may include sympathetic concerns for others, we also have empathetic preferences that reflect our ethical concerns. This means that empathy can be graduated. However, some empathetic identification is crucial to the survival of human societies because society needs commonly understood coordinating conventions that select an equilibrium when many options are available.

When a person empathizes with someone, they do so entirely successfully (the best known historical example of incomplete empathy is quoted from Longinus: "Were I Alexander", said Parmenio, "I would accept of these offers made by Darius". "So, would I too", replied Alexander, "were I Parmenio".). The empathetic utility function of a person simply expresses the fact that the person in question thinks that A and B personal utils can be traded off, so that A's personal utils count the same as a of B's personal utils. In other words, holding an empathetic preference is the same thing as subscribing to a standard for making interpersonal comparisons between A and B.

Under the above assumptions, the Harsanyi's theory can be described using a Von Neumann and Morgenstern utility function.

The power of empathy comes from its association with convergence of opinions. It could be accepted that putting yourself in the place of other favors reaching a common opinion. All that is needed is to verify that increases in empathy lead to convergence of opinions. In this case, participation should be encouraged if empathy among participants improves, and if not, the process should end.

If the overall amount of empathy is high at the end of the participatory process, then the aggregate opinion is representative of the preferences of the people involved. Here, the aggregate opinion will be adopted as the group's joint decision. Otherwise, the group of participants will need to be broken down into homogeneous subgroups (with common opinions within the subgroups) and other aggregation procedures applied to the preferences of the homogeneous communities.

2.1. Calculating Empathy

The methodology described below is applicable to the calculation of empathy in a non-competitive environment. A deeper explanation and the primitive formulation of it can be found in [21]. It is based on calculating empathy for participants connected on social networks [50,51] and in incorporating Adler's developments [52].

To determine the empathy of one person (*A*) towards another (*B*), we must consider the opinions of all those who make up this empathy. At least, *A* and *B* opinions (o_A and o_B), an aggregation of the opinions *A* considers to be closest to *B* in the confidence networks ($o_{n(B)A}$), the history or, at least, the aggregated opinion of the whole group when empathy is calculated (o_g), and an explicit objective opinion formulated by experts (o_{Ob}).

The above opinions can be used to determine the utility that a person believes that a solution has for another person, but the weight of all opinions cannot be the same. Therefore, the utility that A believes that solution X has for B can be calculated as:

$$U_{A \to B}(X) = \sum_{i \in \{A,B,n(B)_A,g,Ob\}} (1 - D_{i \leftrightarrow B}) \times (1 - \|X - X_i\|) \times u_i(X) \tag{1}$$

where: $D_{i \leftrightarrow B}$ is the reciprocal distance between i and B's opinions (calculated using the distance between i and B in the confidence networks) it is the complementary of the probability that the opinion of the first person will align with the opinion of the second person); $\|X - X_i\|$ is the distance between the best solution for i (X_i) [9] and the solution for which we are calculating the utility (X); $u_i(X)$ is the assessment of X made by i; and i refers to all the involved opinions $(o_i = [X_i, u_i(X)])$, $i \in \{A, B, n(B)_A, g, Ob\}$.

However, Equation (1) does not yet indicate the empathy of A relative to B. As mentioned, the empathetic function expresses the equivalence in the utilities of A and B when the aggregate assessment of the group (X_g) is modified. To calculate this, we force L systematic changes in the values of the descriptive variables for the group's opinion (for example, increasing or decreasing the cost and performance of the X_g solution) and we determine the changes that these modifications produce in both A and B's utilities. If each of these changes is represented by $X_{g<k>}$ (for $k = 1, \ldots, L$) then, the modification it induces in the way A believes that B makes their assessments will be represented as:

$$e_{A \to B, k} = U_{A \to B}(X_{g<k>}) - U_{A \to B}(X_g) \tag{2}$$

For $k = 1, \ldots, L$, we get a L coordinates vector representing the empathy function of A relative to B:

$$euf_{AB} = \begin{bmatrix} e_{A \to B, 1} \\ e_{A \to B, 2} \\ \vdots \\ e_{A \to B, L} \end{bmatrix} \tag{3}$$

The set of empathies of A for all participants (N) will be the empathetic utility matrix of A (which will be notated as $\mathbf{em}_A$):

$$\mathbf{em}_A = \begin{bmatrix} euf_{AA_1}, & \ldots, & euf_{AA_N} \end{bmatrix} = \begin{bmatrix} e_{A \to A_1,1} & \cdots & e_{A \to A_N,1} \\ \vdots & \ddots & \vdots \\ e_{A \to A_1,L} & \cdots & e_{A \to A_N,L} \end{bmatrix} \tag{4}$$

This matrix collects all the information referring to the empathy of a person (A) with the other participants. The matrix shows the variations of the utility that A believes that certain changes in the aggregate solution (in the L rows of $\mathbf{em}_A$) induce in the rest of the participants (in the N columns of $\mathbf{em}_A$).

A measure of the separation of empathy between a pair of individuals can be obtained from these matrices. Thus, from general operations with matrices:

$$S_{AB} = 1 - (\|\mathbf{em}_A, \mathbf{em}_B\| / N) \tag{5}$$

where: N is the total number of evaluators,

$$\mathbf{em}_A, \mathbf{em}_B = \sqrt{\mathrm{tr}\left((\mathbf{em}_A - \mathbf{em}_B)^\circ (\mathbf{em}_A - \mathbf{em}_B)^T\right)} \tag{6}$$

$^\circ$ denotes the usual multiplication of matrices and tr( ) represents the trace of a matrix

The average of Equation (5) for all possible participant pairs has been used as a measure of the percentage of empathy among participants:

$$S_T = \frac{1}{N(N-1)/2} \sum_{I=1}^{N-1} \sum_{J=I}^{N} S_{IJ} \tag{7}$$

Obviously, the average of similarities of empathy between pairs of participants does not describe the distribution of empathy among participants. For that, other centralization and dispersion measures would need to be applied to the similarities of empathy, but our objective is to measure the evolution of confidence and variation of average empathy over time, to detect the increase or decrease in total empathy and evolution over time represented by Equation (7) (see Figure 1).

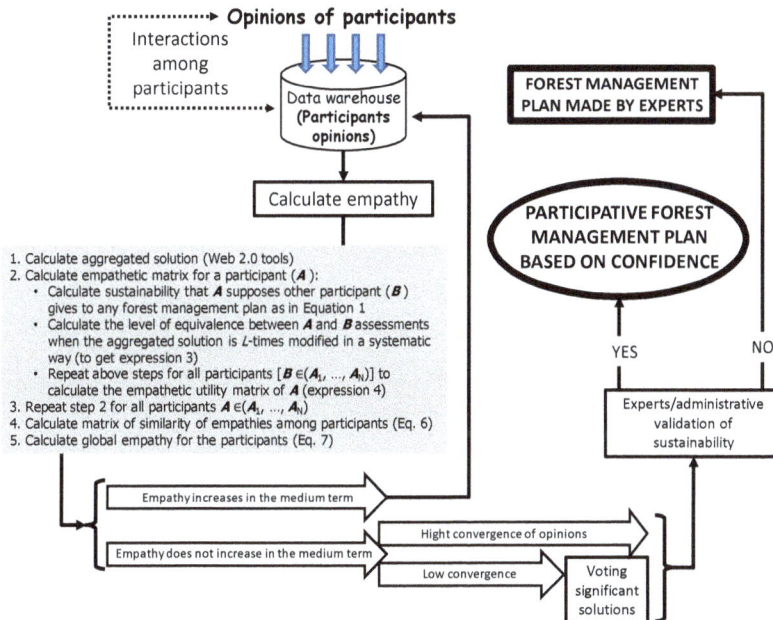

Figure 1. Flow chart to incorporate empathy.

2.2. Simulating Interactions among Participants

Interactions among opinions occur naturally in participatory processes and can be promoted through digital forums and other tools offered by the Internet. Effects of interactions on collective decisions can also be simulated as described below.

The simulation is based on the works of Sayama [51]. Next, let us show the simulation frame with the mathematical notation and simplifications adopted in this paper:

- Let $o_A = [u_A(X), X_A]$ be (as in the rest of the paper) the opinion of participant A.
- Participants do not know the final aggregated assessment (u_{gF}). The goal is to find a joint evaluation, which should be accepted by most participants at the end of the discussion process. However, the current aggregated value (u_g) is known at any given time.
- Each participant has a memory in which they store their personal comprehension of the problem and the history of group discussion. The memory is defined as a list of opinions either they have (or have had) or that other participants have expressed during the discussion. The memory of each participant (M_A) can store up to p of their own opinions (the initial opinion plus the

modifications they have accepted during the discussion) and up to q opinions expressed by each of the other participants ($M_A = \{o_{A,k}\}$, $k = 1, \ldots, p, p + 1, \ldots, p + q$). The total memory of one participant is limited to a certain number of opinions, thus $p + q \leq l$. If the number of opinions participating in the personal understanding of the problem exceeds this number, then some of the opinions in the memory must be removed.

- At least the following opinions always belong to the memory of a participant (A): o_A, o_g, and o_{Ob}.

Interaction does not simply accept changes in view of other opinions. It requires discussion: each participant tries to impose their own solution on the group, but at the same time, the participant accepts changes in their solution if convinced. Discussion is an iterative process with the following phases on each one of the iteration:

1. Each participant examines whether there is an easy way to improve the utility of their current solution.

Every participant examines their neighbors' solution and incorporates the opinion with the highest utility for all the participants belonging to their neighborhood into their memory. We shall denote this plan as $u_{n(A)}$.

2. A randomly selected speaker is chosen.

We shall denote the speaker as S.

3. The speaker suggests the revision of the group plan.

The speaker (S) identifies which aspect of their individual plan has the most significant impact if incorporated into the group's plan. To do this, we built $\mathbf{em}_S$ matrix (Equation (4)), which is now calculated by adding the speaker's opinion to that in the memory of all participants.

First, we have to calculate the utility that the speaker supposes the solution has for another participant (A) as:

$$U_{S \to A}(X) = \sum_{i \in S \cup M_A} (1 - D_{i \leftrightarrow A}) \times (1 - \|X - X_i\|) \times v_i^*(X) \tag{8}$$

where S refers to the speaker's assessments; M_A is the memory of the A-evaluator; $D_{i \leftrightarrow A}$, X and X_i are as in equation 1; and v_i^* is not the assessment of a solution (u_i), it is referred to the expert's opinion expressed as:

$$v_i^* = 1 - |u_i - u_{Ob}| \tag{9}$$

Then, u_g, which belongs to the memory of all participants, is systematically modified to calculate the $\mathbf{em}_S$ matrix (equations 2 to 4). The row of this matrix with the highest sum of its components (j) determines the suggestion of change proposed by the speaker.

The new opinion that the speaker suggests to the group is expressed as $o_{g<j>} = (u_{g<j>}, X_{g<j>})$.

4. The speaker's suggestion is evaluated by all other participants, who respond to the suggestion at the individual level.

Each participant then studies the utility of the speaker's suggestion, which is done by applying the next equation:

$$E_A(o_{g<j>}) = \sum_{p \in M_A} ((1 - D_{p \leftrightarrow A}) \times (1 - \|X_{g<j>} - X_p\|) \times u_p) \tag{10}$$

where p is any opinion in the memory and M_A is the memory of A.

If $E_A(o_{g<j>}) > E_A(o_g)$, then the A-evaluator expresses support for the suggestion, and $o_{g<j>}$ is incorporated to M_A. Otherwise, i.e., $E_A(o_{g<j>}) \leq E_A(o_g)$, then A's individual plan is not affected by the speaker's suggestion. However, $o_{g<j>}$ may still be randomly incorporated into the memory of A with a probability given by:

$$P(o_{g<j>}, o_g) = \exp(-\|E_A(o_{g<j>}) - E_A(o_g)\|/T) \tag{11}$$

where: T is the "temperature" of the participant's cognition (i.e., to what degree participants can tolerate low-utility suggestions).

If incorporating a new opinion exceeds the maximum number of opinions allowed in the memory, then the previous one has to be removed from memory.

5.   Response to suggestion at group level.

The group determines if the speaker's suggestion should be adopted as the new group solution by counting the number of participants that have incorporated the speaker's suggestion into their memories. If this is more than 50%, then the old group opinion will be replaced by the suggestion ($o_g \leftarrow o_{g<j>}$).

Figure 2 summarizes how to simulate iteration among participants. It is assured the convergence to better solutions, but not in a monotone way. Thus, you have to analyze results on all iterations to choose the best aggregated plan.

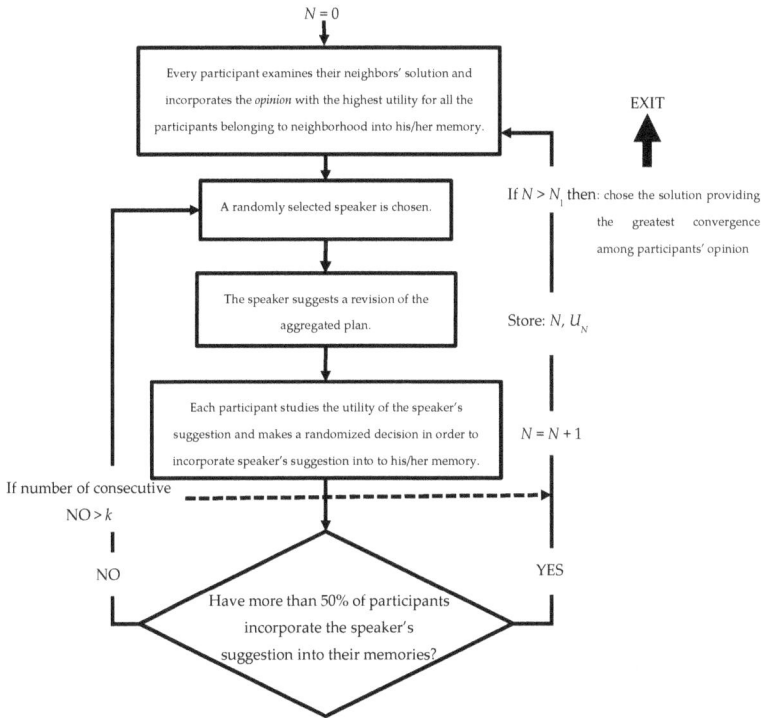

Figure 2. Flow chart to simulate iterations among participants.

2.3. The Strategy for Confidence-Based Collective Decision-Making

From an operational point of view, the proposed strategy consists of the following stages:

a.   Opening of an online information and participation process for anyone to formulate their preferences and to incorporate these preferences into the available information systems.

b. Encouraging participation, through the media or by community managers. Motivation should provide an incentive for both the recurrent participation of individuals already involved (they tend to modify their preferences because of interaction with other participants), and the incorporation of new participants.

c. Automatic representation of participant opinions. This requires determining how a participant assesses a solution (based on their personal preferences) and designing the solution that best fits a preference.

d. On-line calculation of the empathy of each participant with the others and of the total empathy for the opinions among all participants.

e. Supporting the participatory process if total empathy increases and terminating the process when there is no significant increase in empathy among participants.

f. Application, at the time of termination, of Web 2.0 tools, building of interpersonal functions, or both to choose the best aggregate decision. In case of a low convergence of preferences, another social or individual selection process should be used to choose representative and socially acceptable solutions.

It is evident that the confidence-based strategy satisfies the general requirements for participation to a greater degree than the public-participation applications that are currently available (see Table 1).

Table 1. Comparing available public-participation applications.

Capabilities of the Available Application of Public Participation to Manage Collective Decision-Making		Type-A Systems	Type-B Systems	Type-C Systems	Confidence-Based Strategy
			See Section 1.3		
General requirements	To be transparent and allow rebuttal by third parties in all its decisions	Yes	Partially	Yes	Yes
	Adaptive to the available information and to the knowledge that the evaluator has in the participation process.	Partially	Partially	Partially	Yes
	Friendly (use is not affected by the technological, cultural or generational gap).	No	Yes	Partially	Yes
	Optimizer of individual utility and socially intelligent.	No	No	Yes	Yes
	To serve as an instrument for accountability to society.	Yes	Yes	Yes	Yes
Outcome requirements	To provide assessments based on the representation of personal preferences.	Yes	No	Yes	Yes
	Collective decision-making through aggregation of individual preferences	Yes	Yes	Yes	Yes
	Design of the alternative that best fits to any system of preferences (individual or aggregated).	Yes	No	No	Yes
Capacity to solve large combinatorial problems		Not available in current applications			Yes

3. Case Study and Results

The case study focused on verifying the strategy for confidence-based collective decision-making. To do this, we simulated a participative process to design the most sustainable forest management plan at a real forest.

We apply the proposed methodology to "Valle de la Fuenfría" forest (Madrid, Spain). Environmental authorities have approved a sustainable long-term forest management plan with Scots pine (*Pinus sylvestris* L.) as the main vegetation. The case study is a territory within that forest (see Figure 3). It covers an area of 96.04 ha (40°45′ N, 4°5′ W), and has altitudes ranging from 1310 to 1790 m, an average annual temperature of 9.4 °C, and average annual rainfall of 1180 mm. Further information can be found in [53].

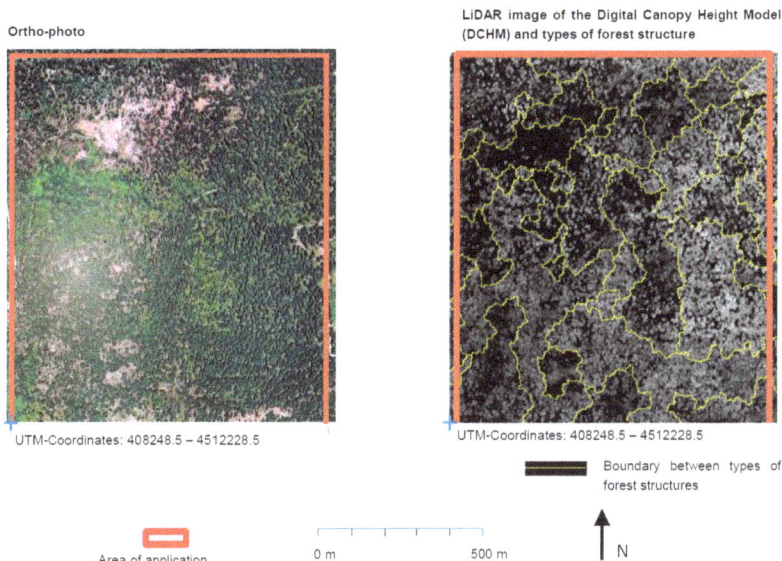

Figure 3. Area selected for the test case.

The strategy consisted of encouraging participation if empathy among participants increased. To verify this strategy, we needed to compare the evolutions of empathy with the convergence of opinions throughout the participative process (Section 3.3). To do that, we needed to measure both empathy and convergence of opinions (Section 3.2). In turn, this calculation requires the representation of the opinions of each of the participants in the process (Section 3.1).

3.1. Representation of Opinions

Sustainability is a complex, multidimensional idea that does not allow an overall method of assessment. It is possible to measure partial aspects—indicators—but each person has their own way of aggregating those indicators for assessing sustainability. In other words, each person has an opinion on sustainability.

For this case study, representing an opinion means:

- Assessing the sustainability on each one of the pixels of the territory according to personal preferences on sustainability.
- Designing the forest management plan that best suits personal preferences on sustainability.

We have adopted pixels (instead of stands) because the accuracy of the inventory allows doing that [54]. Even trees could be decision units if you know the effects that any action carried on tree have in the tress on its neighborhood (see Equation (13)). Additionally, management based on pixels is applicable to analyze sustainability at different scales. In any case, you must remember that we are looking for a management plan of a forest under a long-term management plan based on stands, and any action must fit to this plan.

3.1.1. Personal Assessment of Sustainability at Each Point of the Territory

Sustainability is calculated at each one of the points of a grid superimposed on the study area with a pixel width of one meter.

The methodology for the assessment of sustainability was the one that incorporates SiLVANET [54], a computer application for participatory forest management. It calculates sustainability on hierarchical (and no hierarchical) indicators structuring that must be built by experts [55]. This case study focuses on a partial assessment (an aggregation of attributes to determine wood sustainability) within the larger structure. Nevertheless, the case shows all the challenges that faces sustainability assessment.

The information required came from three different sources: *physical and management parameters of the area*—mainly: height distribution of trees within a 50-m diameter around each point, it is a scale-invariant indicator that can be used in management based on stands or pixels, whatever their size [56], and sustainability indicators such as structural diversity, timber yield, or the biomass at each point; *personal information* on the individual whose opinion was going to be represented (gender, age, level of education, occupation, type of residence—urban or rural—and stakeholder category); and the *personal sustainability preferences* of each participant.

Preferences are captured through a pair-wise comparison (see [30]) of forest management regimens in specific points of the territory. Management regimens must be representative of all the management options for the forest and we shall refer them as meaningful or significant points. Comparison requires the selection of the meaningful points (see [54] for the points in this case study) and, after forming all feasible pairs of points, to ask the participants for their preferences within a pair (they must indicate which of the two points they consider to be more sustainable (see Figure 4). The answers are recorded in a preference matrix and then transformed into a numeric value of the sustainability on each one of the comparison points (see [57]). This value matches the preferences of the participant that has filled the preference matrix. Table 2 shows: geographical coordinates of comparison points, the matrix of preferences filled by a concrete participant, and the sustainability assessment on the comparison points. The number of meaningful points must be small. It is recommended not to choose more than six points, because the number of pairs to compare will be 15. The more comparisons the more decreases of concentration and the judgments tend to be inconsistent with the evaluator's own preferences [58].

Table 2. Data entered (spatial location of comparison points (with meaningful forest management regiments) and pair comparison matrix (columns IDs are the same as rows IDs) filled by a concrete participant) and assessments of sustainability on the meaningful points (1, 2 ,..., 6).

Comparison Points			Individual Preferences (Data Entered by a Participant) [1]						Preferences after Sequel and Transitive Closure [59] [2]						Sustainability at Compared Points	
ID	UTM X-coord.	UTM Y-coord.	1	2	3	4	5	6	1	2	3	4	5	6	Original value	[0,1] Rank
1	408916,5	4512672,5	3	3	3	3	1	2	0	1	1	0	1	1	4.3	0.86
2	408776,5	4513172,5	3	3	1	3	3	3	1	0	1	1	1	1	5	1
3	408916,5	4512272,5	3	2	3	2	2	1	0	1	0	0	1	1	3	0.6
4	408976,5	4512852,5	3	3	1	3	1	2	1	1	1	0	1	0	4.1	0.82
5	408936,5	4512252,5	2	3	1	2	3	1	1	1	1	0	0	1	3.9	0.78
6	408516,5	4513012,5	1	3	2	1	2	3	1	1	1	1	0	0	3.7	0.74

[1] Row more sustainable than column = 1; Column more sustainable than row = 2; No one preferred to another = 3.
[2] Row preferred to column = 1; Column preferred to row = 0.

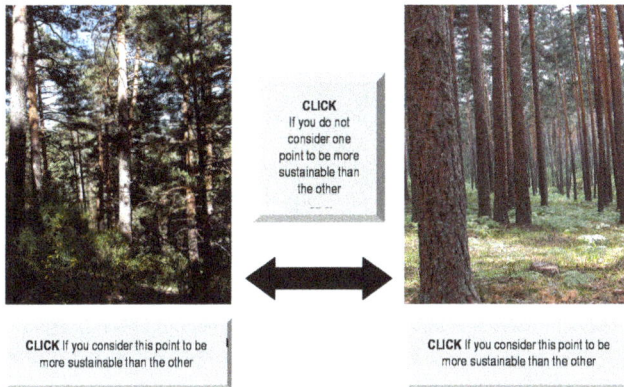

Figure 4. An example of pair comparison.

Also, you must have an aggregation model to extend the above result to all the pixels in the study area. In this case, we have used an objective model that calculates sustainability as the statistical association between indicators on any pixel and indicators on an ideal point [55,56]. Experts must choose the appropriate sustainability indicators (it is possible to use pan-European or Montreal indicators). Although, you always must use an indicator that considers the annual increment of wood under allowable cut standards (we use the timber yield, see [55,56] for a description of how to compute it).

Sustainability in all the territory (Figure 4) is then calculated from information deduced from both procedures: pair-wise comparison and aggregation of criteria (see [55]). The pair-wise comparison value is the framework where the assessment of the pixels that are not subject to direct comparison is fitted. The thoughtful judgments required in the pair-wise comparison set the general pattern of value, and the valuation obtained from the aggregation of criteria incorporates the variability present in all alternatives to the evaluation process.

In the above context, the *SiLVANET* methodology [54,60] was applied to achieve the following outcomes:

1.   To get the personal sustainability value at each one of the compared points (Table 2).
2.   To extend the assessment to the remaining points in the area (Figure 5).
3.   To estimate the mapping from sustainability indicators into the sustainability at each point, Equation (12) via a Multiple Regression Model. Independent variables are constrained to those statistically significant and with absolute values for their parameters greater than 1.5. This expression is used to extend the sustainability to the rest of the forest (the maps of sustainable indicators all over the forest is all you need) and to design the best management plan (Equation (12)).

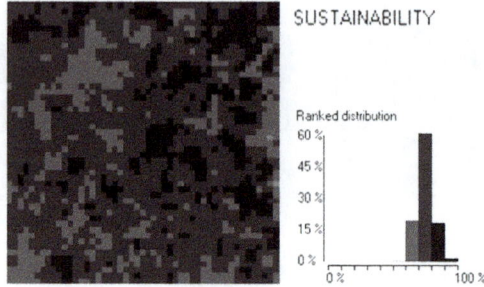

Figure 5. Sustainability assessment throughout the entire the Study Area for the participant who filled Table 2. Map of sustainability (for pixel of 10 m size) calculated by combining pair-wise comparison and objective criteria aggregation procedures.

Analytical equation of sustainability calculated through multiple regression analysis between the value in Figure 5 and the value of the main sustainability indicators at all of the points of the territory is

$$U = -3.166 I_2 + 5.08 I_1 I_3 + 1.519 I_2 I_3 \tag{12}$$

where I_1: Structural diversity; I_2: Timber yield; I_3: Biomass.

The above outcomes are sequential: first, the sustainability on the significant points was obtained (outcome number 1). Then, outcomes 2 and 3 were obtained by adding the physical and management parameters of the territory to outcome 1. Consequently, for specified forest exploitation, the relevant information came from the sustainability in the six points subjected to comparison.

3.1.2. The Management Plan that Best Suits Individual Preferences

A management plan involves formulating a set of temporally-located actions to be carried out at a point in the territory. However, the number of viable plans is so high that it is not feasible to generate all of the possible options to choose the best one. For instance, if there were ten activities per tree and year (cutting, pruning, labor, pest treatment, and any combination) and the number of trees in our study area was 500,000, then the number of potential management plans would be $100^{500,000}$ (for a planning horizon of ten years). To get around this problem, we determined the best management plan by applying combinatorial algorithms for optimization ([58] and [60]).

Specifically, we applied the Simulated Annealing algorithm [61] to optimize the following function:

$$U_A(X) = -\left(\sum_{i=1,\dots,C} \sum_{j=1,\dots,L} \mathbf{1}_{P_i=y_j} u_A\left(P_i = y_j\right) - \beta V(X) \right) \tag{13}$$

where: A is a participant in the decision process. P_i is a point in the territory and its neighborhood (a circle of 50-m diameter around each point); Y_j is a specified set of activities (cut trees, prune lower branches, styling, eliminate feedback, etc.) carried out on P_i; L is the number of feasible activities that can be carried out at any point and in its neighborhood; C is the number of points in the territory (usually the points are pixels, but they can also be trees or homogeneous territorial units); $\mathbf{1}$ is the indicator function (1 if $P_i = y_j$ and 0 otherwise); $\beta V(X)$ is a regularization term; $u_A(P_i = y_j) = k_1 v_A(P_i = y_j) + k_2 B_i(P_i = y_j)$, where: $v_A(P_i = y_j)$ is the A assessment of sustainability for $P_i = y_j$ and $B_i(P_i = y_j)$ is the economic balance for changing the current land use to $P_i = y_j$.

Optimizing expression (13) leads to the most sustainable forest management plan for a specific participant. You can see [54,58,60] for a detailed explanation of the spatial-temporal location of activities given by the management plan. Basically, it determines (on each pixel) the number of trees to be cut (per ha) on each height class, and the labors required to facilitate natural regeneration. To ensure sustainability, the solutions are constrained to feasible sustainable solutions defined by the experts.

Table 3 describes the best forest management plan for the participant in Table 2 and Figure 5. Measures on Table 3 are trees heights because current data collection technologies (combining aerial and terrestrial LiDAR (Light Detection and Ranging), photogrammetry, and remote sensing with measurements in the field) make easy tree-heights measurements [62] in large forest areas. It is also socking to show the number of trees from 0 to 3 m height. It is because a dynamic programming model has been applied to predict the height classes evolution [54,56]. This model generates the future number of trees in all the height classes, which is incorporated in the timber yield indicator. In no case 0 to 3 m trees are harvested by clearing or precommercial thinning, the most correspond to natural self-thinning mortality. In the field, the inventory of trees from 0 to 3 m is made in a sub-plot and its exact height is not recorded, it is only recorded if it is in the range, visually.

Figure 6 shows the flow charge for representation of opinions for multiple participants.

Table 3. Current stock, trees to be cut in the next 10 years and description of the current and of the proposed forest plan for the preferences of the participant in Table 2, for the total study area.

Stock Variables for Height Distribution			
Tree Height (meters)	Timber to Harvest in 10 Years (stems/ha)	Trees after the Management Plan (stems/ha)	Standing Trees (stems/ha)
0–3	30,879	632,880	230,142
3–6	3600	46,407	24,599
6–9	2458	13106	14,741
9–12	630	7176	7508
12–15	210	6155	6774
15–18	165	5032	4869
18–21	64	3515	3171
21–24	264	1459	1493
Economic Variables after the Plan (in Euros for the Entire Area of 96.04 ha)			
Cut & Harvest wood COST			39,549.01
Pruning & Debris elimination COST			54,088.35
Gross Timber INCOME			54,399.77
Bio-fuel & Public subsidies INCOME			41,130.92
Sustainability (as % of the IDEAL FOREST sustainability)			
Current overall sustainability value			74.99
Overall sustainability after Plan			86.51

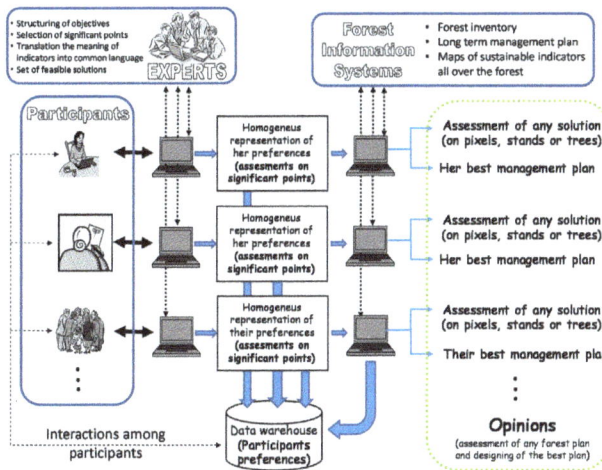

Figure 6. Role of different agents in decision-making for confidence-based participation. Its output is the input of Figure 1.

3.2. Measuring Empathy and Convergence of Opinions

Our approach to collective decisions requires recurrent calculating of two items at any point in the process: the convergence of the expressed opinions, and the overall empathy among participants. This section describes how to calculate both items for a group of twenty participants (the analysis of larger groups is similar, but showing results is more complicated).

The information required to calculate these items is the representation of opinions for all participants, which is calculated from the sustainability at six meaningful points of the territory. Table 4 shows the sustainability for all members of the group at the most meaningful spatial points.

Table 4. Assessments of sustainability for the group of 20 participants.

Participant-ID	Sustainability on the Comparison Points and for All the Territory						
	P-1	**P-2**	**P-3**	**P-4**	**P-5**	**P-6**	**Study Area v(X)**
E-1	0.49	0.41	0.12	0.07	0.12	0.42	0.272
E-2	0.89	0.16	0.37	0.27	0.07	0.42	0.363
E-3	0.37	0.07	0.52	0.57	0.52	0.32	0.395
E-4	0.42	0.34	0.87	0.46	0.42	0	0.418
E-5	1	0	0.22	0.5	0.22	0.59	0.422
E-6	0.59	0.46	0.97	0.27	0.07	0.27	0.438
E-7	0	1	0.62	0.51	0.07	0.02	0.37
E-8	0.85	0.42	0.82	0.62	0	0	0.452
E-9	1	0.62	0.22	0	0	0.59	0.405
E-10	0.75	0.07	0.52	0.47	0.02	0.87	0.45
E-11	0.83	0.62	0.12	0.55	0	0.74	0.477
E-12	0.45	0.62	1	0.47	0.32	0.22	0.513
E-13	0.83	0.67	0.07	0.25	0	0.94	0.46
E-14	0.32	0.22	0.72	0.67	0.92	0.22	0.512
E-15	0.92	0.07	0.72	0.52	0.77	0.07	0.512
E-16	0.57	0.07	0.67	0.52	0.97	0.02	0.47
E-17	0.54	0.66	0.97	0.92	0	0.42	0.585
E-18	1	0.72	0.32	0.7	0.32	0.64	0.617
E-19	0.42	0.07	1	0.72	0.77	0.22	0.533
E-20	1	0.87	0.07	0.8	0.27	0.64	0.608
Group (g)	0.414	0.254	0.341	0.308	0.183	0.238	0.29
Expert (Ob)	0.92	0.75	0.54	0.42	0.26	0.07	0.493

3.2.1. Convergence of Opinions

Calculation of convergence requires knowing the aggregated solution over the entire time. For this case study, the aggregated assessment is the average of the values for all the participants at the six meaningful points. Convergence of opinions (*CoO*) has been calculated as the sum of the variation coefficients at each of the six meaningful points:

$$CoO = 100 \left\{ \sum_{i=1}^{6} (s_i / \bar{x}_i) \right\} / 6 \tag{14}$$

where $\bar{x}_i$ and s_i are, respectively, the average and the standard deviation of the sustainability assessments in the *i*-meaningful point and for all the participants in the group.

3.2.2. Overall Empathy of the Group

All the information necessary to calculate empathy is derived from the data provided in Table 4. Following the methodology described in Section 2.1, the first task to calculate is the utility that a participant (*A*) believes that a solution (*X*) has for another participant (*B*), which involves application of the equation 1. The variables we have used to calculate equation 1 are described below.

- $D_{i\leftrightarrow K}$, which is the reciprocal distance between i and B's opinions, is calculated as:

$$D_{i\leftrightarrow B} = \|u_i(\mathbf{X}_{\text{Real}}), u_B(\mathbf{X}_{\text{Real}})\| = \frac{\sqrt{(u_i - u_B)\Sigma^{-1}(u_i - u_B)^T}}{D_{\max}} \quad (15)$$

where: $D_{i\leftrightarrow B}$ is the reciprocal similarity in the way participants i and B make their sustainability assessments; $u_i(\mathbf{X}_{\text{Real}}) = [v_i(\mathbf{X}_{\text{Real}})_{P1}, \dots , v_i(\mathbf{X}_{\text{Real}})_{P6}]$ are the sustainability values that i assigns to the six points being compared (P1, ... , P6) in the current land use ($\mathbf{X}_{\text{Real}}$); Σ^{-1} is the inverse covariance matrix of the sustainability assessments at the most representative territorial points for the data of all the active participants (Table 5); and $D_{\max}$ is the maximum value of the numerator in Equation (15) for all pairs of active participants.

Table 5. Sample covariance matrix and its inverse for the simulated opinions of 100 participants. Both are calculated for all the pairs of meaningful points (P1 to P6).

Participants	P1	P2	P3	P4	P5	P6
			Sample Covariance Matrix			
P1	0.0865	0.01127	−0.02471	−0.01023	−0.01914	0.03407
P2	0.001127	0.10156	−0.00428	0.00334	0.0266	−0.01522
P3	0.02471	−0.00428	0.08395	0.02746	0.01118	−0.02735
P4	−0.01023	0.00334	0.02746	0.0673	0.01217	−0.01564
P5	−0.01914	−0.0266	0.01118	0.01217	0.0692	−0.01385
P6	0.03407	−0.01522	−0.02735	−0.01564	−0.01385	0.09163
			Inverse of Sample Covariance			
Participants	P1	P2	P3	P4	P5	P6
P1	14.816	−1.6899	2.4984	−0.172	2.1223	−4.7533
P2	−1.6899	11.9061	1.0144	−1.3497	4.8655	3.4124
P3	2.4984	1.0144	15.3689	−5.2802	0.1151	2.9422
P4	−0.172	−1.3497	−5.2802	17.7403	−2.6566	0.8895
P5	2.1223	4.8655	0.1151	−2.6566	17.8146	2.2913
P6	−4.7533	3.4124	2.9422	0.8895	2.2913	14.6224

- $\|X - \mathbf{X}_i\|$ is the distance between the best solution for i ($\mathbf{X}_i$) and the solution to which we are calculating the utility (X):

$$\|X - \mathbf{X}_i\| = |U_i(X) - U_i(\mathbf{X}_i)| \quad (16)$$

where $U_i(X)$ is the utility of X solution for the i participant, calculated like in Equation (12).
- $v_i(X)$ is the assessment of X made by i. It is calculated as the average sustainability at all the points of the study area (see Section 3.1).
- i refers to all the involved opinions ($o_i = [\mathbf{X}_i, v_i(X)]$), $i \in \{A, B, n(B)_A, g, Ob\}$. A and B are the mentioned participants, and:

  - $n(B)_A$ are the opinions that a participant (A) considers to be close to the opinion of each of the other participants (B). These opinions satisfy both: the reciprocal similarities between their sustainability assessments and the A assessment of sustainability are bounded to a fixed value ($D_{A\leftrightarrow K} < K_0$ for $K = 1, \dots , N$ participants; with $D_{A\leftrightarrow K}$ calculated as shown in the Equation (14) and, simultaneously, their opinions belong to stakeholders of the same type as B.
  - o_{Ob} is the expert opinion on sustainability (which is supposed to be more objective than other opinions).
  - o_g is the aggregate opinion, calculated as an average of the assessment for all participants.

The next task is to determine the equivalence of the utilities of A and B when the aggregate assessment of the group (X_g) is modified. The changes we adopted represented a variation of $\pm$ 5% in the values of all the components in $u_g(X_g)$, which is the equivalent of 729 changes. We then carried out the process described in Section 2.1 up to calculate $\mathbf{emf}_A$ for an A-participant (Equation (4)). Each one of these matrices has 729 rows and 20 columns.

The reciprocal similarities of empathy between participants (Equation (5)) are shown in Table 6.

Equation (6) is then applied to the data on Table 6 to calculate the average of the empathetic similarities. The overall similarity for the twenty analyzed participants is $S_T = 80.99\%$.

Table 6. Matrix of similarity of reciprocal empathy (%) between participant opinions for the group of 20 participants.

Group	E-1	E-2	E-3	E-4	E-5	E-6	E-7	E-8	E-9	E-10	E-11	E-12	E-13	E-14	E-15	E-16	E-17	E-18	E-19	E-20
E-1	100	83.27	75.83	72.95	73.79	72.35	73.79	72.37	75.57	71.71	63.08	60.82	52.67	63.2	63.44	59.35	56.78	49.97	55.62	64.3
E-2	83.27	100	88.93	85.39	88.9	87.2	63.2	87.29	71.87	85.57	62.58	75.83	46.9	65.55	74.38	54.14	72.15	66.05	60.56	74.46
E-3	75.83	88.93	100	93.58	88.11	89.97	54.91	89.58	61.35	83.5	52.73	82.4	36.57	72.14	83.11	57.42	75.87	69.08	67.63	68.11
E-4	72.95	85.39	93.58	100	86.34	91.29	54.39	91.33	59.71	82.7	51.6	85.32	35.23	74.45	86.95	57.82	78.19	69.42	71.28	67.35
E-5	73.79	88.9	88.11	86.34	100	90.7	56.55	90.32	66.37	93.83	59.56	82.21	42.41	64.09	79.04	50.24	79.57	73.19	62.53	76.04
E-6	72.35	87.2	89.97	91.29	90.7	100	56.2	97.55	63.9	88.9	57.17	87.9	39.84	66.98	81.22	50.86	83.54	74.08	65.37	74.18
E-7	73.79	63.2	54.91	54.39	56.55	56.2	100	56.36	79.17	57.44	71.43	45.57	68.62	45.21	45.47	43.33	44.42	36.26	40.15	58.97
E-8	72.37	87.29	89.58	91.33	90.32	97.55	56.36	100	64.15	88.18	57.34	87.39	40.03	67.02	81.32	50.98	83.2	73.92	65.28	74.41
E-9	75.57	71.87	61.35	59.71	66.37	63.9	79.17	64.15	100	67.75	86.42	53.09	74.46	61.06	49.74	38.29	53.62	48.19	39.89	73.91
E-10	71.71	85.57	83.5	82.7	93.83	88.9	57.44	88.18	67.75	100	62.35	81.21	44.86	61.06	75.94	47.06	80.96	73.11	60.73	78.33
E-11	63.08	62.58	52.73	51.6	59.56	57.17	71.43	57.34	86.42	62.35	100	48.09	80.3	63.65	81.8	26.53	50.91	46.59	31.29	74.54
E-12	60.82	75.83	82.4	85.32	82.21	87.9	45.57	87.39	53.09	81.21	48.09	100	29.86	63.65	81.8	44.88	90.67	80.79	64.54	68.82
E-13	52.67	46.9	36.57	35.23	42.41	39.84	68.62	40.03	74.46	44.86	80.3	29.86	100	21.63	25.83	18.1	32.06	27.26	17.98	55.26
E-14	63.2	65.55	72.14	74.45	64.09	66.98	45.21	67.02	61.06	61.06	63.65	63.65	21.63	100	60.02	78.02	74.4	56.08	86.73	44.2
E-15	63.44	74.38	83.11	86.95	79.04	81.22	45.47	81.32	49.74	75.94	81.8	81.8	25.83	60.02	100	60.02	65.74	65.74	79.18	58.31
E-16	59.35	54.14	57.42	57.82	50.24	50.86	43.33	50.98	38.29	47.06	26.53	44.88	18.1	78.02	60.02	100	37.49	28.38	69.75	31.03
E-17	56.78	72.15	75.87	78.19	79.57	83.54	44.42	83.2	53.62	80.96	50.91	90.67	32.06	74.4	65.74	37.49	100	85.05	57.97	72.77
E-18	49.97	66.05	69.08	69.42	73.19	74.08	36.26	73.92	48.19	73.11	46.59	80.79	27.26	56.08	65.74	28.38	85.05	100	47.02	71.44
E-19	55.62	60.56	67.63	71.28	62.53	65.37	40.15	65.28	39.89	60.73	31.29	64.54	17.98	86.73	79.18	69.75	57.97	47.02	100	42.45
E-20	64.3	74.46	68.11	67.35	76.04	74.18	58.97	74.41	73.91	78.33	74.54	68.82	55.26	44.2	58.31	31.03	72.77	71.44	42.45	100

3.3. Verification of the Strategy for Confidence-Based Collective Decision-Making

We used simulated data instead of real opinions to verify the strategy because we wanted to analyze the influence of the two main factors affecting collective decisions: the number of well-informed participants and the interactions among them. It is not possible to separate a person's own opinion from the changes due to interactions when dealing with human decision-makers, but it is easier to isolate the effects of these factors with simulated data.

As shown in Sections 3.1 and 3.2, all the relevant information for calculating empathy and convergence comes from the sustainability at the six most meaningful points in the territory. We have therefore simulated the opinions of one hundred participants by assigning a sustainability value to these points, under the following conditions:

– The sustainability for each meaningful point and each stakeholder came from sampling a normal distribution with the mean and standard deviation shown in Table 7. The mean corresponds to the sustainability assessments of a well-informed participant on each type of stakeholder. Standard deviations show variations in data from real participative processes [30,58].
– Table 7 also shows the number of participants for each type of stakeholder.
– The opinions for the simulated participants (one hundred) are summarized in Figure 7.
– To measure the separation between opinions (Equation (4)), we used the inverse of the covariance matrix (for the one hundred simulated participants) between all the possible pairs of significant points. The covariance matrix and its inverse are shown in Table 6.

Table 7. Parameters for generating sustainability of 100 participants, grouped into five stakeholder types (S1 to S5 in Figure 8) and at six meaningful points (P1 to P6 in Table 2). Sustainability assessments are simulated from sampling an $N(\mu; \sigma)$ distribution with the parameters shown in the table. The number of simulations carried out for each type of stakeholder is also shown.

Type of Stakeholder	Parameters: $(\mu; \sigma)$						Number of Simulations
	P1	P2	P3	P4	P5	P6	
S1	(0.88; 0.2)	(0.6; 0.3)	(0.75; 0.4)	(0.55; 0.2)	(0.2; 0.3)	0.25; 0.4)	18
S2	(0.4; 0.4)	(0.92; 0.4)	(0.75; 0.4)	(0.54; 0.2)	(0.1; 0.2)	(0.2; 0.2)	13
S3	(0.5; 0.2)	(0.2; 0.2)	(0.75; 0.2)	(0.6; 0.2)	(0.7; 0.2)	(0.3; 0.2)	14
S4	(0.96; 0.2)	(0.5; 0.3)	(0.25; 0.2)	(0.33; 0.3)	(0.1; 0.2)	(0.72; 0.2)	23
S5	(0.77; 0.3)	(0.69; 0.4)	(0.75; 0.3)	(0.5; 0.4)	(0.2; 0.3)	(0.4; 0.4)	32

We consider the above participants to be well informed because the standard deviation is not too large (means are calculated by consulting experts on each stakeholder group). We consider that the larger the standard deviation, the worse-informed the participants.

The average convergence of opinions (Equation (14)) for the participants in Table 7 is: $CoO = 73.67\%$.

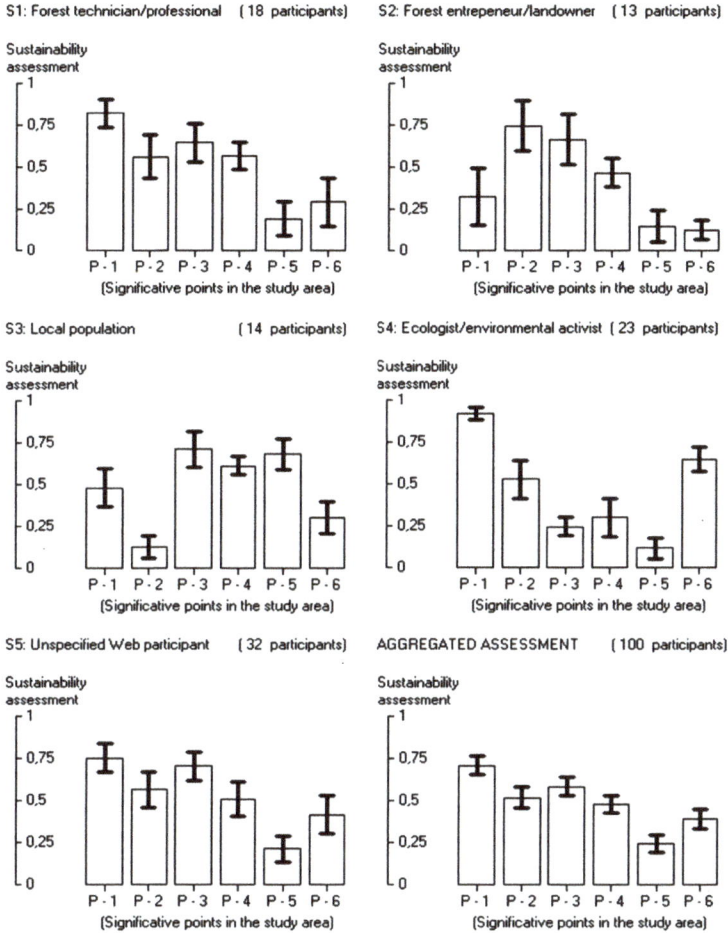

Figure 7. Summary of the sustainability assessments for the participants (5 Stakeholders and the aggregated assessment) under the simulation parameters described in Table 4. Top ends of bars indicate observation means of the group. Line segments represent confidence intervals $(1 - \alpha = 95\%)$ for the individual assessments.

To analyze the evolution of the participative process (Section 3.3.1), we started with a group of five participants (randomly chosen from the set of one hundred). For the participants in this group, we measured both the aggregated empathy among participants and the convergence of their opinions. These calculations were repeated in new groups, formed by randomly adding five new participants to the current group. The expansion process continued until one hundred participant groups were reached.

Later, Section 3.3.2 analyzes the effects of interactions among the simulated participants.

3.3.1. Convergence of Opinions by Adding New Participants

Figure 8 shows the evolution of the measures used to monitor the participative process: the overall empathy between the opinions of the group (S_T, as in Equation (6)), and the convergence of opinions among the participants (CoO, as in Equation (14)).

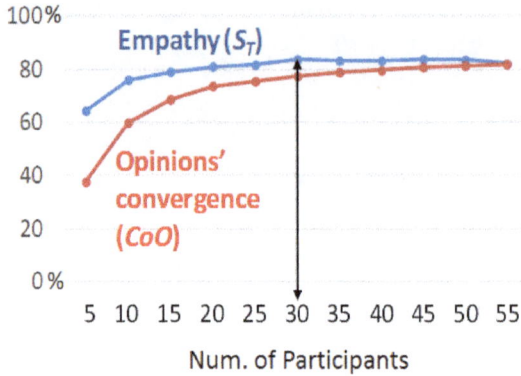

Figure 8. Evolution of the overall empathy and of the convergence of opinions with the number of participants.

As shown in the graph, empathy increased with participation until it reached a local maximum for thirty participants ($S_{T(30)}$ = 83.65%). It then shows very small changes that do not depend on the increase in the number of participants. On the other hand, the convergence of opinions grows rapidly from the group of five participants ($CoO_{(5)}$ = 37.59%) to the group of thirty participants ($CoO_{(30)}$ = 77.41%). It then continues to increase, but at a much lower rate ($CoO_{(55)}$ = 81.89%). The growth rate remains constant in the range between thirty and one hundred participants (for a number of participants not shown in the figure: $CoO_{(80)}$ = 84% and $CoO_{(100)}$ = 86.82%).

The association between empathy and convergence is evident: while empathy increased (up to 30 participants), convergence increased an average of 1.52% for each new participant. Subsequently, it increased at only a rate of 0.13% for each new participant.

The previous result could, on its own, validate the strategy of encouraging participation if empathy increases. However, aside from the number and quality of participants, there is another feature that plays a fundamental role in the convergence of opinions: the interactions among participants.

3.3.2. Influence of the Interactions among Participants on the Collective Solution

In participatory processes, the convergence of opinions due to the interactions among the participants would be expected to be at least as important as the convergence due to the number of well-informed participants. We validated this assertion through a computer simulation that analyzed the influence of interactions on the final decision (see Section 2.2 and Figure 2).

The mathematical simulation was used to calculate and compare the aggregate assessment of sustainability in two scenarios. The first one refers to an aggregate assessment for the participants generated based on the parameters in Table 7. The second scenario incorporates interactions through an agent-based simulation (described on Section 2.2) to explain the interactions among participants for the same participants than in the first case.

To get an aggregated assessment of sustainability without interactions, the joint value was calculated as the average of the individual sustainability for all the participants. We proceeded this way because it was the simplest method of aggregation. In general, there will be a different aggregate assessment, but this does not affect our objective: to see how interactions modify the aggregation of preferences. For this case study, u_g = (0.71, 0.51, 0.58, 0.47, 0.24, 0.39).

We then applied the methodology described in Section 2.2 to trigger interactions among the members of the group of 100 participants. The simulation was parameterized as discussed below:

- We were not interested in artificial acceleration of convergence of opinions. Therefore, only low-utility suggestions of changes were accepted in the random changes of opinion needed to simulate interactions. From a mathematical point of view, this meant adopting a low temperature for Equation (11) that controls the randomized changes of opinions.

- Since we accepted random and synchronized changes of opinion (all participants change, at the same time, to solutions accepted by more than 50% of them), the convergence of this algorithm is not monotonous [63]. This means that each iteration does not have to improve the previous solution, but after a certain number of iterations (in our case we limited this to 300) there are clear improvements.

To measure the convergence for interaction among participants, we used the distance between the last group-opinion accepted by each evaluator into their memory ($o_{g(A)} = (u_{g(A)}, v_{g(A)})$) and the current aggregated group-opinion ($o_g = (u_g, v_g)$). Both opinions change in each iteration of the process (i). We will therefore re-write them as $o_{g(A)-i}$ and o_{g-i} respectively. So:

$$\text{OpC}_i = \frac{1}{N} \sum_{A=1}^{N} \|u_{g(A)-i} - u_{g-i}\| \tag{17}$$

where $\|u_{g(A)-i} - u_{g-i}\|$ as in equation 15; OpC_i is the measure of the convergence of opinions in the i-iteration; and N is the total number of participants ($N = 100$).

In the simulation carried out, the maximum convergence of opinions corresponds to the iteration number 268 (see Figure 9). In this iteration, the aggregate evaluation corresponds to $u_{g\text{-}268} = (0.81, 0.56, 0.53, 0.52, 0.24, 0.44)$, which represents an improvement of 12.07% with respect to the "no interaction" scenario (percentage of improvement between OpC_{268} and OpC_1).

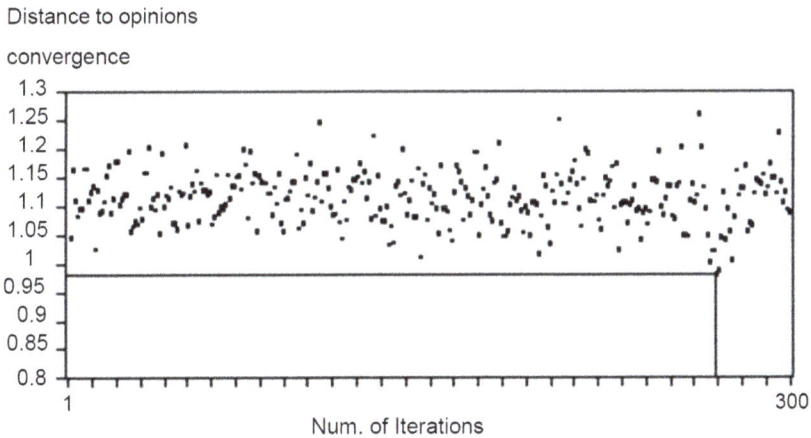

Figure 9. Convergence of aggregated opinion due to interactions among participants.

The reality is that there is no "no interaction" scenario with which to compare. We have simulated opinions around well-informed participants, but no opinion can be formed without interaction. That means all we have is a scenario of well-informed participants who have not interacted after formulating their opinions. On the other hand, we have a scenario with interaction after the initial formulation of preferences in which no new participants have entered. In any case, it is notable that:

- Both scenarios improve convergence.
- The only way we found to simulate participatory processes that do not converge is to avoid the simultaneous emergence of both factors. In other words, to incorporate ill-informed participants (with greater randomness in the opinion of the participants) and not allow post-formulation of opinions.

These outcomes reinforce the idea that participation means both things: the incorporation of a large number of participants and interaction between them.

4. Discussion and Conclusions

The paper validates an Artificial Intelligence methodology applicable to forest management. As in all AI (Artificial Intelligence) process, the algorithms must be trained in the available data, never replicated. This requires a type of manager able to understand the performance of algorithms to propose modifications when they have to apply to real cases. The paper has been written by foresters for these types of forest managers.

The first notable conclusion is that it is possible to measure confidence among participants in collective decisions. This is done through the construction of empathic utility functions.

A second conclusion deals with the association between empathy and the convergence of opinions. This relation has been commonly accepted, but this is the first time that it has been verified: the case study shows that increases in empathy are associated to greater convergence of opinions among participants, and when empathy ceases to increase, the rate of convergence of opinions slows.

The main conclusion of the paper was drawn by combining the two conclusions mentioned above: empathy variations can be incorporated to manage participatory processes, and this can be done by applying a simple strategy. The strategy is to encourage participation while empathy increases and, if not, to stop participation. When the participatory process ends, the results from the *Web* 2.0 and from interpersonal utilities are applied to aggregate opinions.

The strategy always leads to a solution for collective decision making. The aggregated solution will be socially acceptable in the case of high convergence of the opinions. In that case, the aggregate opinion will represent the preferences of almost all the people involved, and it will be the joint solution for the group. Otherwise, the set of participants must be broken down into homogeneous subgroups (with common opinions within each subgroup) and other aggregation procedures must be applied. These procedures may range from voting to an explained personal decision. Even in the case that the final decision was made by an official, the use of this strategy provides the information required for accountability.

It is also notable that incorporation of empathy improves the results obtained so far. When comparing the confidence-based strategy with the available participatory applications, it shows improvements in all the functional requirements required for a participative process.

Other conclusions deal with the power of networking. We confirmed that if participants report their judgments in digital systems, then the representation can be automated for both: the opinions of each participant, and the changes of empathy throughout the participative process. Automation also promotes an increase in the number of participants and facilitates interaction among them.

The influence of the number of participants is well known. Condorcet [12] was the first to relate the number of participants and the quality of the information they have with the final decision: when the components of a group are well-informed and the number of participants expands to incorporate all possible individuals, then the group will then certainly make correct decisions. The *Web* increases the number of people involved because access is free and it also allows "delegation" of the decisions of non-participants to participants close to them in social networks. However, the *Web* also increases the quality of the information, because participatory applications usually explain basic rules for forest management and for understanding natural processes. Besides, this information fits to the characteristics of a participant (for example, *SILVANET*, [54], provides information considering the coherence and depth of everyone's knowledge). All of these factors increase the likelihood that a participant will be well informed, although it does not ensure that everyone to be well-informed (the likelihood of a participant making correct decisions is greater than 50%) as requires Condorcet´s theorem.

The consequences of interaction between participants are usually considered under the umbrella of self-organization (in the terminology of complex systems) [64]. Self-organization is the process by which coordination arises out of interactions between participants in collective decisions (see Sections 2.2 and 3.3.2). A computer simulation has verified that self-organization enhances emergence (the whole—the results of the participatory process—is more than the parties—the mere aggregation of participants) in assessment of sustainability. In fact, the convergence of opinions after simulating interaction was greater than the convergence without self-organization. However, this does not seem to be the result just of the simulation carried out: self-organization has provided great advances in natural evolution, technology, and computation. Consequently, digital discussion forums should be favored to increase interactions among participants.

The confidence-based strategy also provides operative advantages, such as:

- The homogeneous representation of opinions for all participants. This means knowing how a participant assesses a forest plan and which forest management plan is best for them.
- The strategy is easily integrated into the most widely-used information systems. In fact, the information required is obtained from available information systems and from the preferences of each participant (which is new). Information is then processed with using the usual big-data and business analytic tools and, after applying the specific participatory algorithms described in this paper, the outcomes can be transferred to most common decision-support systems.
- Accountability in collective decision-making.

There is another indirect advantage: the proposed strategy counteracts the excesses of new ICT developments that tend to limit universal access to information and propose decisions that are only economically efficient. Specifically, unlike the *Web*, which provides free and global access to results, some of the new ICT developments—such as Artificial Intelligence algorithms or Internet of Things—require large investments and their outcomes are owned by the entities that have made the investments. Most of the time, these entities are for-profit organizations that optimize only their economic returns.

Author Contributions: Conceptualization, E.M.-F.; Methodology, E.M.-F.; Project administration, E.A.-T.; Supervision, E.A.-T.; Visualization, A.G.-A.; Writing—original draft, E.M.-F.; Writing—review & editing, C.G.-G.

Funding: This research received no external funding.

Conflicts of Interest: The authors declare no conflict of interest.

References

1. Bonabeau, E. Decisions 2.0: The power of collective intelligence. *MIT Sloan Manag. Rev.* **2009**, *50*, 45–52.
2. Myers, D.G. *Intuition, its Power and Perils*, 3rd ed.; Yale University Press: New Haven, England, 2002; ISBN 9780300130270.
3. Oztürk, M.; Tsoukiàs, A.; Vincke, P. Preference Modelling. In *Multiple Criteria Decision Analysis: State of the Art Surveys*; Figueira, J., Greco, S., Ehrgott, M., Eds.; Springer: New York, NY, USA, 2005; Volume 78, pp. 27–59. ISBN 978-0-387-23081-8.
4. Debreu, G. Representation of a preference ordering by a numerical function. In *Decision Processes*; Thrall, R., Coombs, C.H., Davies, R., Eds.; John Wiley & Sons: Oxford, UK, 1954; pp. 159–175. Available online: https://archive.org/details/decisionprocesse033215mbp (accessed on 4 March 2018).
5. Arrow, K. A difficulty in the concept of social welfare. *J. Polit. Econ.* **1959**, *58*, 328–346. [CrossRef]
6. Aleskerov, F.; Bouyssou, D.; Monjardet, B. *Utility Maximization, Choice and Preference*, 2nd ed.; Springer: Berlin, Germany, 2007; Volume 16, p. 283. ISBN 978-3-540-34183-3.
7. Suzumura, K. *Rational Choice, Collective Decisions and Social Welfare*; Cambridge University Press: New York, NY, USA, 1983; p. 278. ISBN 0521238625.
8. Sen, A. *On Ethics and Economics*; Blackwell Publishing: Malden, MA, USA, 1987; p. 148. ISBN 978-0-6311-6401-2.
9. Martín-Fernández, S.; Martínez-Falero, E.; Valentín-Gamazo, V. Optimization Methods to Identify the Best Management Plan. In *Quantitative Techniques in Participatory Forest Management*; Martínez-Falero, E.,

Martín-Fernández, S., García-Abril, A., Eds.; CRC Press: Boca Ratón, FL, USA, 2013; pp. 421–498. ISBN 9781466569249.

10. Arrow, K. *Social Choice and Individual Values*, 2nd ed.; Yale University Press: Westford, MA, USA, 1963; p. 124. ISBN 0-300-01364-7.

11. Borda, J.C. Mémoire sur les élections au scrutin. In *Histoire de L'academie Royale des Sciences*; Institut de France Academie des Sciences: Paris, France, 1784; pp. 657–665.

12. Condorcet, M. *Essai sur L'application de L'analyse 'a la Probabilité des Décisius Rendues á la Pluralité des voix*; Imprimerie de Du Pont: Paris, France, 1785; p. 304.

13. Laukkanen, S.; Kangas, A.; Kangas, J. Applying voting theory in natural resource management: A case of multiple-criteria group decision support. *J. Environ. Manag.* **2002**, *64*, 127–137. [CrossRef]

14. Vainikainen, N.; Kangas, A.; Kangas, J. Empirical study on voting power in participatory forest planning. *J. Environ. Manag.* **2008**, *88*, 173–180. [CrossRef] [PubMed]

15. Stensholt, E. Voces populi and the art of listening. *Soc. Choice Welf.* **2011**, *35*, 291–317. [CrossRef]

16. Edgeworth, F.J. *Mathematical Psychics: An Essay on the Application of Mathematics to the Moral Sciences*; C.K. Paul & Co.: London, UK, 1881; p. 168. Available online: https://archive.org/details/mathematicalpsyc00edgeuoft (accessed on 16 April 2018).

17. Marshall, A. *Principles of Economics*, 8th ed.; Palgrave Macmillan: Basingstoke, UK; p. 754. ISBN 978-0-230-24929-5.

18. Bentham, J. *An Introduction to the Principles of Moral and Legislation*; Payne and Son: London, UK, 1789; p. 377. Available online: http://www.koeblergerhard.de/Fontes/BenthamJeremyMoralsandLegislation1789.pdf (accessed on 16 April 2018).

19. Sen, A.K. The possibility of social choice. *Am. Econ. Rev.* **1999**, *89*, 349–378. [CrossRef]

20. Robbins, L. Interpersonal comparison of utility: A comment. *Econ. J.* **1938**, *48*, 635–641. [CrossRef]

21. Ayuga-Téllez, E.; González-García, C.; Martínez-Falero, E. Multiparticipant Decision-Making. In *Quantitative Techniques in Participatory Forest Management*; Martínez-Falero, E., Martín-Fernández, S., García-Abril, A., Eds.; CRC Press: Boca Ratón, FL, USA, 2013; pp. 499–556. ISBN 9781466569249.

22. Rawls, J. *A Theory of Justice*, rev. ed.; Belknap Press of Harvard University Press: Cambridge, MA, USA, 1999; p. 538. ISBN 0-674-00078-1.

23. Hammond, P.J. Interpersonal Comparisons of Utility: Why and how they are and should be made. In *Interpersonal Comparisons of Well-Being*; Elster, J., Roemer, J.E., Eds.; Cambridge University Press: New York, NY, USA, 1991; pp. 200–254. ISBN 0-521-39274-8.

24. Chevaleyre, Y.; Endriss, U.; Lang, J.; Maudet, N. A Short Introduction to Computational Social Choice. In *SOFSEM-2007: Trends in Theory and Practice of Computer Science*; van Leeuwen, J., Italiano, G.F., van der Hoek, W., Meinel, C., Sack, H., Plášil, F., Eds.; Springer: Berlin, Germany, 2007; Volume 4362, pp. 51–69. ISBN 978-3-540-69506-6.

25. Harsanyi, J.C. *Rational Behavior and Bargaining Equilibrium in Games and Social Situations*; Cambridge University Press: New York, NY, USA, 1977; p. 314. ISBN 0 521 20886 6.

26. Harsanyi, J.C. Normative validity and meaning of Von Neumann and Morgenstern utilities. *Stud. Log. Found. Math.* **1995**, *134*, 947–959. [CrossRef]

27. Suppes, P. Some formal models of grading principles. *Synthese* **1966**, *6*, 284–306. [CrossRef]

28. Sen, A. *Collective Choice and Social Welfare*, 1st ed.; Holden-Day: San Francisco, CA, USA, 1970; p. 225. ISBN 9780816277650.

29. Arrow, K. Extended sympathy and the problem of social choice. *Philosophia* **1978**, *7*, 223–237. [CrossRef]

30. Martínez-Falero, E.; Ayuga-Tellez, E.; Gonzalez-Garcia, C.; Grande-Ortiz, M.A.; Sánchez De Medina Garrido, A. Experts' Analysis of the Quality and Usability of SILVANET Software for Informing Sustainable Forest Management. *Sustainability* **2017**, *9*, 1200. [CrossRef]

31. De Sanctis, G.; Gallupe, R. A Foundation for the Study of Group Decision Support Systems. *Manag. Sci.* **1987**, *33*, 589–609. [CrossRef]

32. Bonczek, R.H.; Holsapple, C.W.; Whinston, A.B. *Foundations of Decision Support Systems*; Academic Press Inc.: New York, NY, USA, 1981; pp. 3–25. ISBN 978-0-12-113050-3.

33. Konsynski, B.R.; Kottemann, J.E.; Jay, F.; Nunamaker, J.F.; Stott, J.W. PLEXSYS-84: An Integrated Development Environment for Information Systems. *J. Manag. Inf. Syst.* **1984**, *1*, 64–104. [CrossRef]

34. Stefik, M.; Foster, G.; Bobrow, D.G.; Kahn, K.; Lanning, S.; Suchman, L. Beyond the chalkboard: Computer support for collaboration and problem solving in meetings. *Commun. ACM* **1988**, *30*, 32–47. [CrossRef]

35. Gallupe, R.B.; DeSanctis, G.; Dickson, G.W. The Impact of Computer Support on Group Problem Finding: An Experimental Approach. *MIS Q.* **1988**, *12*, 277–296. [CrossRef]

36. Hale, D.P.; Haseman, W.D.; Munro, D.L. Architectural requirements for integrating group decision support systems into the daily managerial experience. In Proceedings of the Twenty-Second Annual Hawaii International Conference on System Sciences, Kailua-Kona, HI, USA, 3–6 January 1989; IEEE Xplore: Piscataway, NJ, USA; pp. 321–325.

37. Gray, P.; Berry, N.W.; Aronofsky, J.S.; Helmer, O.; Kane, G.R.; Perkins, T.E. The SMU Decision Room Project. In *Transactions of the 1st International Conference on Decision Support Systems*; Young, D., Keen, P.G.W., Eds.; The Institute of Management Sciences: Providence, RI, USA, 1981; pp. 122–129.

38. Aronson, J.E.; Aronofsky, J.S.; Gray, P. An Exploratory Experiment in Group Decision Support Systems Using the SMU Decision Room. In Proceedings of the Decision Sciences Institute, Boston, MA, USA, 1987; Decision Sciences Institute: Houston, TX, USA; pp. 428–430.

39. Davison, R.M. A Survey of Group Support Systems: Technology and Operation. *Sprouts* **2008**, *1*, 1–12.

40. Available online: https://www.loomio.org/about (accessed on 13 April 2018).

41. What is Deliberative Polling®? Available online: http://cdd.stanford.edu/what-is-deliberative-polling/ (accessed on 13 April 2018).

42. Available online: https://www.capterra.com/decision-support-software/ (accessed on 13 April 2018).

43. Introducing Paramount Decisions. Available online: https://paramountdecisions.com/features/ (accessed on 13 April 2018).

44. Available online: https://www.capterra.com/p/104768/Analytica/ (accessed on 13 April 2018).

45. Available online: https://www.capterra.com/p/138255/1000Minds-Decision-Making/ (accessed on 13 April 2018).

46. Bagdasaryan, A. Systems Theoretic Techniques for Modeling, Control and Decision Support in Complex Dynamic Systems. In *Artificial Intelligence Resources in Control and Automation Engineering*; González, E.J., Acosta, L., Hamilton, A.F., Eds.; Universidad de la Laguna: Canary Islands, Spain, 2012; pp. 15–72. ISBN 978-1-60805-589-0.

47. LiquidFeedback. Available online: http://liquidfeedback.org/ (accessed on 13 April 2018).

48. About this project. Available online: https://www.allourideas.org/about (accessed on 13 April 2018).

49. Binmore, K. *Natural Justice*; Oxford University Press: New York, NY, USA, 2005; p. 224. ISBN 9780195178111.

50. Salehi-Abari, A.; Boutilier, C. Empathetic Social Choice on Social Networks. In Proceedings of the 13th International Conference on Autonomous Agents and Multiagent Systems (AAMAS 2014), Paris, France, 5–9 May 2014; Lomuscio, A., Scerri, P., Bazzan, A., Huhns, M., Eds.; 2014; pp. 693–700.

51. Sayama, H.; Farrell, D.L.; Dionne, S.D. The Effects of Mental Model Formation on Group Decision Making: An Agent-Based Simulation. *Complexity* **2011**, *16*, 49–57. [CrossRef]

52. Adler, M.D. Extended Preferences and Interpersonal Comparisons: A New Account. *Econ. Philos.* **2014**, *30*, 123–162. [CrossRef]

53. Pascual, C.; Garcia-Abril, A.; Garcia-Montero, L.G.; Martin-Fernández, S.; Cohen, W.B. Object-based semi-automatic approach for forest structure characterization using lidar data in heterogeneous *Pinus sylvestris* stands. *For. Ecol. Manag.* **2008**, *255*, 3677–3685. [CrossRef]

54. Martínez-Falero, E.; Martín-Fernández, S.; García-Abril, A. *SILVANET. Participación Pública para la Gestión Forestal Sostenible*; Fundación Conde del Valle de Salazar: Madrid, Spain, 2010; p. 106. ISBN 978-84-96442-36-8.

55. Reynolds, K.M. *EMDS Users Guide (Version 2.0): Knowledge Based Decision Support for Ecological Assessment*; General Technical Report PNW-GTR-470; Forest Service, USDA: Washington, DC, USA, 1999; p. 63.

56. Ayuga-Téllez, E.; Mauro-Gutiérrez, F.; Antonio García-Abril, A.; González-García, C.; Martínez-Falero, E. Comparison of estimation methods to obtain ideal distribution of forest tree height. *Comput. Electron. Agric.* **2014**, *108*, 191–199. [CrossRef]

57. Martínez-Falero, E.; Martín-Fernández, S.; Orol, A. Assessment of Sustainability Based on Individual Preferences. In *Quantitative Techniques in Participatory Forest Management*; Martínez-Falero, E., Martín-Fernández, S., García-Abril, A., Eds.; CRC Press: Boca Ratón, FL, USA, 2013; pp. 499–556. ISBN 9781466569249.

58. Martín-Fernández, S.; Martínez-Falero, E. Sustainability assessment in forest management based on individual preferences. *J. Environ. Manag.* **2018**, *206*, 482–489. [CrossRef] [PubMed]

59. Fishburn, P.C. *Interval Orders and Interval Graphs. A Study of Partially Ordered Sets*; John Wiley & Sons Inc: New York, NY, USA, 1985; p. 230.

60. Martín-Fernández, S.; García-Abril, A. Optimisation of spatial allocation of forestry activities within a forest stand. *Comput. Electron. Agric.* **2005**, *49*, 159–174. [CrossRef]

61. Geman, S.; Geman, D. Stochastic Relaxation, Gibbs Distributions, and the Bayesian Restoration of Images. *IEEE Trans. Pattern Anal. Mach. Intell.* **1984**, *6*, 721–741. [CrossRef] [PubMed]

62. Valbuena, R.; Hernando, A.; Manzanera, J.A.; Martínez-Falero, E.; García-Abril, A.; Mola-Yudego, B. Most similar neighbor imputation of forest attributes using metrics derived from combined airborne LIDAR and multispectral sensors. *Int. J. Dig. Earth* **2017**. [CrossRef]

63. Besag, J.E. On the Statistical Analysis of Dirty Pictures. *J. R. Statist. Soc. Series B* **1986**, *48*, 259–302. [CrossRef]

64. Morçöl, G. Self-Organization in Collective Action: Elinor Ostrom's Contributions and Complexity Theory. *Complex. Gov. Netw.* **2014**, *1*, 9–22. [CrossRef]

forests

MDPI

Article

Implementing Participatory Processes in Forestry Training Using Social Network Analysis Techniques

Simone Blanc [1,*], Federico Lingua [1], Livio Bioglio [2], Ruggero G. Pensa [2], Filippo Brun [1] and Angela Mosso [1]

[1] Department of Agricultural, Forest and Food Sciences (DISAFA), University of Torino, 10095 Grugliasco, Italy; federico.lingua@unito.it (F.L.); filippo.brun@unito.it (F.B.); angela.mosso@unito.it (A.M.)
[2] Department of Computer Science, University of Torino, 10149 Torino, Italy; livio.bioglio@unito.it (L.B.); ruggero.pensa@unito.it (R.G.P.)
*   Correspondence: simone.blanc@unito.it; Tel.: +39-011-6708684

Received: 28 June 2018; Accepted: 26 July 2018; Published: 30 July 2018

Abstract: Public participation has become an important driver in increasing public acceptance of policy decisions, especially in the forestry sector, where conflicting interests among the actors are frequent. Stakeholder Analysis, complemented by Social Network Analysis techniques, was used to support the participatory process and to understand the complex relationships and the strong interactions among actors. This study identifies the forestry training sector stakeholders in the Western Italian Alps and describes their characteristics and priorities, in relation to training activities on entrepreneurial topics for forestry loggers. The hierarchy among actors has been identified, highlighting their respective roles and influence in decision-making processes. A lack of mutual communication among different and well-separated categories of actors has been identified, while good connections between stakeholders, operating in different territories, despite the presence of administrative and logistical barriers, have been observed. Training is a topic involving actors with different roles and interests. Nevertheless, all actors consider training about how to improve yields of forest operations and how to assess investments, particularly in innovative machinery, to be crucially important and conducive to a better comprehension of the wood supply chain and the enhancement of the raw material.

Keywords: entrepreneurial education; forestry training; innovative training; participatory process; social network analysis; stakeholder analysis

1. Introduction

Since the UN Conference on Environment and Development, held in Rio de Janeiro in 1992, the idea that the route towards the sustainable exploitation of forestry resources should follow a participatory approach, even in the forest planning phase, has been openly acknowledged [1,2]. In fact, public participation is expected to produce better plans by fostering the exchange of information and views between stakeholders [3]. The new EU forest strategy [4] set out how the participation of different stakeholders has become an important driver in increasing the public acceptance of policy decisions and in creating an inclusive platform for constructive discussions [5].

This participative process is especially important in the forestry sector, which includes a multitude of conflicting interests among the actors [6].

As stated by Ananda and Herath [7], environmental policy cannot be separated from public participation in the decision-making process, however, there is a lack of proven methods that explicitly integrate the beliefs of stakeholders.

One of the fundamental tools for supporting the participatory process is Stakeholder Analysis (SA) [8,9], a technique based on studies of the dynamics of social interaction, introduced for the first time in the 1930s. This technique was developed for the sake of understanding the different roles played by stakeholders with respect to the various interests represented [10]. Stakeholder Analysis is a technique that can be applied in several research areas of the forestry sector and wherever stakeholders are present, i.e., it analyses stakeholders interested in ecosystem services [11], mapping actors who participate in the planning of natural resource management [12] and assessing the participatory process in environmental management and governance [13].

Despite being widely applied, SA has been criticized for several weaknesses regarding analytic, qualitative and academic rigor [8]. Hence, in recent years, SA has frequently been complemented by Social Network Analysis (SNA) techniques [14]. The joint use of these techniques is particularly useful for understanding the complex relationships and strong interactions between actors that are typical of environmental policy processes.

SNA offers a quantitative approach to investigate collaborative processes, given that the analysis unit is the relationship between two entities and not the entity itself [15], considering that the networks consist of mechanisms and patterns of connections [16], measured through communications or exchanges among actors [17].

Using SNA and SA at the same time has been shown to have an impressive potential in generating complimentary results [14], where SNA requires more rigorous data collection and supplies quantitative results, and SA has a qualitative approach, providing fine-grained insights into stakeholders' preferences and characteristics.

Thanks to the positive effects of a combined use of SA and SNA, both were applied in our research to analyze the forestry sector network in the Western Italian Alps, focusing on the actors involved in the training activities of forestry workers.

The need to investigate this sector in depth is also indicated by other authors. For instance, a recent survey [18] has shown that Italian forestry enterprises have a worrying economic instability, mainly due to the low profitability of forestry operations. As stated by Morat [19], one of the suggested strategies to overcome this issue is the introduction of training courses focusing on entrepreneurial topics, such as economics, management, market, performance and quality. In reality, training is the most effective way to increase innovativeness [20], and the development of innovativeness could give loggers competitive advantages [21].

The design of training courses that target forestry workers is at the same time a challenging and an essential task that requires careful consideration.

The study area includes three administrative regions, located in the Western Italian Alps, Piedmont, Liguria and the Aosta Valley, which are representative of Alpine forest resources, as woodlands cover almost 40% (1,359,000 ha) of the territory (in the Alps, forests cover approximately 45% of the terrain) [22].

In this area, there are approximately 3000 forestry workers [23], and training courses have been implemented in recent years. Generally, the enterprises are characterized by small dimensions and small logging volumes [18]. This peculiarity should represent a stimulus for institutions to increase the forestry training offer, as small enterprises show a pronounced ability to adopt new business scenarios [24].

Loggers have shown a high interest in the training courses provided over the last decade, in fact, in this area, between 2007 and 2015, more than 350 courses have taken place, attended by over 2500 participants [25]. Most of the courses focused on technical topics, such as safety and use of exploitation techniques, while only a minor part dealt with managerial and entrepreneurial issues. The gap between the offer of technical and entrepreneurial training courses is evident, and this trend shows no sign of a reversal. Therefore, it is imperative to investigate the stakeholders involved in the forestry sector, with specific reference to those involved in the design and execution of training courses. Specifically, the goals of this study were to identify and to describe: (i) The forestry training sector stakeholders and their characteristics; (ii) the features of their network; (iii) the opinions and priorities of the different stakeholders concerning training activities on entrepreneurial topics for forestry loggers.

2. Materials and Methods

The analyses can be divided into three main steps: (i) Identification and classification of stakeholders involved in the forestry sector with interests in training; (ii) investigation of the stakeholder network and collection of their interest on entrepreneurial forestry training; (iii) gaining an insight into stakeholders' opinions on entrepreneurial forestry training and their mutual cooperation.

2.1. Initial Stakeholder Analysis

The first step of stakeholder analysis is the identification of the actors involved in the network [8]: During this process, all the interested parties, who influence or are influenced by the policies, decisions and actions, are contacted.

In Italy, each administrative region autonomously defines the contents and rules of professional training, in compliance with national guidelines. In order to identify the stakeholders involved in forestry training in the study area, an initial questionnaire was submitted to the Regional Forestry Offices of the three administrative Regions involved in the study. Thanks to this first survey, an initial group, formed by 24 stakeholders who have direct relations with the offices of public institutions, was identified.

Then a snowball sampling technique was applied: This technique consists in asking the actors to list additional people and institutions, involved in the sector, to be interviewed [26,27]. Finally, a list of 54 stakeholders was obtained.

The second step of SA is the stakeholder classification. In this study area, the classification of actors is relatively straightforward. This is because the forestry sector is rather small-scale and not particularly dynamic, where the stakeholders' knowledge of education and training is balanced, and the role they play in the forestry sector is relevant.

The stakeholder classification was carried out by a focus group composed of the authors and officials of the three regional institutions involved in the study. The composition of the focus group was defined, considering the high level of knowledge that each participant has of all the other stakeholders present in the network.

Initially, each participant of the focus group autonomously indicated their classification proposal on the basis of their experience. Subsequently, all the proposals were shared and discussed, in order to obtain the definitive classification.

On the basis of the interest pursued (hereinafter "interest"), the original classification proposed by Lienert et al. [14] was adapted to the studied context, identifying six main stakeholder categories: Economic, education and research, environmental, legal, political and, finally, technical.

On the basis of the stakeholder role in forestry training (hereinafter "class"), the stakeholders were classified, adapting the original classification by Paletto [5] into: Actors of the forest-wood chain, associations, owner associations, public administrations, research institutes, training centers and others.

2.2. Identification of Stakeholder Network and Their Interests in Entrepreneurial Forestry Training

In the second step, a questionnaire was submitted in order to investigate the stakeholders' interest in entrepreneurial forestry training.

This questionnaire was divided into two parts: The first section investigated the collaboration between the different stakeholders, attempting to understand the shape of the network, thanks to their relationships. A list of stakeholders, obtained through the initial SA, was included in the questionnaire in order to investigate the cooperation between actors. The same actors were then asked to indicate with whom they cooperated and to specify the kind of cooperation, according to Coulson [28]: (i) Frequent collaboration, "contacts fortnightly or weekly," considered a strong tie, (ii) occasional collaboration, "one contact per month," considered a weak tie, or (iii) no contact, if less than one contact per month.

Stakeholders characterized by strong ties have similar backgrounds and views, and their communications are effective [29]. For this reason, strong ties are frequent in relationships that contribute to the growth and success of a business sector [30]. On the contrary, weak ties are typical of low emotional intensity relationships, but, on the other hand, weak ties give access to a variety of information and can build bridges between individuals [31].

In the second section, the respondents were asked to indicate the priority level of the entrepreneurial training topics.

To identify the best suited subjects for Italian forestry training, the authors designed a questionnaire, where the management skills were organized into 5 main areas, as proposed by Morat [19]. Then, each area included two specific skills, as indicated as focal for forestry entrepreneurial training by FAO [32] (Table 1).

Table 1. Management skills investigated.

Area	Skills
Economics	Accounting
	Taxation
Performance	Work organization
	Schemes of partnership and associations
Timber & Market	Marketing
	Forestry supply-chain
Quality	Due Diligence in the forestry and wood sector
	Forest certification
Management	Investments
	Information technology

Finally, each stakeholder identified the priority level of entrepreneurial training topics using a three-point Likert scale [33], where the possible answers were: "low interest," "intermediate interest" and "high interest."

2.3. Social Network Analysis

Data obtained from the survey were transferred into a matrix scheme and, subsequently, used for the analysis. In order to describe the general aspects of the social network, the following features were calculated: (i) Diameter: The longest geodesic distance in the network, namely, how many steps are necessary to get from one side of the network to the other; (ii) density: The sum of the ties divided by the number of possible ties; (iii) reciprocity: The ratio of the number of links pointing in both directions to the total number of links (only in directed networks) [34].

Then, two different measures of centrality, degree centrality and betweenness centrality were calculated in order to assess the role and importance of the various stakeholders in the network.

The degree centrality is the number of arches that link one node of the network to other nodes of the network. The Freeman degree centrality formula [35] is shown in Equation (1),

$$Cd(n_i) = \sum_{k=1}^{n} a(n_i, n_k) \, (N-1)^{-1} \tag{1}$$

where:

Cd = degree centrality

a_{ik} = arc between nodes (1 if there is a connection between n_i and n_k, and 0 if there is no connection between n_i and n_k).

Additionally, in directed networks, degree centrality can be divided into outdegree centrality (that only considers the outgoing connections) and indegree centrality (that only considers the incoming connections).

The Freeman betweenness centrality formula [35] is shown in Equation (2),

$$C_b(n_i) = \sum_j^N \sum_k^{N-1} \frac{D_{jk}(n_i)}{D_{jk}} \tag{2}$$

where:

C_b = betweenness centrality

D_{jk} = set of minimum paths between the nodes n_j and n_k

$D_{jk}(n_i)$ = set of minimum paths connecting the node n_j to the node n_k through the node n_i

The stakeholders with high scores in degree centrality are capable of influencing the entire network [30], however, they do not always possess the greatest decisional power [36], thus, to understand and classify the importance of the stakeholders, coupling the two measures of centrality is useful. In fact, the betweenness centrality identifies stakeholders who play an intermediary role in the decision-making process, controlling the spread of information. The two measures of centrality were used to classify the stakeholders into three groups: Key stakeholders, primary stakeholders, secondary stakeholders. This distinction was made on the basis of their importance in the decision-making process. Following the Yamaki [37] methodology, actors who have at least one centrality value in the first quartile were classified as key stakeholders, actors whose two centrality values were in the fourth quartile were classified as secondary stakeholders and the remaining actors were classified as primary stakeholders.

2.4. Combining Social Network Analysis and Stakeholders Analysis

The last phase of our analysis consisted in combining the results of SA and SNA, dividing the stakeholders by "interests" and "classes" in order to understand which groups were the most influential and which were marginal in the network.

Next, two adjacency tables, reporting the density of connections among the different "interests" and "classes," were calculated, enabling the identification of the groups who are more inclined to cooperate and where, conversely, there is a lack of communication. The density of contacts was evaluated, considering frequent contacts with a value of 1, while occasional contacts were considered to have a value of 0.5.

The final analysis performed on the survey regarded the training preferences expressed by the stakeholders and assessed the importance of each proposed area, weighted on the role of the actors in the network. This calculation was performed by multiplying the responses of the stakeholders as follows: 0.5, if the stakeholder was secondary; 1, if the stakeholder was primary; and 2, if the stakeholder was key.

3. Results

The main features of the network are displayed in Table 2, where 43 of the 54 stakeholders initially contacted responded to the questionnaire (80%). This response rate is high, compared to others obtained in previous studies [5] and may be interpreted as a sign of interest in the issues investigated in the study area. Among the 43 stakeholders, 634 connections are present, showing a low-density value (0.351), a result that is consistent with values found in similar contexts by Paletto et al. [38]. Despite the low density, the network is efficiently connected, as we can deduce from the diameter parameter value, which is only 4. On the basis of the very high score of reciprocity (0.779), it is clear that this network is mainly characterized by bidirectional links.

Table 2. The main features of network.

Feature	Value
nodes	43
edges	634
av. degree	14.744
density	0.351
diameter	4
reciprocity	0.779

As shown in Table 3, the three most influential stakeholders represent different interests—"political," "technical" and "education and research"—and belong to two different classes—"public administrations" and "research institutes". On the one hand, there is a clear political effort (RegPie) to guide choices in the field of training, with the support of both the university (UniTo) and a research center with technical functions (Ipla), in defining the contents of the courses and providing refresher courses for trainers.

Table 3. Social Network Analysis results.

ID	Name	Label	Interest	Class	Degree	Betweenness	Role
1	Piedmont Region	RegPie	POL	PUAD	**71**	**0.050**	KEY
8	Institute for Wood Plants and the Environment	Ipla	TEC	REIN	**68**	**0.067**	KEY
5	University of Turin—DISAFA	UniTo	E&R	REIN	**52**	**0.031**	KEY
24	Association of Forestry Instructors	Aifor	TEC	OTHS	49	0.023	KEY
17	Forest Consortium Upper Susa Valley	Cfavs	ECO	OWNS	**48**	**0.059**	KEY
11	Farmers' association—Coldiretti	Coldir	TEC	AFWC	46	0.008	KEY
16	Forest Association Rosa Valleys	AsRosa	ECO	OWNS	40	0.012	KEY
20	Canavese Forest Consortium	Cfc	ECO	OWNS	**39**	**0.136**	KEY
2	Liguria Region	RegLig	POL	PUAD	**38**	0.001	KEY
14	National Confederation of Artisans	CNA	TEC	AFWC	**36**	0.010	KEY
15	Confederation of cooperatives—FEDAGRI	Confco	TEC	AFWC	**35**	**0.038**	KEY
34	Centre for Agricultural Education and Technical Assistance	Cipaat	E&R	TRCE	35	0.019	PRIMARY
10	Italian Confederation of Farmers	Cia	TEC	AFWC	34	0.008	PRIMARY
38	Managers of regional protected areas and Natura 2000 sites	Rn2000	ENV	PUAD	34	0.023	PRIMARY
13	Artisanal enterprises	Confar	TEC	AFWC	33	0.006	PRIMARY
23	Piedmont's Regional Foresters Association	Areb	TEC	AFWC	32	**0.024**	KEY
32	National Board of Vocational Education—Acli	Enaip	E&R	TRCE	31	**0.037**	KEY
4	Carabinieri—Unit command for forestry, environmental and Agri-food protection	CarFor	LEG	PUAD	30	0.017	PRIMARY
22	Mountain Union Lanzo Valleys	Umvl	ECO	OWNS	30	0.006	PRIMARY
25	Orders of the Agronomic and Forest Doctors	Odaf	TEC	OTHS	30	0.008	PRIMARY
33	Training Consortium Innovation and Quality	Cfiq	E&R	TRCE	30	0.006	PRIMARY
3	Autonomous region Valle d'Aosta	RegVda	POL	PUAD	28	**0.028**	KEY
12	Farming Confederation	Confag	TEC	AFWC	28	0.009	PRIMARY
29	Cebano Monregalese Professional Training Centre	Cfp	E&R	TRCE	27	**0.070**	PRIMARY
30	Training Agency FOCUS Piedmont	Focus	E&R	TRCE	27	0.008	PRIMARY
31	Farmers' Federation Consortium GESTCOOPER	Gestco	E&R	TRCE	27	0.007	PRIMARY
7	Agricultural Research Council	Crea	E&R	REIN	26	0.016	PRIMARY
39	ProNature	Pronat	ENV	ASSO	26	0.006	PRIMARY
18	Villar Fioccardo Forestry Consortium	Cfvf	ECO	OWNS	25	0.007	PRIMARY
27	Giuseppini del Murialdo National Body	Engim	E&R	TRCE	25	0.003	PRIMARY
28	Artisan and Trades Charity Association	Cdcam	E&R	TRCE	25	0.016	PRIMARY
19	Biella Mountain Forest Consortium	Cfmb	ECO	OWNS	24	0.014	PRIMARY
21	Monte Armetta Forestry consortium	Cfma	ECO	OWNS	22	0.009	PRIMARY

Table 3. *Cont.*

ID	Name	Label	Interest	Class	Degree	Betweenness	Role
26	Federation wood industry	Federl	TEC	AFWC	22	0.009	PRIMARY
40	WWF	Wwf	ENV	ASSO	15	0.013	PRIMARY
41	Ligurian training institution—Elfo	Elfo	E&R	TRCE	14	0.009	PRIMARY
43	Gran Paradiso National Park Authority	Epngp	ENV	PUAD	12	0.002	SECONDARY
6	University of Genoa	UniGe	E&R	REIN	11	0.019	PRIMARY
9	Agricultural Experimentation and Assistance Centre	Cersaa	E&R	REIN	10	0.010	PRIMARY
37	Training Services Agency	Asf	E&R	TRCE	9	0.002	SECONDARY
35	Provincial Centre of Vocational Training G. Pastor	Cpfp	E&R	TRCE	8	0.002	SECONDARY
36	San Salvatore Youth Centre	Vrss	E&R	TRCE	8	0.002	SECONDARY
42	Arbores Domi	Arbore	E&R	REIN	8	**0.040**	KEY

Note: Numbers in bold in columns Degree and Betweenness refer to values in the first quartile. KEY: ECO: Economic, E&R: Education and research, ENV: Environmental, LEG: Legal, POL: Political and TEC: Technical; AFWC: Actors of forest-wood chain, ASSO: Associations, OWNS: Owners associations, PUAD: Public administrations, REIN: Research institutes, TRCE: Training centers and OTHS: Others.

The majority of the stakeholders (67%) are characterized by high similarity scores (over 70%) of incoming and outgoing links (Figure 1). This means that cooperation and communication in the forestry training sector are almost always mutual and are not dictated by some stakeholders to others. Furthermore, similarity is almost the same for key and secondary stakeholders. Only 1% of the actors involved in the study have a value of similarity lower than 50. Hence, it is possible to assume that the majority of forestry training stakeholders have a precise idea of their own collocation in the network.

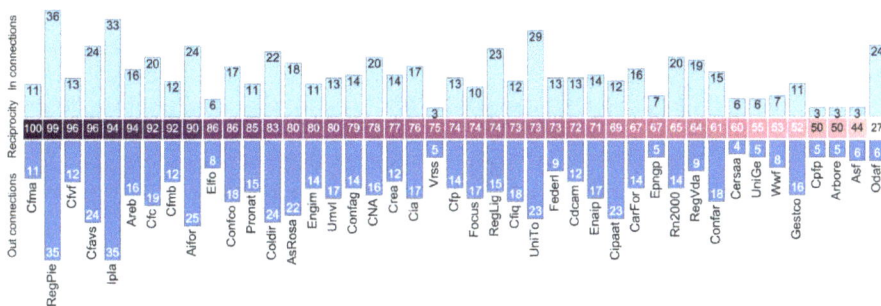

Figure 1. Similarity analysis among incoming and outgoing links.

In the two adjacency tables (Figure 2a,b), the density of connections between stakeholders who belong to the different groups is reported.

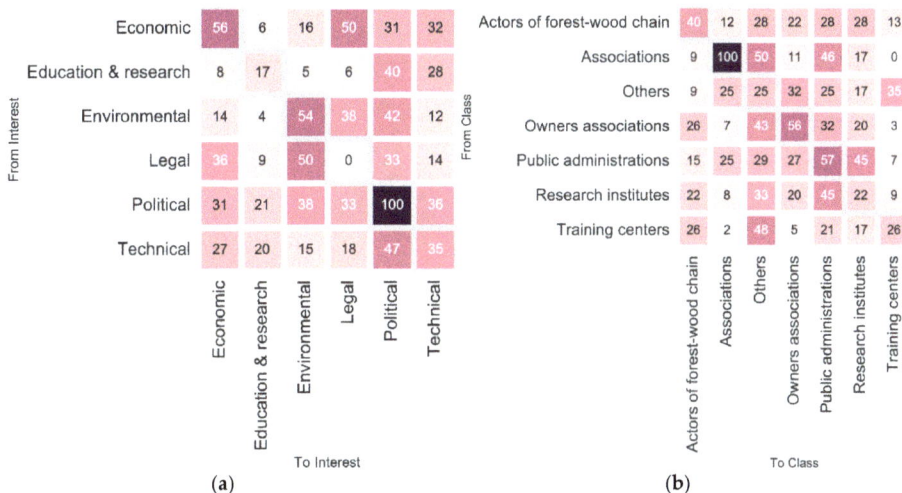

Figure 2. Connections between interests (**a**) and between classes (**b**) (in percentages).

The values of adjacency of the stakeholders, divided by interest, are displayed in Figure 2a. Following this classification, the highest values lie on the bisector, meaning that stakeholders focusing on the same aspect of the forestry training sector are more inclined to cooperate (e.g., all the stakeholders that pursue a political interest cooperate among themselves). There are some exceptions. In fact, the internal cooperation among actors whose pursued interest is education and research (17%) and technical (35%) is quite low. At the same time, the cooperation among the education and research groups, as well as the economic (8%) and environmental (5%) groups, is very low, whereas the network between stakeholders pursuing technical and political interests (47%) has a high level of cooperation between these two categories, which, as highlighted by the SNA analyses, revealed the highest grades for both centrality indicators.

As shown in Figure 2b, the highest values lie on the bisector, meaning that the highest levels of cooperation are between stakeholders belonging to the same class, e.g., associations are characterized by a density of 100%. Interestingly, research institutes are the least connected (22%), while there is a high level of cooperation between public administrations (57%) and owner associations (56%). On the contrary, the classes characterized by internal economic competition, such as training centers and actors of the forest–wood chain, present low levels of cooperation. Finally, the density of the connections is high between public administrations and research institutes (45%), and between associations and public administrations (46%), while the lowest value is between training centers and associations.

The level of interest in entrepreneurial training is displayed in Figure 3, and we can observe how "Work organization," "Investments" and "Supply chain" are the items that show the highest scores for the actors of the forestry sector. On the contrary, "Accounting," "Marketing" and "Taxation" are topics considered unnecessary for forestry workers by the majority of the stakeholders.

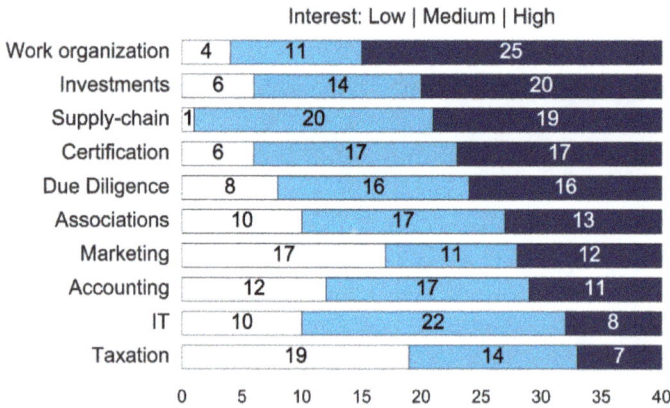

Interest: Low | Medium | High

	Low	Medium	High
Work organization	4	11	25
Investments	6	14	20
Supply-chain	1	20	19
Certification	6	17	17
Due Diligence	8	16	16
Associations	10	17	13
Marketing	17	11	12
Accounting	12	17	11
IT	10	22	8
Taxation	19	14	7

Figure 3. Stakeholder preference regarding entrepreneurial training.

With regard to management skill classes, as proposed by Morat [19], Italian stakeholders located in the Western Alps consider those skills aimed at improving the performance of companies and those relating to knowledge of the wood market and marketing tools to be fundamental. There is also a high level of interest in the qualification of the company with regard to the quality of production and to the compliance with regulations and certification. On the other hand, as far as management is concerned, the responses are divergent. It is considered necessary to train entrepreneurs in the opportunities provided by investments, although digitalization is not perceived as a binding necessity.

Only the specific accounting aspects reported a low general interest, probably because these are services that the entrepreneur does not manage directly but relies on third party professionals or professional organizations.

The preferences, expressed by the various stakeholders in relation to the issues presented, divided by interest and class, are shown in Figure 4.

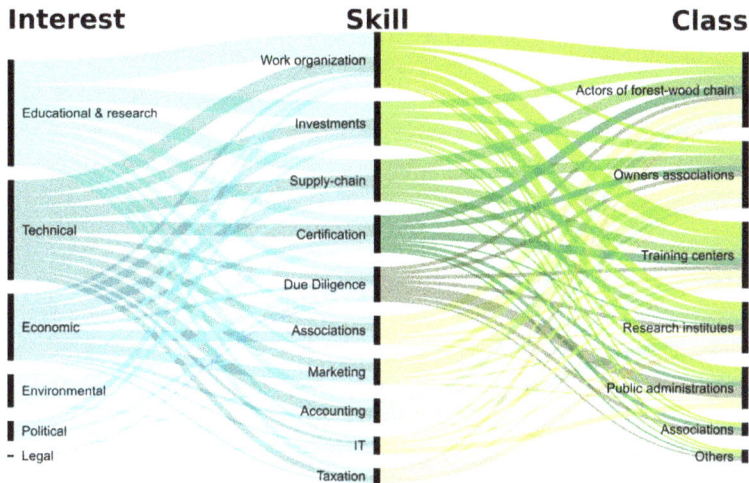

Figure 4. Stakeholder preferences regarding entrepreneurial training, divided by interest and class.

These analyses were carried out using only the preferences of the stakeholders who indicated a "high" level of interest for each topic.

It is clear that, by analyzing information with an overall vision, the network expresses a high degree of coherence in the choices expressed. In fact, each of the categories that allow the network to be segmented express an interest in all the issues proposed. This absolute lack of polarization in the choices expressed is configured as an important indicator of internal consistency between the various actors in the network, but, at the same time, it expresses the need for training on many issues, highlighting the criticality of the sector from multiple management perspectives.

Here we find a reconfirmation of what emerged in Figure 3, i.e., it is clear that the actors representing the "education and research" and "technical" interests place the issues of "work organization" and "investments" in the foreground, followed by the "supply chain". In light of the findings, these three issues need to be prioritized when defining new training programmes.

4. Discussion

The importance of the stakeholder analysis, together with that of their networks and the influence of different categories in decision-making processes, has been highlighted by many authors [39,40], although the use of network analysis in the forestry sector is still in its infancy. In fact, only in the recent past have some authors applied Social Network Analysis to forestry stakeholders, for instance, Paletto et al. [38] and Yamaki [37], and this technique is deemed to be appropriate for describing small yet highly interconnected networks, such as those typically found in the forestry sector.

This study combined the Social Network Analysis with the Stakeholder Analysis, exploring the relationships among the actors involved in the training sector. The combined use of these two tools, making the best use of the specific aspects of each, made it possible to identify the networks' properties in the participatory process and, as stated by Paletto et al. [5], to recognize and measure, in an objective way, the presence of strong relationships characterizing the actors involved in the network.

In further researches, in addition to being complemented by SNA, the SA can be developed with Multi-criteria analysis in order to more accurately represent decision-making problems and help decision-makers to define priorities and best solutions. Another aspect that should be developed is the challenge related to communication, technology transfer and dissemination, providing policy-makers with effective tools that are directly applicable in defining sector policies [41]. As a general result, the hierarchy among actors has undoubtedly been identified, highlighting their so-called "social power," namely, their respective roles and influence in decision-making processes, and clarifying the political role that each actor has in an often opaque context, where, in addition to structural positions, direct relations between individuals are also significant. Thanks to this approach, it has been possible to identify a preliminary network of actors, and the response rate has shed light on both their high level of interest in belonging to the network and their own interest in the training topics. The choice to investigate specific topics, however, represents a research limit, as these results cannot be automatically applied to other contexts.

Another interesting result, worthy of further in-depth discussion, is the lack of mutual communication between different and well-separated categories of actors. In fact, some stakeholders are excluded from the decision-making process, confirming the findings of other studies [7,42].

Currently, several "secondary" stakeholders find no place in the decision-making process at all and, even when formally included in the network, are relegated to a marginal position. This is the case, for instance, in training agencies, which, even though are fully part of the network, are evidently not able to condition it. The consequence may be the creation of a dangerous short circuit, since the failure to meet mutual expectations may lead to ineffective training courses. This study provided several initial results, which need to be examined in greater depth through subsequent investigations. The network and the position of each actor, with respect to the various indicators, are relevant only for the professional training topics. The position and importance of the actors could in fact change if analyzed with respect to other topics of investigation.

The next steps of our research could continue by analyzing stakeholders with reference to other topics, studying which factors lead to the confinement of some stakeholders to gregarious and accessory positions and the elements which cause the exclusion of some of them from the decision-making process, deepening the examination of the social and political needs defined by each actor and possibly also including in the survey those who have not provided their feedback in this first analysis. The research should also systematically investigate any conflicts or synergies and trade-offs between actors who belong to different classes or occupy different roles.

A focus on similarities and differences expressed in other fields (e.g., ecosystem services and the wood supply chain) will make it possible to identify a more stable and stronger network with an "absolute" value.

Additionally, the marginality of another actor, the associations, is to be considered a risk because the training proposals may not reflect their needs and, therefore, may be opposed by the loggers themselves, as also highlighted by Egan [43]. It is worth underlining how very few stakeholders declined to participate in the survey, justifying their decision by citing confidentiality concerns about their contacts, which confirms the importance of informal individual links.

Regarding the results obtained in the stakeholder classification, it appears evident that all groups, respectively divided by interest and by class, are represented with a strong social connection and depict a sort of general balance in the network. Within these groups, good levels of connection between stakeholders, belonging to the same class and pursuing the same purpose, have been shown by the combined use of the two techniques. On the contrary, connections among stakeholders of different groups are lower, as stated by Bruña-García and Marey-Pérez [44]. However, despite the presence of administrative and logistical barriers, a good connection was found even among stakeholders operating in different territories. In fact, the physical or administrative distance is cancelled out by the desire to establish common guidelines on a crucial and delicate issue, such as the professional training of forestry workers, beginning with the high interest shown by the political decision-makers.

As stated by other authors [18], who have investigated the training needs expressed directly by forestry loggers, we find confirmation that issues related to the supply chain aspects, work organization and investments are considered the most important ones. This is a view that is also held by other stakeholders who express different interests and occupy different roles in the forestry sector. This is in line with other researches that have shown that the main concern of forestry companies is linked to the low profitability of logging operations [24].

5. Conclusions

The availability of wood resources in the Western Italian Alps would potentially allow for the existence of a widespread network of business enterprises, guaranteeing employment and social benefits in fragile environments and, at the same time, allowing for the sustainable management of the resource itself. However, for this to happen, it is essential that businesses are efficient and have the appropriate level of professional knowledge.

The effort shown by political institutions in this area, in setting up training courses for forestry workers, has increased in recent years, thanks to continuous investment in this sector, with over € 7 million allocated in the 2014–2020 period. For several years now, political institutions and research centers have been working together to define the contents of future training courses. The associations, representing the various professional figures that work in this sector, are also particularly active, and training is considered a fundamental aspect for the professional growth of entrepreneurs.

It should be noted that training centers have a marginal role in the network, and this weakness should be considered in future investigations, bearing in mind the important contribution that trainers can make in defining the contents of such training courses and also considering that they have direct contact with workers and their needs.

Training is a topic that involves actors with different roles, with different interests, but the need expressed by the majority of actors involved in the survey is clear: To increase the offered training on managerial issues.

Interpreting the results of the survey, it is clear that all stakeholders are concerned about the economic performance of the forestry sector. As a result, being able to provide the system with the tools to implement profitability is of paramount importance, and the actors have confirmed this result, as they consider training on how to improve the yield of forest operations and how to assess investments in innovative machinery to be crucially important. In the same manner, gaining more knowledge about the wood supply chain and the enhancement of the raw material is requested.

On the other hand, stakeholders do not have confidence in the opportunities provided by information technologies, probably because they are still linked to traditional market forms, mostly local, based on direct contacts among operators of the supply chain.

In conclusion, the forestry sector stakeholders in the Western Italian Alps have highlighted the need for the training of a "modern woodsman," identifying the profile of an entrepreneur who, in addition to expressing an excellent knowledge of forestry operations and the profitable management of the site, seeks to better comprehend the timber market and the opportunities it offers.

Author Contributions: S.B. and F.L. contributed equally to the conceptualization and design of the work; Writing—Original Draft Preparation, F.L.; Writing—Review & Editing, S.B.; Software, L.B.; Visualization, L.B.; Validation, R.G.P.; Methodology and Formal Analysis, L.B., F.L.; Investigation, F.L.; Supervision and critical revision of the manuscript, F.B. and A.M.; Project Administration, A.M. and F.B.; and Funding Acquisition, A.M. All the Authors approved the version to be published.

Funding: This research was funded by the EU cross-border territorial cooperation programme INTERREG V-A Italy-France-ALCOTRA 2014-2020, contribution FESR, within the INFORMA PLUS research project (No. 1574).

Conflicts of Interest: The authors declare no conflict of interest.

References

1. Appelstrand, M. Participation and societal values: The challenge for lawmakers and policy practitioners. *For. Policy Econ.* **2002**, *4*, 281–290. [CrossRef]
2. Teder, M.; Kaimre, P. The participation of stakeholders in the policy processes and their satisfaction with results: A case of Estonian forestry policy. *For. Policy Econ.* **2018**, *89*, 54–62. [CrossRef]
3. Kangas, A.; Saarinen, N.; Saarikoski, H.; Leskinen, L.A.; Hujala, T.; Tikkanen, J. Stakeholder perspectives about proper participation for Regional Forest Programmes in Finland. *For. Policy Econ.* **2010**, *12*, 213–222. [CrossRef]

4.   European Commission. The New EU Forest Strategy. Available online: https://ec.europa.eu/agriculture/forest/strategy_en (accessed on 26 June 2018).

5.   Paletto, A.; Giacovelli, G.; Pastorella, F. Stakeholders' opinions and expectations for the forestbased sector: A regional case study in Italy. *Int. For. Rev.* **2017**, *19*, 68–78. [CrossRef]

6.   Kleinschmit, D.; Pülzl, H.; Secco, L.; Sergent, A.; Wallin, I. Orchestration in political processes: Involvement of experts, citizens, and participatory professionals in forest policy making. *For. Policy Econ.* **2018**, *89*, 4–15. [CrossRef]

7.   Ananda, J.; Herath, G. The use of Analytic Hierarchy Process to incorporate stakeholder preferences into regional forest planning. *For. Policy Econ.* **2003**, *5*, 13–26. [CrossRef]

8.   Reed, M.S.; Graves, A.; Dandy, N.; Posthumus, H.; Hubacek, K.; Morris, J.; Prell, C.; Quinn, C.H.; Stringer, L.C. Who's in and why? A typology of stakeholder analysis methods for natural resource management. *J. Environ. Manag.* **2009**, *90*, 1933–1949. [CrossRef] [PubMed]

9.   Peterson St-Laurent, G.; Hoberg, G.; Sheppard, S.R.J. A Participatory Approach to Evaluating Strategies for Forest Carbon Mitigation in British Columbia. *Forests* **2018**, *9*, 225. [CrossRef]

10.   Grimble, R.; Wellard, K. Stakeholder methodologies in natural resource management: A review of principles, contexts, experiences and opportunities. *Agric. Syst.* **1997**, *55*, 173–193. [CrossRef]

11.   Barnaud, C.; Corbera, E.; Muradian, R.; Salliou, N.; Sirami, C.; Vialatte, A.; Choisis, J.-P.; Dendoncker, N.; Mathevet, R.; Moreau, C.; et al. Ecosystem services, social interdependencies, and collective action: A conceptual framework. *Ecol. Soc.* **2018**, *23*, 15. [CrossRef]

12.   Krupa, M.; Cenek, M.; Powell, J.; Trammell, E.J. Mapping the stakeholders: Using social network analysis to increase the legitimacy and transparency of participatory scenario planning. *Soc. Nat. Resour.* **2018**, *31*, 136–141. [CrossRef]

13.   Raum, S. A framework for integrating systematic stakeholder analysis in ecosystem services research: Stakeholder mapping for forest ecosystem services in the UK. *Ecosyst. Serv.* **2018**, *29*, 170–184. [CrossRef]

14.   Lienert, J.; Schnetzer, F.; Ingold, K. Stakeholder analysis combined with social network analysis provides fine-grained insights into water infrastructure planning processes. *J. Environ. Manag.* **2013**, *125*, 134–148. [CrossRef] [PubMed]

15.   Robertson, P.J.; Lewis, L.B.; Sloane, D.C.; Galloway-Gilliam, L.; Nomachi, J. Developing networks for community change: Exploring the utility of network analysis. *Commun. Dev.* **2012**, *43*, 187–208. [CrossRef]

16.   Burt, R.S. The Network Structure of Social Capital. *Res. Organ. Behav.* **2000**, *22*, 345–423. [CrossRef]

17.   Powell, W.M. Neither market nor hierarchy; Network forms of organization. In *Research in Organizational Behavior*; Staw, B.M., Cummings, L.L., Eds.; JAI Press: Greenwich/London, UK, 1990; Volume 12, pp. 295–336. Available online: http://www.scirp.org/(S(czeh2tfqyw2orz553k1w0r45))/reference/ReferencesPapers.aspx?ReferenceID=1751733 (accessed on 26 June 2018).

18.   Spinelli, R.; Magagnotti, N.; Jessup, E.; Soucy, M. Perspectives and challenges of logging enterprises in the Italian Alps. *For. Policy Econ.* **2017**, *80*, 44–51. [CrossRef]

19.   Morat, J. *Skills Needed by Forest Contractors*; Report Project ConCert—Programme Leonardo 2009–2011; La Bastide des Jourdans, France, 2011.

20.   Nybakk, E.; Crespell, P.; Hansen, E.; Lunnan, A. Antecedents to forest owner innovativeness: An investigation of the non-timber forest products and services sector. *For. Ecol. Manag.* **2009**, *257*, 608–618. [CrossRef]

21.   Wiklund, J.; Shepherd, D. Entrepreneurial orientation and small business performance: A configurational approach. *J. Bus. Ventur.* **2005**, *20*, 71–91. [CrossRef]

22.   Elmi, M.; Streifeneder, T. *The Alps in 25 Maps*; Permanent Secretariat of the Alpine Convention: Bozen, Italy, 2018; ISBN 978-88-97500-43-8.

23.   Motta Fre, V.; Tondeur, V.; Marchaison, P.; Blin, L.; Dubois, C.; Haudemand, J.-C. Vocational forestry training in the transalpine area between Italy and France (in Italian). In Proceedings of the Summary of Studies and Analyses Carried Out in the Framework of the EUROPEAN Cooperation Project in Forma, Torino, Italy, 2 March 2015.

24.   Accastello, C.; Blanc, S.; Mosso, A.; Brun, F. Assessing the timber value: A case study in the Italian Alps. *For. Policy Econ.* **2018**, *93*, 36–44. [CrossRef]

25.   Magagnotti, N.; Smidt, M.; Spinelli, R. Characteristics of the students and appreciation of the courses: Each one his own? (in Italian). In Proceedings of the Seminar: Forest Training and Information: Where We Come from and Where We Are Going, Turin, Italy, 11 January 2018.

26. Harrison, S.R.; Qureshi, M.E. Choice of stakeholder groups and members in multicriteria decision models. *Nat. Resour. Forum* **2000**, *24*, 11–19. [CrossRef]

27. Noy, C. Sampling Knowledge: The Hermeneutics of Snowball Sampling in Qualitative Research. *Int. J. Soc. Res. Methodol.* **2008**, *11*, 327–344. [CrossRef]

28. Coulson, J. *The Strength of Weak Ties in Online Social Networks: How Do Users of Online Social Networks Create and Utilize Weak Ties to Amass Social Capital?* LAP LAMBERT Academic: Columbia, SC, USA, 2010; ISBN 978-3-8383-6965-5.

29. Prell, C.; Hubacek, K.; Reed, M. Stakeholder Analysis and Social Network Analysis in Natural Resource Management. *Intern. Soc.* **2009**, *22*, 501–518. [CrossRef]

30. Paletto, A.; Ferretti, F.; De Meo, I. The role of social networks in forest landscape planning. *For. Policy Econ.* **2012**, *15*, 132–139. [CrossRef]

31. Mark, S. Granovetter The Strength of Weak Ties. *Am. J. Sociol.* **1973**, *78*, 1360–1380.

32. FAO. *Guide to Good Practice in Contract Labour in Forestry*; Report of the UNECE/FAO Team of Specialists on the Best Practices in Forest Contracting; Kastenholz, E., Dyduch, C., Fitzgerald, R., Hudson, B., Jaakkola, S., Lidén, E., Monoyios, K., Morat, J., Pasek, F., Sachse, M., et al., Eds.; Food and Agriculture Organization of the United Nations: Rome, Italy, 2011; ISBN 978-92-5-106877-9.

33. Likert, R. A technique for the measurement of attitudes. *Arch. Psychol.* **1932**, *22*, 55.

34. Scott, J. Social Network Analysis. Available online: https://uk.sagepub.com/en-gb/eur/social-network-analysis/book249668 (accessed on 22 June 2018).

35. Freeman, L.C. Centrality in social networks conceptual clarification. *Soc. Netw.* **1978**, *1*, 215–239. [CrossRef]

36. Sinclair, P.A. Network centralization with the Gil Schmidt power centrality index. *Soc. Netw.* **2009**, *31*, 214–219. [CrossRef]

37. Yamaki, K. Applying social network analysis to stakeholder analysis in Japan's natural resource governance: Two endangered species conservation activity cases. *J. For. Res.* **2017**, *22*, 83–90. [CrossRef]

38. Paletto, A.; Hamunen, K.; Meo, I.D. Social Network Analysis to Support Stakeholder Analysis in Participatory Forest Planning. *Soc. Nat. Resour.* **2015**, *28*, 1108–1125. [CrossRef]

39. Bryson, J.M.; Cunningham, G.L.; Lokkesmoe, K.J. What to Do When Stakeholders Matter: The Case of Problem Formulation for the African American Men Project of Hennepin County, Minnesota. *Public Adm. Rev.* **2002**, *62*, 568–584. [CrossRef]

40. Dos Muchangos, L.S.; Tokai, A.; Hanashima, A. Stakeholder analysis and social network analysis to evaluate the stakeholders of a MSWM system—A pilot study of Maputo City. *Environ. Dev.* **2017**, *24*, 124–135. [CrossRef]

41. Blanc, S.; Accastello, C.; Bianchi, E.; Lingua, F.; Vacchiano, G.; Mosso, A.; Brun, F. An integrated approach to assess carbon credit from improved forest management. *J. Sustain. For.* **2018**, *1–15*, 1–15. [CrossRef]

42. Ananda, J. Implementing participatory decision making in forest planning. *Environ. Manag.* **2007**, *39*, 534–544. [CrossRef] [PubMed]

43. Egan, A.F. Training preferences and attitudes among loggers in northern New England. *For. Prod. J.* **2005**, *55*, 19–26.

44. Bruña-García, X.; Marey-Pérez, M.F. The challenge of diffusion in forest plans: A methodological proposal and case study. *Forests* **2018**, *9*, 240. [CrossRef]

forests

MDPI

Article

Qualitative Assessment of Forest Ecosystem Services: The Stakeholders' Point of View in Support of Landscape Planning

Isabella De Meo [1,*], Maria Giulia Cantiani [2], Fabrizio Ferretti [3] and Alessandro Paletto [3]

[1]   CREA, Research Centre for Agriculture and Environment, 50125 Firenze, Italy
[2]   Department of Civil, Environmental and Mechanical Engineering, University of Trento, 38123 Trento, Italy; maria.cantiani@unitn.it
[3]   CREA, Research Centre for Forestry and Wood, 52100 Arezzo, Italy; fabrizio.ferretti@crea.gov.it (F.F.); alessandro.paletto@crea.gov.it (A.P.)
*   Correspondence: isabella.demeo@crea.gov.it; Tel.: +39-055-2492238

Received: 4 July 2018; Accepted: 30 July 2018; Published: 1 August 2018

Abstract: In the last decades, the ecosystem services (ES) concept has become one of the main challenges of study and discussion in the scientific community. The quantitative and qualitative assessment of ES is as a tool to address forest management planning on a local scale. Forest landscape management planning is the most suitable level for integrating social needs and demands in the enhancement of different forest ES. Some regions in Italy have developed forest landscape management plans taking into account the social preferences for the different ES. In this paper, we refer to five case studies in three pilot areas in Italy. A survey collected and analyzed the opinions and preferences, from 362 stakeholders, for ten ES included in three categories (provisioning, regulating and cultural services). The main aim of this study is to understand what type of variables (study area, the groups of interest and socio-demographic characteristics of respondents) most influence stakeholder preferences for ES. The results show that for the sample of stakeholders involved in the survey, the most important ES category is regulating services followed by cultural services. In addition, the results show that the group of stakeholders' interest is the most important variable influencing their preferences for ES.

Keywords: social assessment; forest multifunctionality; stakeholders' involvement; forest planning; questionnaire survey

1. Introduction

Following the United Nations Conference on the Human Environment held in Stockholm in 1972, the scientific community started to focus on the importance of the benefits provided by natural ecosystems for human society [1]. At that time, a growing number of authors stressed the role of natural ecosystems for human well-being in order to increase society's interest in biodiversity conservation. This utilitarian approach aimed at improving policy decisions and was mainly focused on raising public interest in biodiversity conservation, highlighting the fact that loss of biodiversity directly affects benefits provided by natural ecosystems to society [2–4].

About two decades later, the concept of ecosystem services (ES) was definitively and firmly placed into the policy agenda by the Millennium Ecosystem Assessment (MEA) [5]. The MEA [5] defines ecosystem services as "the benefits people obtain from ecosystems", focusing on human dependency on ecosystem services and on the importance of biodiversity and ecological processes for human well-being [6,7]. MEA classified ecosystem services into four categories: (1) Provisioning, (2) regulating, (3) cultural, and (4) supporting services. The first three categories directly affect people and human well-being, whereas supporting services are the underlying ecosystem processes that maintain the others [8,9].

The Economics of Ecosystems and Biodiversity (TEEB) classification [10] substituted supporting services with habitat services, which include lifecycle maintenance and gene pool protection. Subsequently, ES have been reclassified into three categories (provisioning, regulating and maintaining, cultural), adopting the Common International Classification of Ecosystem Services (CICES) [11]. The CICES classification is widely used when ecosystems and economic accounts have to be linked, and is useful when the outputs of ecosystems have to be turned into benefits for human well-being. Moreover, the CICES classification is able to overcome 'double counting' problems in valuation studies [11], and it can be considered a suitable approach for integrating ES into analytical models that support landscape planning [12].

Many other definitions and classifications of ES have also since been provided with the aim of integrating the concept or to analyze it from a different point of view [13,14].

With regard to the forest sector, forests generate a variety of important ES for human well-being, such as timber and non-timber forest products among provisioning services; water regulation, protection from natural hazards and carbon sequestration among regulating services; and, scenic beauty and recreation opportunities among cultural services [15,16].

In order to maintain and improve the ES available from natural, semi-natural or planted forests, the assessment of the ES provided by forests at different spatial and temporal scales is required. These data are fundamental in order to define priorities and forest management strategies. In the literature, there are two main approaches aimed to assess the ES [17–19]: The first approach is based on a quantitative assessment of ES provided by forests through field measurements of biophysical outcomes; the second approach is based on a qualitative assessment of ES using expert or stakeholders' opinions [20]. In some studies, the biophysical assessment of ES is integrated with a monetary evaluation in order to provide policy makers with information and to facilitate communication with public opinion and citizens [21].

Recently, policy makers have also emphasized the importance of forest ES at local and global levels, taking into account the documents and prescriptions provided by the scientific community in the previous period. In 2013, the European Commission (EC) in "A new EU Forest Strategy: For forests and the forest-based sector" included in the main social objectives of Sustainable Forest Management (SFM), the protection of forest ecosystems in order to maintain and improve the ES. This new EU Forest Strategy considers forest management plans as the main instrument for implementing the principles of SFM and increasing the supply of multiple forest ES.

In "European Forests 2020"—elaborated during the 6th Ministerial Conference on the Protection of Forests in Europe (MCPFE) held in Oslo in 2011—the role of ES provided by European forests is repeatedly emphasized. The first two goals of "European Forests 2020" are focused on sustainable management of European forests to enhance lasting provision of goods and services and on their role in contributing to a green economy, including increased provision of timber, other forest products and ecosystem services from sustainable sources. Subsequently, the Madrid Ministerial Resolution 1 (MCPFE, Madrid, 20–21 October 2015) emphasizes the importance of promoting a forest sector and its related value chain that provides society with increasing opportunities for green jobs, which ideally should be connected to the management and use of forests, and to environmentally friendly production processes, based on goods and services from sustainably managed forests. The regulations of European policy makers have been implemented in Italy in two main political documents: The new Italian Forest Law (Testo Unico per le Foreste 2018) and the Italian Bio-economy Strategy (2017). Recently, the Action Plan of the Italian Bio-economy Strategy considers as a priority moving from the existing economy to a sustainable bio-economy valorizing the provision of ES [22].

The new Italian Forest Law emphasizes the importance of protecting the forest by promoting actions to prevent natural and anthropogenic risks, to contrast biotic and abiotic adversities, and to assure hydrogeological and forest fire protection, adaptation to climate change, recovery of degraded or damaged areas, carbon sequestration and the provision of other ES generated by SFM. In addition, the regions can promote systems for the Payment for Ecosystem Services (PES) generated by SFM activities [23].

Despite the huge literature on ES and all the applications at the policy level, there is still a lack of research and knowledge in how to integrate the concept of ES in planning and management and especially at the level of decision making [6,24–27]. Various studies contribute to the integration of the concept of ES into landscape scale planning and provide information about the spatial distribution of ES and their value. These studies help to improve the understanding of the effect of planning decisions on ES and to avoid resulting trade-offs when planning focuses on only one ES at time [28], and has proved useful when decision-making processes are concerned with environmental matters at a regional level [6,29–31].

Regarding the forest sector, it is important to define the proper scale when considering the relationship between forest planning and ES assessment. The Forest Unit Management Plan (FUMP), which refers to a forest area managed on a medium/short-term basis, takes into consideration the technical and managerial aspects of each individual forest ownership [32]. When considering ES provision, the fact of focusing on individual forest ownership can lead to trade-offs or cause the undervaluation of synergies between ES, because ES transcend borders between different ownerships [33–35]. The need for taking into account the multiple ES in forest planning has led to a profound revision of Italian forest planning [32,36–38] and a hierarchical approach has been consolidated, which moves from a detailed level of planning (FUMP) to a medium scale. At medium scale, the Forest Landscape Management Plan (FLMP) has been introduced. The FLMP has the main scope of providing management guidelines for the subordinate FUMP, and integrating and coordinating with other types of plans [32,37,39–41].

FLMP is based on the principles of SFM and public participation, and this scale is the most suitable for taking into account ES assessment issues and for developing management guidelines that consider the sustainability of the relationship between the local community and nearby forests [42–46]. However, while implementing FLMPs, concerns are being raised as to how to consider stakeholders' preferences. Indeed, various groups of stakeholders have different needs and social demands concerning the various forest ES that must be included in the objectives and actions of the FLMP [32,47].

Starting from these preliminary considerations, the main aim of this study was to analyze stakeholders' opinions and preferences for different categories of forest ES. The secondary objective was to identify which variables influence the stakeholders' point of view in the qualitative evaluation of forest ES. The study was applied in three pilot areas in Central and Southern Italy (Matese, Arci-Grighine and Alto Agri forest districts) where we identified five case studies characterized by marked differences from an environmental, socio-economic and cultural point of view.

2. Materials and Methods

2.1. Study Areas

The research was developed in three pilot areas where FLMPs were carried out. They are located in three Italian regions (Molise, Sardinia and Basilicata), which are rural areas with similar socio-economic features but different ecological characteristics [32] (Figure 1).

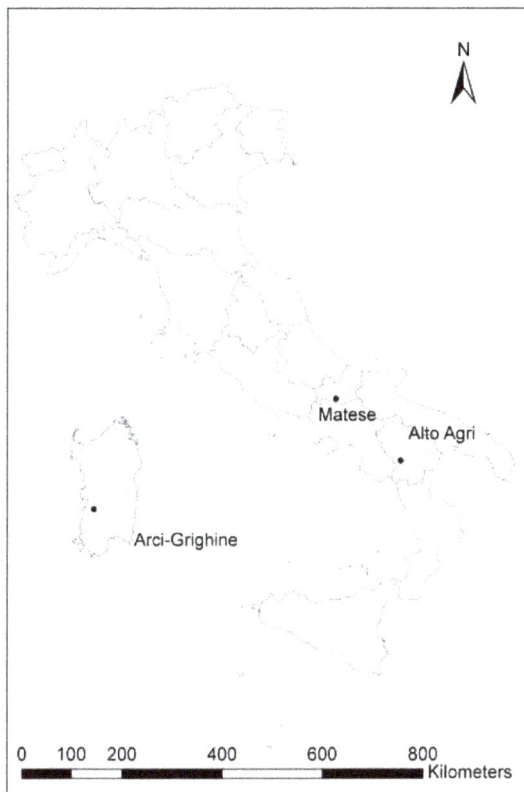

Figure 1. Location of study areas in Italy.

From a geographic point of view, the Matese district (41°29′12″ N, 14°28′26″ E), located in central Italy, in the Molise region, covers about 365 km². It has a population of 21,022 inhabitants and a density of 58 inhabitants per km². The Arci-Grighine district (39°42′7″ N, 8°42′4″ E), located in the central-eastern area of the island of Sardinia, covers a land area of 552 km². It has a population of 26,207, with a density of about 47 inhabitants per km². The Alto Agri district (40°20′25″ N, 15°53′52″ E) is located in southern Italy (Basilicata region) and occupies a land area of 726 km², with a resident population of 33,739 inhabitants (a population density of 46.5 inhabitants per km²). The pilot areas are characterized by very low population densities compared to the national situation (201 inhabitants per km²), the result of slowly declining populations, which started in the 1950s and has continued to present times, driven by mass unemployment and, in the 1990s, by young adults moving away from rural areas to find work elsewhere [48,49].

The forest surface covers 44.1% of the territory in the Matese district, 51.2% in Arci-Grighine and 58.4% in Alto Agri. Other main land uses are agriculture and grazing. Regarding ownership, most of the forests are privately owned in the Matese and Alto Agri districts (66% and 73%, respectively). In Arci-Grighine, privately owned forests comprise approximately 34%; there is an important percentage of common forests (53%), whereas the remaining 13% are public forests. These differences in ownership have important effects on forest management strategies [32,40].

In the Matese and Alto Agri districts, the main forest types are Turkey oak (*Quercus cerris* L.) and European beech (*Fagus sylvatica* L.) forests, mainly coppices. Other forest types are mixed broadleaved formations and both evergreen and broadleaved plantations [40,50]. In these districts, forests are characterized by a high level of multifunctionality. Conversely, Arci-Grighine is characterized by a high vocation in terms of non-timber forest products (cork and myrtle). In this district, the major forest types are Mediterranean forests (*Ericoarboreae-Arbutetumunedonis* and *Pistaciolentisci-Calicotometumvillosae* associations) and evergreen forests with holm oak (*Quercus ilex* L.) and cork oak (*Quercus suber* L.).

The agricultural sector plays an essential role in the economic structure of the three districts; conversely, the industrial sector is, in general, weak, and the infrastructure is poor [51].

In Alto Agri, 66% of the study area is located in protected areas (part in a national park and part in the Natura 2000 sites, which partially overlap) [40,52]. For this reason, we decided to consider two different case studies within the Alto Agri district. In fact, the sector which does not fall into protected areas has to deal with the economic and management problems typical of rural areas, while the national park is faced mainly with nature conservation and recreation issues. For different reasons we also considered separately data concerning Arci and Grighine, within the Sardinian district. Actually, Grighine has an agricultural connotation and a vocation for agri-forestry activities, in particular cork production, whereas Arci is characterized by a forested landscape and, more importantly, forestry activities.

2.2. Survey Methodology

The survey in the three FLMPs was conducted by means of a two-stage approach. Such an approach was based on a preliminary stakeholder analysis followed by the investigation of stakeholders' opinions and preferences through the administration of questionnaires.

In the FLMPs, a three-step model of stakeholder analysis was adopted [32] to identify the key stakeholders, their roles and their relationships. This approach is based on an iterative process: In the first step, some stakeholders are identified based on information from institutional actors and experts, and then, in the second step, based on the indications of this first group, also previously unknown stakeholders are identified. In the third step, after being identified, the stakeholders are categorized not only on the basis of their power (the ability of a stakeholder in a relationship to exert influence on another stakeholder in order to obtain the expected outcomes), but also considering their proximity to the issues and interests at stake, and their capacity to control information and to influence the decision-making process. The result of this step is a classification into three types of stakeholders [32,53]: (a) Key stakeholders—the main actors in terms of social role and power, and actively involved in the decision process; (b) primary stakeholders—the beneficiaries of the FLMP, only partly involved in the decision process; and, (c) secondary stakeholders—the public at large, marginally involved in the issue but constantly kept informed. In each study area, approximately 20% of total stakeholders were classified as key stakeholders, 30% as primary stakeholders, and the remaining 50% as secondary stakeholders. Key and primary stakeholders were involved in the present survey (FLMPs consultation), while secondary stakeholders were simply informed about the planning process.

In the present survey, 380 stakeholders were identified as key and primary stakeholders in the three study areas. The non-response rate was equal to 5%, mainly concentrated in the Matese district. Ultimately, 362 stakeholders completed the questionnaire. The distribution of the sample by study area is as follows: 39 stakeholders in the Matese district (Molise); 212 stakeholders in the Alto Agri district (Basilicata); and, 111 stakeholders in the Arci-Grighine district (Sardinia).

In addition, four groups of interest were identified: (i) Public administrations (i.e., regions, provinces, municipalities or other public agencies); (ii) actors of the forestry industry (private forest owners, forest enterprises, and sawmills); (iii) representatives of environmental Non-Governmental Organizations (NGOs); and, (iv) actors of the tourism sector (i.e., tourism promotion bodies and hotel owners). The group of environmental NGOs includes a broad range of associations with different aims such as the World Wide Fund for Nature (WWF), Legambiente and Greenpeace, but also small local associations with restricted and specific objectives. With regard to the distribution of the stakeholders by groups of interest, 168 are representatives of public administrations, 131 are actors of the forestry industry, 44 are actors of the tourism sector and the remaining 19 are representatives of environmental NGOs.

The differences in the number of stakeholders and their distribution in the case studies are due in part to the different extension of the areas, in part to the specific socio-economic features and in part to the diversity of forest contexts, in particular concerning the different relevance of forest resources among the study areas. Matese is the smallest study area (36,500 ha) and has a limited number of stakeholders with regard to the other areas. The number of stakeholders is the highest in Alto Agri, which has an area twice that of Arci-Grighine.

During the development of FLMPs, the stakeholders were firstly contacted by telephone to set up a location and a time for a face-to-face interview based on a semi-structured questionnaire. Before conducting the interview, informed consent was obtained. Consenting stakeholders participated in this interview, which lasted an average of 30 min to 90 min.

The final version of the questionnaire was produced after a test phase on a sample of stakeholders used to verify question effectiveness, clarity of the language and completeness of the information required; then, the final questionnaire was administered through personal interviews with the 362 stakeholders. As generally reported in questionnaire-type approaches, the questionnaire was divided into thematic sections in order to maintain a high level of attention in the interviewees [54].

During the interviews, respondents were given the opportunity to expand the conversation and besides ticking the given answer, there was discussion and exchange of information. The time for discussion was also important in order to support some respondents who found the significance of some concepts or technical words difficult to understand.

The questionnaires were firstly used to investigate individual preferences attributed by stakeholders to forest ES. They were also used to consider other aspects of the social perception of forests, which could be usefully employed to support decision making in forest planning and to define the priority actions of the FLMP.

In the present study, we have analyzed data from the section of the questionnaire regarding the forest ES and the main forest goods and services. Indeed, the main objective of the study was to determine stakeholders' preferences for ES so that forest management priorities—in terms of forest ES—can be aligned based on the values expressed by respondents from different study areas or groups of stakeholders. Respondents were asked questions about the importance of the various ES. They had to express their opinion about the importance of each ecosystem service, on a five-point Likert scale format, ranging from 0 = "very low importance" to 4 = "very high importance". The forest ES considered were: Timber production, fuelwood production, grazing and non-timber forest products (NTFP) for the provisioning services; natural hazard protection, water and air quality, and nature conservation for the regulating services; and, hunting, sporting and tourism recreation for the cultural services. The attribution of the ES to the different categories follows the main internationally acknowledged classifications. With regard to hunting, according to Balkan and Kahn (1988) [55], Forster (1989) [56], and Bissell et al. (1998) [57], hunting was considered a cultural service, while the products obtained by hunting (i.e., meat, trophy and skin) that are provisioning services have not been considered.

2.3. Data Processing

The survey responses were analyzed statistically with respect to the following variables: Study area, group of interest, and socio-demographic characteristics of respondents (gender, age and level of education). The main descriptive statistics (mean, standard deviation, min. and max.) were provided for all ES and the chi-square (χ^2) test–using the standard $\alpha = 0.01$ cutoff—was used to test the differences between the groups of respondents based on the above-mentioned variables. All statistical analyses were carried out using R software (R version 3.4.4—"Someone to Lean On" Copyright ©2018 The R Foundation for Statistical Computing; R Studio Version 1.1.442—©2009–2018 RStudio, Inc., Vienna, Austria).

3. Results

The socio-demographic characteristics of the sample show that the majority of respondents are males (91.4%), whereas only 8.6% are females. With regard to age, the results show that the majority of the respondents are between 35 and 55 years old (67.7%), while the distribution of respondents in other age groups are the following: 9.9% of respondents are less than 35 years old, 19.1% are between 55 and 65 years old and the remaining 3.3% are more than 65 years old. Concerning the level of education, the respondents with a high school qualification comprise 43.4%, while the respondents with a university and post-university degree comprise 28.2%. The remaining 28.5% are respondents with an elementary school certificate.

The results show that for our sample of stakeholders the three most important ES are: Air and water quality (mean value = 3.54), tourism/recreation (mean value = 3.54) and fuelwood production (mean value = 3.52). Conversely, timber production has a marginal role for the stakeholders involved in the survey (mean value = 1.80) (Figure 2). Such a result is strictly related to the forest stand characteristics (mainly coppices) and potential wood assortments (mainly poles or woodchips) of the five case studies involved in the survey.

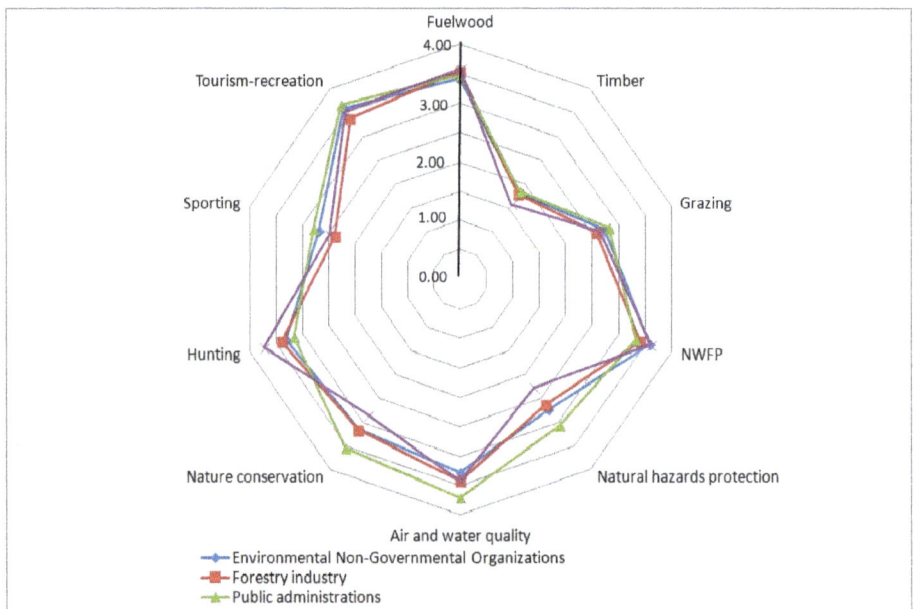

Figure 2. Stakeholders' preferences for ten ecosystem services (ES) by group of interest (mean value).

The ranking of importance for the three ES categories assigned by stakeholders was the following: Regulating services (mean value = 3.21) > cultural services (mean value = 3.16) > provisioning services (mean value = 2.85).

Below we illustrate the results of the analysis carried out in order to understand the influence of the different variables on stakeholders' preferences.

3.1. Study Areas

As we highlighted in Section 2, for two of the three pilot areas (Alto Agri and Arci-Grighine districts), we analyzed the data keeping separate the two different case studies for each pilot area. Concerning the Alto Agri district, we distinguished Alto Agri (with 113 stakeholders) and the National

Park Apennines Lucano-Val d'Agri-Lagonegrese (with 99 stakeholders). Concerning the Arci-Grighine district, we considered separately Arci (with 72 stakeholders) and Grighine (with 39 stakeholders).

The results show that the ranking of importance for ES changes from case study to case study. Hunting is considered the most important ES in two cases (Arci and Grighine), fuelwood production in one case (Matese), air and water quality in one case (National Park Apennines Lucano-Val d'Agri-Lagonegrese), and NTFPs in another case (Alto Agri) (Table 1).

Table 1. Stakeholders' preferences for ten ES by study area (mean value).

ES/Study Area	Alto Agri (*n* = 113)	NP Lucano-Val d'Agri-Lagonegrese (*n* = 99)	Arci (*n* = 72)	Grighine (*n* = 39)	Matese (*n* = 39)
Provisioning services					
Fuelwood	3.49 ± 0.78	3.42 ± 0.82	3.64 ± 0.63	3.56 ± 0.64	**3.59 ± 0.72**
Timber	1.98 ± 0.77	2.42 ± 0.94	1.13 ± 0.41	1.10 ± 0.31	1.59 ± 0.72
Grazing	2.73 ± 0.89	2.85 ± 0.71	2.67 ± 1.03	2.51 ± 0.88	2.56 ± 1.05
Non-timber forest products (NTFP)	**3.62 ± 0.57**	2.87 ± 0.69	**3.78 ± 0.48**	**3.64 ± 0.54**	3.18 ± 0.79
Regulating services					
Natural hazards protection	2.32 ± 1.08	3.69 ± 0.58	2.25 ± 1.21	2.21 ± 1.15	3.46 ± 0.85
Air and water quality	3.38 ± 0.83	**3.86 ± 0.38**	3.60 ± 0.73	3.13 ± 0.95	**3.51 ± 0.79**
Nature conservation	2.93 ± 1.01	**3.83 ± 0.41**	3.14 ± 1.03	2.97 ± 1.01	**3.51 ± 0.79**
Cultural services					
Hunting	**3.60 ± 0.70**	2.57 ± 0.98	**3.90 ± 0.34**	**3.92 ± 0.27**	2.72 ± 0.97
Sporting	2.06 ± 0.97	2.98 ± 0.89	3.03 ± 0.90	2.64 ± 0.87	2.26 ± 0.99
Tourism-recreation	3.35 ± 0.72	**3.77 ± 0.47**	**3.76 ± 0.49**	**3.64 ± 0.54**	3.03 ± 0.96

In bold are the three most important ES by study area.

NTFPs are considered important not only in Alto Agri but also in two other cases (Arci and Grighine in the region of Sardinia). However, it is interesting to note that in Alto Agri the main NTFPs are mushrooms and black truffles, while in Arci and Grighine the most important NTFPs are the cork extracted from cork oak (*Quercus suber* L.) and the myrtle fruit (*Myrtus communis* L.) used for the preparation of a local liqueur.

Tourism-recreation activities are of particular importance in the protected area of the National Park Lucano-Val d'Agri-Lagonegrese (third position in the ranking with a mean value of 3.77) and in the two Sardinian cases.

In accordance with the characteristics and peculiarities of the study area, the stakeholders of the National Park Lucano-Val d'Agri-Lagonegrese emphasize the role of this protected area for nature conservation (second position in the ranking of importance). Also, in the Matese district, the stakeholders highlight the relevance of nature conservation. Such a result is probably due to the fact that in this territory there are six areas falling into the Natura 2000 network. This, over time, has increased sensitivity, at least in some local actors, towards issues of nature conservation.

Observing the results by ES category (Table 2), in three case studies (Alto Agri, Arci and Grighine) the most important ES category is cultural services with 3.00, 3.56 and 3.40, respectively, as mean values. Conversely, in the National Park Apennines Lucano-Val d'Agri-Lagonegrese and in the Matese district, the stakeholders consider regulating services as a more important ES category (with mean values of 3.79 and 3.50, respectively).

Table 2. Stakeholders' preferences for three ES categories by study area (mean value).

ES Category/Study Area	Alto Agri (*n* = 113)	NP Lucano-Val d'Agri-Lagonegrese (*n* = 99)	Arci (*n* = 72)	Grighine (*n* = 39)	Matese (*n* = 39)
Provisioning services	2.96 ± 0.38	2.89 ± 0.52	2.80 ± 0.36	2.71 ± 0.32	2.73 ± 0.51
Regulating services	2.88 ± 0.80	**3.79 ± 0.36**	3.00 ± 0.66	2.77 ± 0.86	**3.50 ± 0.73**
Cultural services	**3.00 ± 0.52**	3.10 ± 0.57	**3.56 ± 0.42**	**3.40 ± 0.38**	2.67 ± 0.61

In bold is the most important ES category by study area.

The chi-square test shows no statistically significant differences for all ES categories. Despite this, it is highlighted that in Grighine and Alto Agri the regulating services category has less importance than in the other three case studies (National Park Apennines Lucano-Val d'Agri-Lagonegrese, Arci and Matese).

3.2. Groups of Interest

The results show that two groups of interest (actors of the forestry industry and actors of the tourism sector) have a similar order of priorities in the ranking of importance for the ES. The three most important ES for these two groups of interest are the following (Table 3): Hunting (ranked first by actors of the tourism sector and ranked third by actors of the forestry industry); NTFP (placed in third position by actors of the forestry industry and in second position by actors of the tourism sector); and, fuelwood (in first position for actors of the forestry industry and third position for actors of the tourism sector). The only substantial difference is that actors of the forestry industry emphasize the importance of air and water quality, while for the actors of the tourism sector this ES is only in fifth position.

Table 3. Stakeholders' preferences for ten ES by groups of interest (mean value).

ES/Group of Interest	Environmental NGOs (*n* = 19)	Forestry Industry Actors (*n* = 131)	Public Administrations (*n* = 168)	Tourism Actors (*n* = 44)
Provisioning services				
Fuelwood	**3.41 ± 0.80**	**3.54 ± 0.70**	3.49 ± 0.81	**3.58 ± 0.62**
Timber	1.82 ± 0.66	1.80 ± 0.90	1.84 ± 0.90	1.58 ± 0.81
Grazing	2.73 ± 1.08	2.58 ± 0.95	2.82 ± 0.81	2.64 ± 0.96
Non-timber forest products (NTFP)	**3.59 ± 0.73**	**3.40 ± 0.72**	3.32 ± 0.72	**3.60 ± 0.58**
Regulating services				
Natural hazards protection	2.73 ± 1.20	2.63 ± 1.24	3.07 ± 1.09	2.27 ± 1.05
Air and water quality	3.27 ± 0.98	**3.42 ± 0.82**	**3.71 ± 0.63**	3.42 ± 1.03
Nature conservation	3.14 ± 0.99	3.15 ± 1.03	**3.53 ± 0.78**	2.84 ± 1.09
Cultural services				
Hunting	3.32 ± 0.89	**3.40 ± 0.84**	3.16 ± 1.01	**3.73 ± 0.69**
Sporting	2.68 ± 1.13	2.38 ± 0.95	2.79 ± 0.97	2.49 ± 1.10
Tourism-recreation	**3.59 ± 0.50**	3.39 ± 0.75	**3.67 ± 0.56**	3.53 ± 0.76

In bold are the three most important ES by groups of interest.

The representatives of environmental NGOs emphasize the relevance of fuelwood production and NTFP, but they consider tourism-recreation in forests in first position in the ranking of importance, with a mean value of 3.59.

For the representatives of public administrations, the three most important forest ES are air and water quality (first position with a mean value of 3.71), tourism-recreation in forests (second position with a mean value of 3.67), and nature conservation (third position with a mean value of 3.53). This ES ranking should not be surprising if we consider that public administration has as its objective the valorization of common goods such as environmental externalities. Consequently, the productive aspects with positive effects in the private sphere—e.g., timber and fuelwood production, NTFP—have a secondary importance for public administrations.

Observing data by ES categories (Table 4), the results show that for public administrations the most important category of ES is regulating services (mean value = 3.44), emphasizing the importance of nature conservation and the quality of air and water. This fact is presumably due to the public administrations' aim to pursue the well-being of their community; consequently, the valorization of the air, water and environment quality is a point of strategic importance. For the other three groups of interest (environmental NGOs, actors of the forestry industry and actors of the tourism sector) the most important category of ES is cultural services, with mean values of 3.20, 3.06 and 3.25, respectively. It is interesting to highlight that the high value assigned to cultural services is due to different preferences for single ES by group of interest: Actors of forestry industry and tourism emphasize the importance of hunting, while the representatives of environmental NGOs and public administrations assign a high importance to tourism-recreation. In addition, all groups of interest, apart from the actors of the tourism sector, consider provisioning services as the least important ES category with the following mean values: 2.89 for the environmental NGOs, 2.83 for the actors of the forestry industry, 2.87 for the public administrations.

The chi-square test shows statistically significant differences for the following ES categories: Regulating services for representatives of actors of the tourism sector (p-value = 0.0080) and cultural services always for actors of the tourism sector (p-value < 0.00004).

Table 4. Stakeholders' preferences for three ES categories by groups of interest (mean value).

ES Category/Group of Interest	Environmental NGOs (n = 19)	Forestry Industry Actors (n = 131)	Public Administrations (n = 168)	Tourism Actors (n = 44)
Provisioning services	2.89 ± 0.56	2.83 ± 0.41	2.87 ± 0.45	2.85 ± 0.41
Regulating services	3.05 ± 0.77	**3.06 ± 0.79**	**3.44 ± 0.70**	2.84 ± 0.84
Cultural services	**3.20 ± 0.54**	3.06 ± 0.61	3.21 ± 0.54	**3.25 ± 0.62**

In bold are the most important ES by groups of interest.

3.3. Socio-Demographic Characteristics of Respondents

The results show that socio-demographic characteristics of respondents have an influence on preferences for some ES with special regard to two ES categories (Tables 5 and 6): Regulating services and cultural services.

Table 5. Stakeholders' preferences for ten ES by socio-demographic characteristics of respondents (mean value).

Socio-Demographic Characteristics/ES	Fuelwood	Timber	Grazing	NTFP	Natural Hazards Protection	Air and Water Quality	Nature Conservation	Hunting	Sporting	Tourism -Recreation
Age										
Less than 35 years old	3.53 ± 0.81	1.65 ± 0.82	2.71 ± 0.88	3.47 ± 0.56	2.71 ± 1.27	3.56 ± 0.81	3.32 ± 1.12	3.21 ± 0.94	2.79 ± 1.00	3.53 ± 0.65
36–45 years old	3.61 ± 0.65	1.85 ± 0.98	2.65 ± 0.91	3.39 ± 0.74	2.80 ± 1.18	3.52 ± 0.87	3.18 ± 1.03	3.24 ± 1.02	2.55 ± 1.04	3.52 ± 0.74
46–55 years old	3.45 ± 0.78	1.91 ± 0.84	2.76 ± 0.87	3.35 ± 0.73	2.92 ± 1.16	3.66 ± 0.61	3.45 ± 0.84	3.27 ± 0.89	2.55 ± 0.99	3.54 ± 0.66
56–65 years old	3.44 ± 0.74	1.56 ± 0.74	2.80 ± 0.93	3.38 ± 0.71	2.88 ± 1.14	3.53 ± 0.88	3.35 ± 0.90	3.39 ± 0.86	2.68 ± 1.03	3.67 ± 0.60
More than 65 years old	3.58 ± 0.79	1.83 ± 1.19	2.33 ± 0.78	3.75 ± 0.62	2.17 ± 1.34	3.42 ± 1.00	3.08 ± 1.16	3.75 ± 0.62	3.17 ± 0.72	3.58 ± 0.79
Gender										
Male	3.51 ± 0.74	1.82 ± 0.90	2.72 ± 0.89	3.40 ± 0.72	2.82 ± 1.17	3.58 ± 0.75	3.34 ± 0.92	3.31 ± 0.93	2.63 ± 1.00	3.56 ± 0.66
Female	3.47 ± 0.67	1.47 ± 0.67	2.63 ± 0.88	3.32 ± 0.68	2.89 ± 1.26	3.53 ± 1.07	3.05 ± 1.21	3.115 ± 0.85	2.47 ± 1.04	3.42 ± 0.74
Level of education										
Elementary school	3.58 ± 0.63	1.69 ± 0.92	2.56 ± 0.85	3.40 ± 0.76	2.49 ± 1.24	3.38 ± 0.90	3.06 ± 1.01	3.49 ± 0.81	2.61 ± 1.00	3.50 ± 0.73
High school	3.44 ± 0.84	1.75 ± 0.85	2.74 ± 0.89	3.41 ± 0.68	2.87 ± 1.11	3.61 ± 0.77	3.41 ± 0.91	3.26 ± 0.96	2.63 ± 1.05	3.59 ± 0.63
University or post-University degree	3.55 ± 0.67	1.98 ± 0.89	2.84 ± 0.91	3.35 ± 0.71	3.10 ± 1.17	3.72 ± 0.64	3.44 ± 0.94	3.14 ± 0.96	2.62 ± 0.97	3.57 ± 0.68

Table 6. Stakeholders' preferences for categories of ES by socio-demographic characteristics of respondents (mean value).

Socio-Demographic Characteristics/Category of ES	Provisioning Services	Regulating Services	Cultural Services
Age			
Less than 35 years old	2.84 ± 0.44	3.20 ± 0.89	3.18 ± 0.56
36–45 years old	2.88 ± 0.51	3.16 ± 0.89	3.10 ± 0.63
46–55 years old	2.87 ± 0.40	3.34 ± 0.69	3.12 ± 0.55
56–65 years old	2.80 ± 0.37	3.25 ± 0.71	3.25 ± 0.56
More than 65 years old	2.88 ± 0.42	2.89 ± 0.72	3.50 ± 0.50
Gender			
Male	2.86 ± 0.44	3.24 ± 0.75	3.17 ± 0.58
Female	2.72 ± 0.35	3.16 ± 1.02	3.00 ± 0.53
Level of education			
Elementary school	2.81 ± 0.39	2.96 ± 0.81	3.19 ± 0.65
High school	2.84 ± 0.44	3.27 ± 0.76	3.16 ± 0.54
University or post-University degree	2.92 ± 0.47	3.35 ± 0.76	3.12 ± 0.56

With regard to age, it is interesting to highlight that for one ES (i.e., hunting) the level of importance increases with the age of respondents (from respondents who are less than 45 years old to respondents more than 65 years old: 3.24 < 3.27 < 3.39 < 3.75). In addition, respondents over 65 consider cultural services more important than respondents under 65 (mean values of 3.25 and 3.50, respectively); conversely, respondents below 65 assign a higher value to regulating services than respondents above 65 (mean values of 3.25 and 2.89, respectively).

The chi-square test shows a statistically significant difference for regulating services regarding respondents younger than 35 (p-value = 0.0044).

Observing the data by gender, the results show that females consider natural hazards protection more important than males (mean values of 2.89 for females and 2.82 for males). The stakeholders' preferences for ES categories confirm that males assign the highest value to all three categories, but the chi-square test shows no statistically significant differences.

With regard to the level of education, respondents with a lower level assign a higher importance to one ES than respondents with a higher level, i.e., hunting (mean values of 3.49 for respondents with an elementary school certificate, 3.26 for respondents with a high school degree, and 3.14 for respondents with a university or post-university degree). In contrast, respondents with a higher level of education assign a higher importance to four ES than respondents with a lower level of education (mean values from elementary school certificate to university degree): Timber production (1.69 < 1.75 < 1.98); grazing (2.56 < 2.74 < 2.84); natural hazards protection (2.49 < 2.87 < 3.10); and, air and water quality (3.38 < 3.61 < 3.72). The chi-square test shows statistically significant differences only for two ES: Grazing for persons with a university degree (p-value < 0.0009) and Natural hazards protection for persons with a high school degree (p-value < 0.008). Considering the ES categories, the results show that respondents with a higher level of education prefer two of these (provisioning services and regulating services), while respondents with a lower level of education prefer cultural services provided by forests. However, the chi-square test shows no statistically significant differences between ES categories.

4. Discussion

The results of the present study provide evidence of the preferences of stakeholders for ten forest ES and three main ES categories. Results show the preferences considering both the various stakeholders' groups of interest and the socio-demographic characteristics of the respondents. The information collected during the interviews is usually used in the next phase of the FLMP for the definition of objectives and priority actions. In particular, the overall ranking can be used to identify a priority order of actions to be implemented to enhance the most important forest ES. In addition, the rankings of each group of interest can be used in order to take into account all interests. Our experience shows that the ES concept is very useful in forest planning, despite being developed in a much wider context than the forest sector itself. It allows the integration of different actors in the decision-making process related to natural resources management and makes it possible to also take into account the new needs that are being expressed by society towards ecosystems. It provides decision makers useful information to be incorporated into the planning process and a common language that facilitates communication with and between researchers/technicians, decision makers, stakeholders and the public at large, placing at the center of all considerations a discourse regarding quality of life.

In our research, we constantly faced two important questions: At what scale can the whole range of ES be appropriately taken into account, and what is the best approach for ES identification?

In forest planning the missing link was actually the landscape-scale planning. This scale proved to be the most suitable for taking into account issues of public interest, such as soil protection or nature, and landscape conservation and, at the same time, harmonizing them with needs and targets expressed by the different local actors. Once the objectives and the management strategies to realize them have been identified, the FUMP can acknowledge them and contextualize them on the territory.

As for the approach to identify ES, the qualitative assessment experimented by means of our case studies have given satisfactory and promising results.

We had a challenge of stakeholders' identification and categorization, which was successfully overcome, thanks to the three-step model of stakeholders' analysis we adopted [32]. After that, the administration of questionnaires, by means of face-to-face interviews, allowed for the fruitful involvement of stakeholders. In particular, the relaxed and informal atmosphere intentionally created during the interviews, not only permitted one to elicit people's needs and opinions, but also allowed for a climate of trust and reciprocal understanding. Inclusiveness is actually acknowledged as a critical requisite for an effective participatory planning process. When planning in rural areas, it may be especially difficult to involve those stakeholders from the primary sector who often have a marginal role in the social system, despite being directly in charge of the management of local natural resources. Such a situation may entail two main consequences: The loss of valuable information during the elaboration of the plan and the origination of possible conflicts during the implementation stage [42]. Strictly connected to the issue of inclusiveness is the need for a large and effective representation of different views and interests in the planning process [58,59]. Indeed, a fair and balanced representation is difficult to attain. Since we were aware of this problem, in our case studies we made great efforts in order to give different voices the chance to be heard and to represent different interests in a satisfactory way, focusing in particular on both the key and primary stakeholders.

When observing the results of the ES assessment in our case studies, we are faced with some expected and predictable outcomes, but also some surprising ones. As to regulating services, it is evident that in Mediterranean areas we find less attention is paid to natural hazards compared to alpine or mountainous regions. It is striking, on the other hand, to find a keen sensitivity towards issues such as air and water quality and nature conservation, which one would not expect in highly rural areas, such as the districts investigated. It is a clear sign that people's perception is changing and this may be in part due to the existence of protected areas which, despite having initially caused conflicts with local populations, are now producing a slight evolution of their awareness [60].

Concerning provisioning services, it is interesting to note that not only fuelwood production, but also NTFP, are considered very important. This means that it is not just the forest that provides people with goods, but the entire landscape, in all its components. The fact that timber is considered less important is largely due to the prevalence of coppices. However, factors related to local traditions play an important role, as has been also highlighted in other European studies [61–63]. The Matese district may be considered emblematic in this regard: Here, in fact, despite beech high forests being present, which are also potentially very productive, most of the wood production is utilized for energy purposes. Actually, many of the present high forests previously were coppices intensively utilized for fire-wood production and systematically subjected to grazing. Today we do not find in this territory a well-structured forestry industry, able to suitably valorize the wood assortments coming from high forests. This is a typical problem that has to be dealt with at the FLMP scale. Considering the importance of the primary sector in the districts investigated, we would expect greater importance attributed to grazing in line with what has emerged in other studies carried out for the realization of other FLMPs in southern Italy [32,64]. This is probably a sign of the socio-economic change in progress in the territories of the districts investigated, which affects the priorities expressed towards the ES.

With regard to cultural services, the results of this study show that in various contexts tourism-recreation is considered an important ecosystem service. In this sense, several authors have highlighted that key forest attributes to increase recreational attractiveness are naturalness of forest landscape, tree species composition (e.g., mixed forests vs. pure broadleaved of conifer forests), horizontal and vertical stand structure, and tourism facilities (e.g., hiking paths, parking and picnic facilities) [65–67]. In addition, Czajkowski et al. [68] highlighted that according to public opinion the most important biodiversity and recreation attributes of Polish forests are the amount of litter (garbage and rubbish) in forests, and the level of recreational infrastructure. Concerning the aesthetic aspects of tourism-recreation, De Valck et al. [69] investigated people's preferences for nature restoration scenarios showing that the conversion of forest plantations can increase landscape diversity and species richness. Similarly,

Valasiuk et al. [67] showed that Swedish and Norwegian citizens are willing to pay for forest landscape restoration of the transboundary Fulufjället National Parks.

Concerning hunting, we placed it in the cultural services category, following most acknowledged classifications. However, considering the highly rural context, we could have included it in the category of provisioning services. In fact, the actors of the forestry industry attribute a high value to cultural services also due to the fact that hunting has been included in this category. This should lead us to reflect on the importance of considering specific local contexts in categorizing ES. Speaking more generally, as highlighted also for alpine areas, the role of some specific ES should be analyzed and clarified better [29].

Taking into account the influence of the groups of interest variable, it is apparent that there is certain sensitivity on the part of the actors of the forestry industry towards regulating services. A further interesting aspect is that the actors of the forestry industry and those of the tourism sector have a similar order of priorities in the ranking of importance for the ES. In particular, the actors from the tourist sector show great interest in some provisioning services. All this opens up to interesting prospects related to opportunities of making the most of synergies between different ES. Furthermore, positive interaction could be stimulated between local actors that would enable a sustainable and self-propelling development of areas long considered economically marginal. Obviously this would necessitate being adequately supported during the implementation stage of the plan.

5. Conclusions

The present study analyzed stakeholders' opinions and preferences for different categories of forest ES in order to provide new information to the decision makers (forest planners and managers). The data were collected through the face-to-face administration of a semi-structured questionnaire. This method has the main advantage of being able to collect a large number of standardized information in a short time. In addition, the data are an overview of social preferences to be included in the landscape planning in accordance with the principles of FLMP. Conversely, the main disadvantage is that the method is not able to deepen the data collected with qualitative information. The future steps of the study are to collect also qualitative information through in-depth interviews with a sub-sample of stakeholders. Besides, stakeholders' preferences for ES and categories of ES should be mapped in order to facilitate the inclusion of social preferences in forest planning and management choices.

Finally, we think the exercise we made in the present research—to support landscape planning with stakeholders' point of view—can help design the handling of a FLMP, where multiple stakeholders and conflicting criteria could be involved. We are also conscious that a lot of work still remains, because preferences for ES are influenced by a combination of factors deeply rooted in the local context and related to places and people (including attitudes; cultural, social, and human values; and biophysical differences in forest features).

Author Contributions: Conceptualization, I.D.M., A.P., F.F. and M.G.C.; Data curation, A.P. and F.F.; Formal analysis, A.P.; Investigation, I.D.M. and F.F.; Methodology, I.D.M., A.P., F.F. and M.G.C.; Software, F.F.; Supervision, A.P.; Writing—original draft, A.P.; Writing—review and editing, I.D.M. and M.G.C.

Funding: The Alto-Agri research was coordinated by the National Institute of Agricultural Economics Research (INEA) Basilicata Head-quarters, with the financial support of the Basilicata Region—"Operational Programme Val d'Agri Melandro-Sauro-Camastra" and the National Park of Lucano Apennine-Val d'Agri-Lagonegrese. The Matese research was funded by the Region of Molise (Accordo di programma quadro pluriennale per attività di ricerca e sviluppo nel settore forestale) and by Progetto Collezioni E-A-OR MIPAF (Ministry for Agricultural, Food and Forestry Policies). The Arci-Grighine research was funded by the Regione Autonoma della Sardegna, Assessorato della Difesa dell'Ambiente, Servizio Tutela del suolo e Politiche Forestali and by Progetto Collezioni E-A-OR MIPAF (Ministry for Agricultural, Food and Forestry Policies).

Acknowledgments: The authors wish to acknowledge the participation and collaboration of the stakeholders who provided useful information and suggestions.

Conflicts of Interest: The authors declare no conflict of interest.

References

1. Daily, G.C. *Nature's Services: Societal Dependence on Natural Ecosystems*; Island Press: Washington, DC, USA, 1997.
2. Westman, W. How much are nature's services worth. *Science* **1977**, *197*, 960–964. [CrossRef] [PubMed]
3. Ehrlich, P.R.; Ehrlich, A. *Extinction: The Causes and Consequences of the Disappearance of Species*; Random House: New York, NY, USA, 1981.
4. De Groot, R.S. Environmental Functions as a Unifying Concept for Ecology and Economics. *Environmentalist* **1987**, *7*, 105–109. [CrossRef]
5. MEA. *Millennium Ecosystem Assessment Ecosystems and Human Well-Being: Biodiversity Synthesis*; World Resources Institute: Washington, DC, USA, 2005.
6. De Groot, R.S.; Alkemade, R.; Braat, L.; Hein, L.; Willemen, L. Challenges in integrating the concept of ecosystem services and values in landscape planning, management and decision making. *Ecol. Complex.* **2010**, *7*, 260–272. [CrossRef]
7. Gómez-Baggethun, E.; De Groot, R.; Lomas, P.L.; Montes, C. The history of ecosystem services in economic theory and practice: From early notions to markets and payment schemes. *Ecol. Econ.* **2010**, *69*, 1209–1218. [CrossRef]
8. Brauman, K.A.; Daily, G.C.; Duarte, T.K.E.; Mooney, H.A. The nature and value of ecosystem services: An overview highlighting hydrologic services. *Annu. Rev. Environ. Resour.* **2007**, *32*, 67–98. [CrossRef]
9. Dominati, E.; Patterson, M.; Mackay, A. A framework for classifying and quantifying the natural capital and ecosystem services of soils. *Ecol. Econ.* **2010**, *69*, 1858–1868. [CrossRef]
10. TEEB. *The Economics of Ecosystems and Biodiversity: Mainstreaming the Economics of Nature: A Synthesis of the Approach, Conclusions and Recommendations of TEEB*; The Economics of Ecosystems and Biodiversity: Geneva, Switzerland, 2010.
11. Haines-Young, R.; Potschin, M. *Proposal for a Common International Classification of Ecosystem Goods and Services (CICES) for Integrated Environmental and Economic Accounting*; EEA: Copenhagen, Denmark, 2010.
12. Antognelli, S.; Vizzari, M. Ecosystem and urban services for landscape liveability: A model for quantification of stakeholders' perceived importance. *Land Use Policy* **2016**, *50*, 277–292. [CrossRef]
13. Boyd, J.; Banzhaf, S. What are ecosystem services? The need for standardized environmental accounting units. *Ecol. Econ.* **2007**, *63*, 616–626. [CrossRef]
14. Costanza, R. Ecosystem services: Multiple classification systems are needed. *Biol. Conserv.* **2008**, *141*, 350–352. [CrossRef]
15. Busch, M.; La Notte, A.; Laporte, V.; Erhard, M. Potentials of quantitative and qualitative approaches to assessing ecosystem services. *Ecol. Indic.* **2012**, *21*, 89–103. [CrossRef]
16. Deal, R.L.; White, R. Integrating forest products with ecosystem services: A global perspective. *For. Policy Econ.* **2012**, *17*, 1–2. [CrossRef]
17. Baral, H.; Guariguata, M.R.; Keenan, R.J. A proposed framework for assessing ecosystem goods and services from planted forests. *Ecosyst. Serv.* **2016**, *22*, 260–268. [CrossRef]
18. Burkhard, B.; Kroll, F.; Nedkov, S.; Müller, F. Mapping ecosystem service supply, demand and budgets. *Ecol. Indic.* **2012**, *21*, 17–29. [CrossRef]
19. Campagne, S.C.; Roche, P.K. May the matrix be with you! Guidelines for the application of expert-based matrix approach for ecosystem services assessment and mapping. *One Ecosyst.* **2018**, *3*, e24134. [CrossRef]
20. Paudyal, K.; Baral, H.; Burkhard, B.; Bhandari, S.P.; Keenan, R.J. Participatory assessment and mapping of ecosystem services in a data-poor region: Case study of community-managed forests in central Nepal. *Ecosyst. Serv.* **2015**, *13*, 81–92. [CrossRef]
21. Häyhä, T.; Franzese, P.P.; Paletto, A.; Faath, B.D. Assessing, valuing, and mapping ecosystem services in Alpine forests. *Ecosyst. Serv.* **2015**, *14*, 12–23. [CrossRef]
22. Paletto, A.; Giacovelli, G.; Pastorella, F. Stakeholders' opinions and expectations for the forest-based sector: A regional case study in Italy. *Int. For. Rev.* **2017**, *19*, 68–78. [CrossRef]
23. Gatto, P.; Pettenella, D.; Secco, L. Payments for forest environmental services: Organisational models and related experiences in Italy. *iForest* **2009**, *2*, 133–139. [CrossRef]
24. Egoh, B.; Reyers, B.; Rouget, M.; Richardson, D.M.; Le Maitre, D.C.; van Jaarsveld, A.S. Mapping ecosystem services for planning and management. *Agric. Ecosyst. Environ.* **2008**, *127*, 135–140. [CrossRef]

25. Geneletti, D. Reasons and options for integrating ecosystem services in strategic environmental assessment of spatial planning. *Int. J. Biodivers. Sci. Ecosyst. Serv. Manag.* **2011**, *7*, 143–149. [CrossRef]

26. Hauck, J.; Görg, C.; Varjopuro, R.; Ratamäki, O.; Maes, J.; Wittmer, H.; Jax, K. "Maps have an air of authority": Potential benefits and challenges of ecosystem service maps at different levels of decision making. *Ecosyst. Serv.* **2013**, *4*, 25–32. [CrossRef]

27. Von Haaren, C.; Albert, C. Integrating ecosystem services and environmental planning: Limitations and synergies. *Int. J. Biodivers. Sci. Ecosyst. Serv. Manag.* **2011**, *7*, 150–167. [CrossRef]

28. Bennett, E.M.; Peterson, G.D.; Gordon, L.J. Understanding relationships among multiple ecosystem services. *Ecol. Lett.* **2009**, *12*, 1–11. [CrossRef] [PubMed]

29. Cantiani, M.G.; Geitner, C.; Haida, C.; Maino, F.; Tattoni, C.; Vettorato, D.; Ciolli, M. Balancing economic development and environmental conservation for a new governance of Alpine areas. *Sustainability* **2016**, *8*, 802. [CrossRef]

30. Grêt-Regamey, A.; Walz, A.; Bebi, P. Valuing Ecosystem Services for Sustainable Landscape Planning in Alpine regions. *Mt. Res. Dev.* **2008**, *28*, 156–165. [CrossRef]

31. Hermann, A.; Schleifer, S.; Wrbka, T. The Concept of Ecosystem Services regarding Landscape Research: A review. *Living Rev. Landsc. Res.* **2011**, *5*, 1–26. [CrossRef]

32. Paletto, A.; Cantiani, M.G.; De Meo, I. Public Participation in Forest Landscape Management Planning (FLMP) in Italy. *J. Sustain. For.* **2015**, *34*, 465–482. [CrossRef]

33. Howe, C.; Suich, H.; Vira, B.; Mace, G.M. Creating win-wins from trade-offs? Ecosystem services for human well-being: A meta-analysis of ecosystem service trade-offs and synergies in the real world. *Global Environ. Chang.* **2014**, *28*, 263–275. [CrossRef]

34. Goldman, R.L.; Thompson, B.H.; Daily, G.C. Institutional incentives for managing the landscape: Inducing cooperation for the production of ecosystem services. *Ecol. Econ.* **2007**, *64*, 333–343. [CrossRef]

35. Hendee, J.; Flint, C.G. Incorporating Cultural Ecosystem Services into Forest Management Strategies for Private Landowners: An Illinois Case Study. *For. Sci.* **2014**, *60*, 1172–1179. [CrossRef]

36. Cantiani, P.; De Meo, I.; Ferretti, F.; Paletto, A. Forest functions evaluation to support forest landscape management planning. *For. Ideas* **2010**, *16*, 44–51.

37. Ferretti, F.; Di Bari, C.; De Meo, I.; Cantiani, P.; Bianchi, M. ProgettoBosco: A Data-Driven Decision Support System for forest planning. *Int. J. Math. Comput. For. Nat.-Resour. Sci.* **2011**, *3*, 27–35.

38. Paletto, A.; Ferretti, F.; Cantiani, P.; De Meo, I. Multifunctional approach in forest management land plan: An application in Southern Italy. *For. Syst.* **2011**, *2*, 66–80.

39. Bettinger, P.; Lennette, M.; Johnson, K.N.; Spies, T.A. A hierarchical spatial framework for forest landscape planning. *Ecol. Model.* **2005**, *182*, 25–48. [CrossRef]

40. De Meo, I.; Ferretti, F.; Frattegiani, M.; Lora, C.; Paletto, A. Public participation GIS to support a bottom-up approach in forest landscape planning. *iForest* **2013**, *6*, 347–352. [CrossRef]

41. De Meo, I.; Ferretti, F.; Paletto, A.; Cantiani, M.G. An approach to public involvement in forest landscape planning in Italy: A case study and its evaluation. *Ann. Silvicul. Res.* **2017**, *41*, 54–66.

42. Cantiani, M.G. Forest planning and public participation: A possible methodological approach. *iForest* **2012**, *5*, 72–82. [CrossRef]

43. Kant, S. Extending the boundaries of forest economics. *For. Policy Econ.* **2003**, *5*, 39–56. [CrossRef]

44. Focacci, M.; Ferretti, F.; De Meo, I.; Paletto, A.; Costantini, G. Integrating Stakeholders' Preferences in Participatory Forest Planning: A Pairwise Comparison Approach from Southern Italy. *Int. For. Rev.* **2017**, *19*, 413–422. [CrossRef]

45. Cubbage, F.; Harou, P.; Sillsa, E. Policy instruments to enhance multi-functional forest management. *For. Policy Econ.* **2007**, *9*, 83–85. [CrossRef]

46. Schmithüsen, F. Multifunctional forestry practices as a land use strategy to meet increasing private and public demands in modern societies. *J. For. Sci.* **2007**, *53*, 290–298. [CrossRef]

47. Paletto, A.; Hamunen, K.; De Meo, I. Social network analysis to support stakeholder analysis in participatory forest planning. *Soc. Nat. Resour.* **2015**, *28*, 1108–1125. [CrossRef]

48. Paletto, A.; Giacovelli, G.; Grilli, G.; Balest, J.; De Meo, I. Stakeholders' preferences and the assessment of forest ecosystem services: A comparative analysis in Italy. *J. For. Sci.* **2014**, *60*, 472–483. [CrossRef]

49. Kelly, C.; Ferrara, A.; Wilson, G.A.; Ripullone, F.; Nolè, A.; Harmer, N.; Salvati, L. Community resilience and land degradation in forest and shrubland socio-ecological systems: Evidence from Gorgoglione, Basilicata, Italy. *Land Use Policy* **2015**, *46*, 11–20. [CrossRef]

50. Di Salvatore, U.; Ferretti, F.; Cantiani, P.; Paletto, A.; De Meo, I.; Chiavetta, U. Multifunctionality assessment in forest planning at landscape level. The study case of Matese Mountain Community (Italy). *Ann. Silvicul. Res.* **2013**, *37*, 45–54.

51. Paletto, A.; De Meo, I.; Ferretti, F. Social network analysis to support the forest landscape planning: An application in Arci-Grighine, Sardinia (Italy). *For. Ideas* **2010**, *1*, 28–35.

52. Paletto, A.; Ferretti, F.; De Meo, I. The role of social networks in forest landscape planning. *For. Policy Econ.* **2012**, *15*, 132–139. [CrossRef]

53. Reed, M.S.; Graves, A.; Dandy, N.; Posthumus, H.; Hubacek, K.; Morris, J.; Prell, C.; Quinn, C.H.; Stringer, L.C. Who's in and why? A typology of stakeholder analysis methods for natural resource management. *J. Environ. Manag.* **2009**, *90*, 1933–1949. [CrossRef] [PubMed]

54. Adamowicz, W.L.; Louviere, J.; Swait, J. *Introduction to Attribute-Based Stated Choice Methods*; NOAA—National Oceanic Atmospheric Administration: Washington, DC, USA, 1998.

55. Balkan, E.; Kahn, J.R. The value of changes in deer hunting quality: A travel cost approach. *Appl. Econ.* **1988**, *20*, 533–539. [CrossRef]

56. Forster, B.A. Valuing Outdoor Recreational Activity: A Methodological Survey. *J. Leis. Res.* **1989**, *21*, 181–201. [CrossRef]

57. Bissell, S.J.; Duda, M.D.; Young, K.C. Recent studies on hunting and fishing participation in the United States. *Hum. Dimensions Wildl.* **1998**, *3*, 75–80. [CrossRef]

58. McCool, S.F.; Guthrie, K. Mapping the dimensions of successful public participation in messy natural resources management situations. *Soc. Nat. Resour.* **2001**, *14*, 309–323.

59. Bruña-García, X.; Marey-Pérez, M.F. The Challenge of Diffusion in Forest Plans: A Methodological Proposal and Case Study. *Forests* **2018**, *9*, 240. [CrossRef]

60. De Meo, I.; Brescancin, F.; Graziani, A.; Paletto, A. Management of Natura 2000 sites in Italy: An exploratory study on stakeholders' opinions. *J. For. Sci.* **2016**, *62*, 511–520. [CrossRef]

61. Dhubháin, Á.N.; Fléchard, M.C.; Moloney, R.; O'Connor, D. Stakeholders' perceptions of forestry in rural areas—Two case studies in Ireland. *Land Use Policy* **2009**, *26*, 695–703. [CrossRef]

62. Hiedanpää, J. The edges of conflict and consensus: A case for creativity in regional forest policy in Southwest Finland. *Ecol. Econ.* **2005**, *55*, 485–498. [CrossRef]

63. Matilainen, A.; Koch, M.; Zivojinovic, I.; Lähdesmäki, M.; Lidestav, G.; Karppinen, H.; Didolot, F.; Jarsky, V.; Põllumäe, P.; Colson, V.; et al. Perceptions of ownership among new forest owners—A qualitative study in European context. *For. Policy Econ.* **2018**, in press. [CrossRef]

64. De Meo, I.; Cantiani, M.G.; Ferretti, F.; Paletto, A. Stakeholders' perception as support for forest landscape planning. *Int. J. Ecol.* **2011**, *1*, 1–8. [CrossRef]

65. Paletto, A.; De Meo, I.; Cantiani, M.G.; Maino, F. Social perceptions and forest management strategies in an Italian Alpine community. *Mt. Res. Dev.* **2013**, *33*, 152–160. [CrossRef]

66. Abildtrup, J.; Garcia, S.; Olsen, S.B.; Stenger, A. Spatial preference heterogeneity in forest recreation. *Ecol. Econ.* **2013**, *92*, 67–77. [CrossRef]

67. Valasiuk, S.; Czajkowski, M.; Giergiczny, M.; Żylicz, T.; Veisten, K.; Landa Mata, I.; Halse, A.H.; Elbakidze, M.; Angelstam, P. Is forest landscape restoration socially desirable? A discrete choice experiment applied to the Scandinavian transboundary Fulufjället National Park Area. *Restor. Ecol.* **2018**, *26*, 370–380. [CrossRef]

68. Czajkowski, M.; Budzinski, W.; Campbell, D.; Giergiczny, M.; Hanley, N. Spatial Heterogeneity of Willingness to Pay for Forest Management. *Environ. Resour. Econ.* **2017**, *68*, 705–727. [CrossRef]

69. De Valck, J.; Vlaeminck, P.; Broekx, S.; Liekens, I.; Aertsens, J.; Chen, W.; Vranken, L. Benefits of clearing forest plantations to restore nature? Evidence from a discrete choice experiment in Flanders, Belgium. *Landsc. Urban Plan.* **2014**, *125*, 65–75. [CrossRef]

![forests logo] *forests*

MDPI

Article

Orchestrating Forest Policy in Italy: Mission Impossible?

Laura Secco [1], Alessandro Paletto [2], Raoul Romano [3], Mauro Masiero [1], Davide Pettenella [1], Francesco Carbone [4] and Isabella De Meo [5,*]

[1]   Department of Land, Environment, Agriculture and Forestry (TESAF), University of Padua, 35020 Padova, Italy; laura.secco@unipd.it (L.S.); mauro.masiero@unipd.it (M.M.); davide.pettenella@unipd.it (D.P.)
[2]   CREA, Research Centre for Forestry and Wood, 38123 Trento, Italy; alessandro.paletto@crea.gov.it
[3]   CREA, Research Centre for Agricultural Policies and Bioeconomy, 00198 Roma, Italy; raoul.romano@crea.gov.it
[4]   Department for Innovation in Biological, Agro-Food and Forest Systems (DIBAF), Tuscia University, 01100 Viterbo, Italy; fcarbone@unitus.it
[5]   CREA, Research Centre for Agriculture and Environment, 50125 Firenze, Italy
*   Correspondence: isabella.demeo@crea.gov.it; Tel.: +39-055-2492238

Received: 30 May 2018; Accepted: 30 July 2018; Published: 1 August 2018

Abstract: In the Italian political and economic agenda the forest sector occupies a marginal role. The forest sector in Italy is characterized by a high institutional fragmentation and centralized decision-making processes dominated by Public Forest Administrations. Public participation in forest policy processes has been implemented since the 1990s at national, regional and local levels in several cases. However, today no significant changes have been observed in the overall governance of the forest sector and stakeholders' involvement in Italian forest policy decision-making is still rather limited. The aims of this paper are to describe the state of forest-related participatory processes in Italy at various levels (national, regional and local) and identify which factors and actors hinder or support the establishment and implementation of participatory forest-related processes in the country. The forest-related participatory processes are analyzed adopting a qualitative-based approach and interpreting interactive, complex and non-linear participatory processes through the lens of panarchy theory.

Keywords: panarchy theory; national forest policy framework; stakeholders; participatory forums; interviews survey

1. Introduction

In Italy, forest policy issues are not included in public debates. The forest sector has a marginal role in the political and economic agenda due to its low contribution to GDP [1,2]. Forests and forest-related issues are interconnected with other more important sectors, such as agriculture, rural development, biodiversity conservation, climate change and tourism, but even the role of these sectors in national economic growth and societal challenges has been systematically underestimated or disregarded in spite of their collective importance to the nation. The only reasons for forests becoming visible to public opinion are forest fires during summer.

This situation is due to various factors. First of all, the highly fragmented institutional framework, with 21 regional or provincial public authorities (regions or autonomous provinces) which includes forest sector tasks, has remained decentralized since the 1970s. Decentralization has produced a strong fragmentation in the national forest system: According to the different socio-economic and environmental conditions, there is a heterogeneous framework of regional forest policies with priorities and issues that vary from region to region [3]. Secondly, there is no single national political institution that efficiently coordinates and links international, national, and regional sectoral policies. Three Ministries are directly involved in the forest sector—"Agriculture, Food and Forestry Policies", "Environment and Protection of Land and Sea", "Culture and Tourism", plus the Ministry for "Economic Development" (for the relevance of the timber industry, which is mainly based on imports). Often, overlaps and unclear distribution of roles and responsibilities among these Ministries result in conflicts between vision statements and incongruence among policies. Recently, in accordance with Law no. 124/2015, the National Forest Service, an important historical institution specialized in protection of the national forestry heritage, environment, landscape and territory, has been merged into another national unit, a national police force that is military in nature (*Arma dei Carabinieri*), not focused on the forest sector. This has created additional conflicts without significantly solving the problem of fragmentation, because the local offices and units are basically maintained even if the hierarchical structure is changed. Recently, a new national Forest Law that tries to address the sector as a whole was approved in 2018 (Law No. 34/2018). Lastly, there is an absence of national private and public forest owners' associations, which are large enough to be considered representative of the whole sector. Only a few and highly fragmented forest-related groups of interest can be identified as organized stakeholders, with an almost complete absence of active private owners.

Public participation in forest policy processes might help to understand and deal with the social and ecological complexity that characterizes the Italian forest sector [4,5]. Participation—in its simpler form of "stakeholders consultation"—has been implemented in Italy in several cases, at national, regional and local levels, since the first experiences connected with the introduction of forest certification mechanisms in the country after 1995 [3]. In the last 20 years, there have been several initiatives based on participation, at various levels, with different drivers such as some European rules (e.g., EC Regulations 1698/2005 and 1305/2013 on the Rural Development Programme for 2007–2013 and 2014–2020, respectively; Agenda 21 Local Programmes; Nature 2000 network), some voluntary market-based policy instruments (e.g., the development of Sustainable Forest Management standards and procedures for forest certification schemes; the development and implementation of schemes for Payments for Ecosystem Services and network- and non-timber forest products based territorial marketing initiatives), and some specific local factors (e.g., the existence of community-based ancient institutions in mountainous areas). However, all the above-mentioned factors create a difficult context for effective and successful participation. They contribute to maintaining the status quo, limiting entrepreneurial innovation and retaining latent conflicts, thus ultimately leading to still centralized decision-making processes dominated by forest public administrations, while other stakeholders seem to have minor roles [3].

Italy suffers from a large number of ineffective and failing participatory processes in the forest sector [3], one of the reasons being the high fragmentation of participatory forums that can be an effect of the fragmentation of the institutional framework, and consequently of objectives and means but also actors and interests. Italian participatory forums often overlap, replicate efforts, and do not interact or create synergies with one another. As a consequence, despite the various participatory-oriented initiatives launched in the country, no significant changes can be observed in the overall governance of the sector.

Starting from these considerations, the aims of this paper are to describe the state of forest-related participatory processes in Italy at various levels (national, regional and local), to provide elements for understanding whether "orchestrating" forest policy [6] in such a context is a "mission impossible" and, regardless, to make recommendations as to what actions could be done in the future that could result in effective national-level forest policy and processes. Crucial guiding questions are: (i) which factors are able to hinder or support the establishment and implementation of effective participatory forest-related processes involving scientists, stakeholders and the general public in the country; and (ii) whether we have (or not) learned lessons from participatory-oriented experiences in the Italian forest sector in the last 20 years that can now be used to orchestrate policy making.

These questions are particularly challenging if we consider the emerging and stimulating societal changes that participatory forest policy and decision-making processes should take into account at global (e.g., the increasing use of new communication instruments to exchange information, such as social networks and massive open science initiatives) and domestic (e.g., the complex, multi-level and inefficient institutional and legal framework, the absence of organized interested groups, the neglected economic role of forests) levels. Deliberative politics are generative, i.e., they create and animate actors and organizations, define problems and knowledge, and form new institutions for collective action [7], but the processes for reaching these outputs are definitely not linear [8]. This is especially true today, where we must consider the growing complexity of interactive decision-making processes which involve not only policy makers, experts and scientists and other target stakeholders (e.g., forest owners and their associations), but where also the general public is gaining a new role, not only in decision-taking, but also in knowledge co- constructing (e.g., mass-science initiatives).

Concepts of Forest Policy Participatory Processes

To meet our aim of describing participatory processes in Italy, a background on concepts of these processes is needed. Participation is defined by the Food and Agriculture Organization (FAO) of the United Nations [9] as various forms of direct public involvement where people, individually or through organized groups, can exchange information, express opinions and articulate interests, and have the potential to influence decisions or the outcome of specific forestry issues. For the World Bank [10], participation is a process through which stakeholders influence and share control over development initiatives, and the decisions and resources that affect them.

In the last decades, the role of public participation in the development and implementation of forest policy has been recognized at international level. In the 1990s, the Intergovernmental Panel on Forests [11] recommended the development of National Forest Programmes (NFPs) through a participatory process using appropriate mechanisms to involve all interested parties [12,13]. Subsequently, the first EU Forestry Strategy (1998) stressed the relevance of collaboration with all stakeholders in the implementation of international commitments, principles and recommendations through national or sub-national forest programmes, while the new EU Forestry Strategy (2013) emphasized the role of stakeholders to address the three dimensions of sustainable development (social, economic and environmental) in an integrated way.

Recently, in the forest sector new competences and demands are being asked of decision makers, due to the new forest functions that have progressively appeared alongside timber production (i.e., nature conservation, climate change mitigation, protection against natural hazards, recreational use of forests), in order to satisfy the needs of post-modern society [14,15]. The international debate on sustainable development recognizes the importance for forest policies to promote the consideration of the "full value" of forests [16]. An important element in this new approach to policy definition is the shift from a one-dimensional target related to forest productivity to multifunctionality [17].

In this framework, participatory policy processes are a necessity, because the definition of a forest-related policy must take into account a wide range of information related to various settings, involving local communities, adjusting and implementing decisions across different levels with a continuous interplay between top-down and bottom-up governance [18], promoting exchange of knowledge and collaborative learning about problems and their solutions [19], and approaching decisions that are both legitimate and sustainable [20]. Despite these fundamental characteristics, the specific aims of participation may differ according to historical period and the geographical and socioeconomic context [5]. Nevertheless until lately only a few national or regional contexts exist where the participation process for the development of forest policies followed a rigorous procedure and there is still no agreement on a unique, appropriate participation process [21].

2. Materials and Methods

Because participatory-oriented decision-making processes in forestry are systems characterized by a very high degree of complexity, and by multidimensional and non-linear processes and structures [7], we adopt a framework for this study that combines panarchy theory and participatory processes. First, panarchy theory is described. The panarchy model differs from the more static view of traditional hierarchy and "controlled top-down rationalist procedures" [22], so it seems particularly useful for understanding complex multi-level, multi-actor and multi-sector and non-linear decision-making processes typical of participatory forums. The situation in Italy is then described, followed by results based on the use of the panarchy model as a qualitative-based predictive analytical tool to grasp at what stage of development a forum is in, as well as what actions could be used to improve the participatory process.

2.1. Our Framework: Combining Panarchy Theory and Forest Policy Participatory Processes

Panarchy is a model of linked, hierarchically arranged adaptive cycles representing the cross-scale dynamic interactions among the levels of a complex self-organizing system [8]. Panarchy considers the interplay between change and persistence, evaluating both the spatial and time scales [7,8]. Originating from ecological sciences, in its applications to social sciences this theory has been defined as a metaphor used to describe four commonly occurring phases of change in complex systems [8], as a framework to understand the institutional and organizational change needed to enhance resilience [23], to underline discontinuities in urban and regional economic systems [24], and to identify legal reforms that allow for resilience-based governance [25,26]. We argue that panarchy can also be used to interpret forest-related participatory policy making processes and policy changes. Specifically, the four recurrent phases of the adaptive cycle described by panarchy [7] can be used as a qualitative-based predictive analytical tool, to understand how a participatory forum is evolving, its stage of development and, above all, what can be done in advance to prevent failures or reduce problems. The four recurrent phases (identified as r, k, Ω and α in the general theory) are described hereafter as adjusted to forest-related participatory processes. Panarchy theory phases and their complex dynamic interactions are represented in Figure 1.

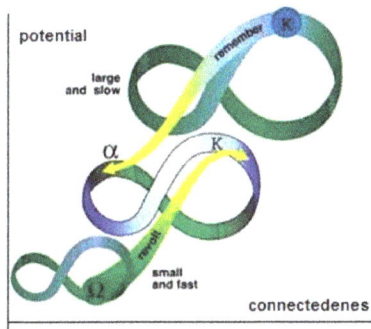

Figure 1. Panarchy theory: phases and dynamic cross-scale interactions. Source: Shannon 2015 [8].

2.1.1. Phase 1: Growth (r)

The first phase is considered not only a growth phase, but also an exploitation phase, where a rapid increase of system components (resources) and scramble competition occurs. It can be assimilated to the first, enthusiastic phase of participatory processes, where human and financial resources are allocated to support the launch and future expected management of the participatory policy initiative. During this phase, a large number of stakeholders are contacted and activated to become involved, preparatory meetings are organized to set the scene, and plans for further development are made. The policy process is typically rapid and quite flexible, many different solutions are still possible, and participants' perception of the potential of their own and others' participation is still open and positive. To be effective, participatory processes should optimize this phase, taking the maximum advantages from the initial enthusiasm (e.g., using the phase to build reciprocal trust between participants as well as between participants and the governance system [27,28], and consolidate relationships that are crucial for collaboration).

2.1.2. Phase 2: Conservation (k)

The second phase, Conservation, typically lasts longer than the first phase: Growth rates are slower, context limits and potential appear. Besides, resources start to be conserved and tend to be bound up in the system's structure, which becomes more rigid. In our field of application, it corresponds to the phase where policy problems are debated, latent conflicts may first become visible, discussions are long and contribute to slowing the process. Furthermore, resources are being progressively reduced and soon become scarce, the system becomes more rigid, with fixed and strict deadlines and appear the first signs of stakeholders becoming demotivated and/or disappointed. In participatory processes, options for preventing the problems connected to this phase include, for example, allocating enough resources (time, funds) [28,29] for long discussions to be satisfactorily completed and defining a clear step-by-step plan where discussions are followed by taking intermediate decisions.

2.1.3. Phase 3: Release (Ω)

The phase of Release is a rapid phase of collapse or creative destruction, where tightly bound resources become increasingly fragile and are finally dispersed. In our field of application, it corresponds to a stage in which one or more stakeholders decide to leave the process (e.g., being dissatisfied by the results or procedures, or not feeling involved enough) and the process itself seems to have failed. Energy, efforts and time voluntarily dedicated to the process by single participants become scarce, while financial resources allocated for supporting participation are almost finished and cannot be quickly re-instated. In order to prevent potential problems connected to this phase from arising, the previous phases have to be kept as short as possible, and accurately and professionally managed. At the same time enough resources have to be allocated from the beginning. In other words, the organizers should be aware that a good participatory process needs time, qualified soft-skills and money [29–31].

2.1.4. Phase 4: Reorganization (α)

In this last phase, the system innovates and re-structures itself, with two possible solutions: (i) persistence of the original regime, or (ii) a shift to a new one with a different set of processes and structures. In the first case, the participatory process is abandoned and the decision-making procedures remain the same as usual, as defined in the previous regime, often based on a predominant role of the public forest authority, which in the end takes a rational-based decision by primarily engaging experts and scientists, and only marginally considering ideas and proposals expressed by non-experts or the general public. In the second case, the participatory initiative continues but with new procedures, new power distribution among those stakeholders who remained onboard, and with re-defined objectives and preferably with new resources allocated.

Adaptive cycles have the potential to affect both smaller and larger scales. Consequently, the theory emphasizes "cross-scale linkages whereby processes at one scale affect those at other scales to influence the overall dynamics of the system" [7]. This means that, sometimes, small-scale variables and bottom-up processes, which act faster than larger-scale processes, can control and determine the overall system dynamics. Cross-scale or multi-level interactions between more timely policy processes occurring at lower levels (i.e., local, where policies are implemented through projects) and more slower processes occurring at higher and larger hierarchical levels (i.e., national or global, where policies are conceptualized and designed as guiding principles) have been mentioned in the literature [32,33]. In general, many authors have highlighted that participatory processes are more effective at local level [31,34]. Two other principles of panarchy, Remember and Revolt, are connected with scale and cross-scale issues [7]. The first one, Remember, is based on the idea that larger scale socio-ecological systems regulate lower levels by providing institutional arrangements for resilience i.e., forms of memory that encourage reorganization around the same structures and processes rather than a new set. In our field of application, this principle can be understood as the conservative attitude of forest public administrations, resisting changes [35] and external influences. The second principle, Revolt, is based on the idea that smaller scale socio-ecological systems foster change, experimentation and institutional revolution. In our field of application, it can be interpreted as the chance of inducing institutional revolution typical of pioneers, innovators and bottom-up promoters of new concepts and initiatives (e.g., Rametsteiner 2002 [36] in relation to Sustainable Forest Management (SFM) standards setting processes).

2.2. The Italian Context

The complexity of panarchy adaptive cycles and their phases is exacerbated by their application to the Italian forest context, which is characterized by variegated ecological, economic and social conditions throughout the country (see Table 1).

In this context, several participatory processes have been implemented, with different results, over the spatial and administrative scales [37].

Table 1. Key concepts to understand the forest sector in Italy.

Key Concept	Description
Forest cover	From the National Forestry Inventory data, the area covered by forests is 10.9 M ha (more than 30% of the total land) and has increased by 0.6 M ha since 2005. 95% of forest is distributed in the mountains and hilly areas, mainly the Alps and Apennines. The majority of forest area is under coppice regime (41.8%), especially in Central and Southern Italy, while high forests of coniferous and broadleaves (36.1%) are predominant in the North. The percentage of "not defined" forest types is significant (20.8%).
Property regime	More than 63.5% of forest area is private, 32.4% is public and 4.1% is not qualified. There are positive forms of management association in some areas, but at national scale forest firms are managed by forest-owners. Results of the Census 2011 have counted 328,358 forest firms and the forest area included in the firms is 2.9 M ha, with an average size of 8.9 ha. A few bigger firms (>100 ha), 3.7%, manage the largest amount of forest (64.7%) with an average firm of more than 150 ha; while the largest number of firms (<100 ha), 96.4%, manage only 35.3% of forest area, with an average of less than 3 ha/firm. The most relevant figure, however, is the amount of forest not included in active forest firms i.e., 3 M ha.
Forest management	It is developed according to the rules dictated by the regional forest code and regional forest laws. Frequently firms in the North manage forest according to the Forest Plan forecasts, and thanks to the support of EU funds; the firms adopting a forest plan are increasing even in the other regions and especially for public property.
Production	In the period 2010–2014 average timber production was in the range 7–6 M m^3/year. This is mainly for energy use (63.7%), while 31.9% is classified as roundwood. Other (non-timber) forest products (NTFPs), chestnuts, mushrooms, strawberries and plants for food are gathered. A large amount is for self-consumption, but companies that use NTFPs for economic activities are increasing.
Forestry economic performance	The added value of the forest sector has been estimated as €1.2–1.5 million in the last 5-years, with a contribution to the total value of national economic activity of ca. 0.05% while its contribution to GDP at national level is 0.09%.

2.3. Survey Methodology and Analytical Matrix

A qualitative-based research approach has been adopted in the present research. The survey methodology was developed with a 3-step approach including: (i) semi-structured interviews with policy makers, experts and stakeholders; (ii) identification and description of explicatory case studies; and (iii) definition of a set of criteria for assessing the success level of selected case studies' participatory processes.

2.3.1. Semi-Structured Interviews

A set of semi-structured interviews was conducted to find out the main participatory-based arenas/forums dealing with forest-related policies in Italy. The interviews were held from March to June 2015. A total of 7 interviewees were selected subjectively by the authors and included: (i) professors and researchers at universities or research institutes ($n = 2$); (ii) public decision makers ($n = 3$); (iii) professionals of private associations and representatives of NGOs ($n = 2$). An attempt to have a higher number of respondents was made but without success. The main reason for this is the limited number of experts and other types of actors directly dealing with forest policy issues in Italy, so that the 7 selected interviewees constitute a rather representative sample [38].

A list of questions suitable for a face-to-face interview was drawn up and pre-tested. It comprised 13 questions (11 open-ended and 2 closed-ended questions), which were chosen to keep the structure simple. In the 2 closed-ended questions the respondents had to choose from a list of pre-set responses or from a ranking scale with n options. Themes included: (i) position and number of years interviewees had been involved in their organizations; (ii) experience and personal participation concerning forest-related participative forums and decision making processes in Italy; (iii) main forms of participation used in the processes; and (iv) own opinion and level of satisfaction with the results of the participatory forums and processes. All the interviews were conducted by the authors through a face-to-face meeting and lasted from 20 to 40 min. The order in which questions were asked remained the same, and the questions were read together by the interviewer and interviewees. For each question, there was discussion and an exchange of information, useful to obtain explanations for responses and interpret the results [38].

2.3.2. Case Studies Selection

On the basis of interview results, a list of case studies of forest and forest-related participatory forums and processes in Italy was identified. These participatory processes, mentioned as the most relevant by respondents, have been categorized by level (national, regional or local) and described in terms of period, name, story (in brief), main promoter(s) and managers, and participants (Table 2). While the list at national level is almost complete, the table is not exhaustive for all the forums launched and managed at regional and local levels throughout the country. To create a full catalogue of all existing processes at these two administrative levels is out of the scope of this study.

Four cases were selected from the listed forums and included in the study: Two developed at national level and were reported by all respondents as the most relevant and successful, and two reported by key-informants at local/regional level. These four participatory forums have been described as "explanatory" cases according to Yin (2009) [39]: (1) the Framework Program for the Forest Sector (PQSF); (2) the Table on Forest-Wood Chain; (3) the partnership for the process of building the Model Forest "Montagna Fiorentina" in Tuscany (Central Italy), and (4) the formulation of regional regulation of the forest sector in Piedmont (Northern Italy).

The case studies analysis has been integrated through the direct knowledge of and observations by the paper's authors. Some of them have more than 30 years of direct experience with forest policy processes in the country, some are currently involved in forest policy reforms at national level, and others have published papers on participatory-oriented processes in forestry in Italy [3,5,38,40–46].

Table 2. Examples of forest-related participatory forums in Italy, at different levels.

Political Level	Period	Participative Forums or Decision-Making Processes	Description/Policy Field/Goals	Responsible for Launching and/or Managing the Forum/Process
National	1996–ongoing	State-Regions Conference-Forest Division	To coordinate state and regions debate on various forest policy issues: Law contents, emergency plans for pests and diseases management, market strategies, forest management and policy reforms guidelines.	Italian government
	1996–2000	National Working Group on Sustainable Forest Management (SFM) standards for Northern Italy (Milano Forum)	To develop a commonly agreed set of SFM standards to guide forest management and create a basis for forest certification implementation in Northern Italy, with no references to any specific forest certification scheme.	Department of Land, Environment, Agriculture and Forestry (TESAF)-University of Padova, Department DEIAFA-University of Turin
	2001–ongoing	Forest Stewardship Council (FSC)-Italy working group on SFM standards for the Italian Alpine regions	To develop, approve and periodically update with stakeholders participation a set of national standards for SFM consistent with the international set of FSC Principles and Criteria.	FSC-Italy (National Secretariat)
		Programme for the Endorsement of Forest Certification (PEFC)-Italy working group on SFM national standards	To develop and approve with stakeholders participation a set of standards for SFM consistent with the Pan-European set of Criteria and Indicators specific to the Italian context for PEFC forest certification processes.	PEFC-Italy (National Secretariat)
	2002–ongoing	Scientific Committee of Services Consortium wood-cork (CONLEGNO)	To promote the quality of companies and their products in timber and related sectors. To promote the provision of services related to stages of production of the member's undertakings.	CONLEGNO Monitoring Organization (National Committee)
	2003–2005	National Working Group on SFM standards for Apennine and Mediterranean forests (SAM)	To develop a commonly agreed set of SFM standards to guide forest management and create a basis for forest certification in Central and Southern Italy, with no references to any specific forest certification scheme.	Italian Academy of Forest Sciences
	2007–2009	Framework Program for the Forest Sector (PQSF)	To identify general strategies and policy guidelines for the forest sector. No funds or other resources have been allocated for implementation.	Ministry of Agriculture, Food and Forestry Policies
	2012–ongoing	Table on Forest-Wood chain	To define new tools, strategies and networks to increase the supply of domestic timber. The table is structured into working groups, including key forest sector stakeholders. Other sectors and interests (e.g., environmentalists) not involved.	Ministry of Agriculture, Food and Forestry Policies. The National Rural Network manages the process
	2007–ongoing	Carbon Monitoring Nucleon	To provide updated information on forest carbon sequestration/stocks, and develop common guidelines for monitoring, forest carbon investments, etc.	National Institute of Agricultural Economics (INEA) (now transformed into CREA)

Table 2. *Cont.*

Political Level	Period	Participative Forums or Decision-Making Processes	Description/Policy Field/Goals	Responsible for Launching and/or Managing the Forum/Process
Regional	2005–2006	Rural Development Plans (RDP) 2007–2013	The RDPs at regional level were developed by adopting a participatory approach, as formally required by the EU rules.	Regional Administrations (e.g., Veneto, Piedmont)
	2012–ongoing	Sub-sectoral forum on poplar plantations	To develop strategies for improving the production of timber from poplar plantations in Northern Italian regions. In this sense, it is an interregional initiative.	Ministry of Agriculture, Food and Forestry Policies through Consulta Nazionale Pioppo, jointly with the National Research Council in Agriculture and Poplar Plantation Owners Association
	2014–ongoing	Regulations for implementing the new regional forest law in Piedmont (approved in 2009)	To define specific rules and regulations for a recently approved forest reform in Piedmont.	Regional Administration (Piedmont)
	2010 (2 months)	Regulation on harvesting allocation in the Monte Rosa Foreste Association	To decide how to allocate harvesting on public stands to private logging companies on the basis of public rules.	Monte Rosa Foreste Association
Local	2009–2012	Partnership for the process of building the Model Forest "Montagna Fiorentina"	To improve the integration and sustainability of forest and land management, increasing the cohesion and awareness of the network of all the social-economic components that directly or indirectly belong to that territory.	Tuscany Region in coordination with Mediterranean Model Forest Network secretariat. Union of Municipalities Valdisieve Valdarno (Florence) is the manager

2.3.3. Set of Criteria for Assessing the Level of Success of Participatory Processes

The success of a participatory process is a multi-dimensional concept strictly linked to the local policy and governance context, and it depends on the motivation and perspective adopted in the approach [28,46].

The set of success criteria adopted in the present study was selected by the authors from a previous research conducted for the evaluation of the success of a participatory process in the framework of forest planning. Among the criteria reported by De Meo et al. (2017) [46], and based on the information collected with the interviews, two criteria suitable for the present study were selected: inclusiveness and transparency. An additional criterion specifically introduced in this study is representativeness. In the present case, particular attention was paid to both the normative and substantive rationale. From the normative perspective, people's empowerment deriving from participation is a measure of success; from the substantive one, success derives from the inclusion of a multiplicity of concerns and values [47,48].

Representativeness (criterion 1) depends on the stakeholders involved in the participatory process and their capacity to be representative of their specific interests. The identification, characterization and analytical categorization of stakeholders in order to determine the extent of their future involvement in the decision making process is typically realized with a stakeholder analysis [49,50]. In this study, the level of representativeness in each case study was assessed through a stakeholder analysis based on two steps: in the first, the involvement of three macro-categories of social actors (citizens, organized groups of interest, and scientists) was assessed; in the second step, for the different organized groups of interest, their involvement was assessed.

Inclusiveness (criterion 2) is the degree of involvement in the decision-making process of the different categories of stakeholders willing to participate. This characteristic of the participatory process is strictly linked to the power of stakeholders to influence the final decisions. In this study, the inclusiveness of the participatory process was assessed using four participation methods with an increasing level of inclusiveness: participation as information; participation as consultation; participation as advocacy; participation as decision-making. For each case study the corresponding level of participation was indicated using the information provided by respondents.

The transparency of the process (criterion 3) is the possibility for participants to understand what is going on and how decisions are made [46–48,51] and it refers to stakeholder satisfaction, in the sense as to whether the outcomes and process are accepted [47]. In our survey, a specific question assessed transparency through respondents' satisfaction with the results of the participatory-oriented forums and processes. Respondents were asked to assess their satisfaction using a 5-point Likert scale format (from 1 not at all satisfied to 5 extremely satisfied).

3. Results and Discussion

3.1. Key-Examples of Forest-Related Participatory Processes in Italy

The results of the survey show that only a few participatory forums are still ongoing in Italy, active in the medium-long period and able to influence policy making. Four of them—two at national level and two at sub-national level (Figure 2)—are those analyzed for the purposes of this study.

The two national forums—i.e., the PQSF and the Table on Forest Wood Chain—were both launched by the Ministry of Agriculture, Food and Forestry Policies and deal with overall forest policy at the national level. The main goal of the PQSF 2007–2009 was to provide a general, modern framework to enhance forestry activities in the country and innovate the sector. This forum is typically an institutional forum, with public institutions as the main targets.

The second forum is more specifically oriented towards enhancement of the forest-wood chain, creating new networks among actors and is known as the Table on Forest-Wood Chain. This Table, launched in 2012, is still ongoing and targets stakeholders in the forest-wood chain.

In both national forums, the participatory process was not open and only invited representatives of most relevant stakeholder categories were consulted and allowed to take part in discussions, but the final decisions were taken through a centralized process conducted by the ministry in collaboration with the regional authorities, through the so-called State-Regions Conference. Invited stakeholders were representatives of forest owners associations, farmers unions, scientists, chartered foresters, officials representative of the regional public forest services nominated by the State-Regions Conference, environmental non-governmental organizations (NGOs), pulp and paper companies, forest workers unions and cooperatives. Stakeholders were invited to some meetings and took part in consultation sessions where a discussion took place about a document that had been prepared by a steering committee composed of a restricted number of actors (representatives of the ministry, regions and professional organizations). Participation of scientists in these forums was not particularly relevant for problems such as lack of time and limited interest and no advanced tools were used to increase participation, such as scenarios building or online survey.

A specific aspect of the forum for the PQSF is that a draft version of the program was shared and discussed by a wide audience of interested stakeholders during a forum with open access to documents; this aspect of the PQSF development is crucial, as it represents a model of governance in which a participatory decision making approach is developed.

For the development of the third forum analyzed in our study, regional regulation of the forest sector in Piedmont Region, an ad hoc committee was created approximately 10 years ago as a permanent consultation body with relevant stakeholders, a large variety of stakeholder categories are involved in this committee, including academics/scientists, harvesting companies associations, public officials, environmentalists, etc. The committee typically meets 2–4 times per year, for stakeholders' consultation on specific issues (e.g., the regional forest plan and its environmental impacts assessment procedures). The committee is quite effective and, in the last 10 years, all decisions taken at regional level in the forest sector have been consistent with the recommendations emerging from such consultations, without significant conflicts between the public authority in charge of the decisions and the stakeholders participating in the forum. On the basis of the committee's work, the implementation rules of several important pieces of legislation have been approved, the most relevant example being approval of the regional forest regulation.

The fourth forum is the partnership for the process of building the Model Forest "Montagna Fiorentina", in Tuscany Region. This Model Forest is a non-profit association that aims to improve the integration and sustainability of forest and land management, increasing the cohesion and awareness of the network of all the social-economic components of the Tuscany Region territory. The partnership for the process of building the Model Forest "Montagna Fiorentina" included among its partners local authorities, associations, companies and citizens, under the responsibility of the Tuscany Region in coordination with the Mediterranean Model Forest Network secretariat (Junta Castilla y Leon and the Canadian Government). The foundation process of the Model Forest began in 2009 when the Tuscany Region joined the Mediterranean Model Forest Network. The participatory process was officially launched in December 2010, and continued with public meetings open to various groups focusing on the different forestry aspects. The participation process was considered very satisfactory, primarily by reason of the relationships established among local citizens, decision makers and scientists through interaction.

Figure 2. Location of participatory forums at sub-national level.

3.2. Factors Influencing the Effectiveness of Forest-Related Participatory Processes in Italy

This part of the analysis refers to our first guiding question, i.e., which factors hinder or support effective participatory forest-related processes in the country.

3.2.1. High Fragmentation of Forums

A first factor evidenced in our study is the high fragmentation of participatory forums. Forums often overlap one another, different authorities start different processes thus replicating efforts, and forums neither interact nor create synergies. As a consequence, many processes stopped after 2–3 years, with no significant results. In several cases, during the processes, stakeholders lost their motivation: without any resources available, their interest decreased and they had the perception of not having influenced any decision. These dynamics seem aligned with the release phase of panarchy that leads in the end to a failing reorganization scenario, where the participatory process is abandoned and decision-making does not change. Respondents also reported that highly fragmented interests and knowledge (e.g., different fields of expertise of invited scientists) led in several cases to long discussions, without being able to agree on a common point of view. In some cases, each participant was trying to defend his/her interests, and was not aware of the others. This is consistent with the conservation phase of the panarchy model, where competition appears and the system becomes more rigid. These dynamics, observed in a number of meetings, contributed to the extreme fragmentation of the forums, in terms of both actors and interests, which were unable to aggregate around a few common components. Being fragmented forums, neither coordinated nor structured around key points of common interest, they probably do not represent components of the same system, they are disconnected and not able to positively influence one another. In other words, the stakeholders did not adopt collaborative approaches in their dialogues or corrective actions to prevent failures (e.g., by pre-defining an agreed timetable for decisions to be taken after discussion sessions).

However, concerning national forums, the State-Regions Conference is reported to have positive results in increasing the capacity of the regional forest administrations to aggregate around common goals and advocate with respect to the European Commission and its rules and programmes. Even if not effective in guaranteeing stakeholders and scientists' participation in forest policy making in terms of inclusiveness—they were consulted (see Table 3), but did not take part in any decision–the conference has an important role in increasing interactiveness of the process, stimulating a constructive long-lasting face-to-face interaction [52,53] and enforcing the horizontal coordination among organizations while respecting the autonomy of each region. In this sense, the cross-scale linkages and influences of one scale on the others described by Allen et al. 2014 [8] appear quite evident.

Table 3. Qualitative-based evaluation of stakeholders' participation in forest-related policy processes in selected cases studied in Italy.

Criteria		National Level		Local Level	
		Forum 1	Forum 2	Forum 3	Forum 4
	Stakeholders				
	State authorities (i.e., Ministries)	X	X	X	
	Regional authorities	X	X	X	X
	Local authorities	X	X	X	X
	Scientists/experts	X	X	X	X
1. Representativeness	Tourist associations	X			X
	Forest owner associations	X	X	X	X
	Private timber enterprises	X	X	X	X
	Farmers associations	X	X	X	X
	Hunting associations	X			X
	Citizens				
	Levels				
	Information				
2. Inclusiveness	Consultation	X	X		
	Advocacy				
	Decision-making			-	X
	Levels				
	Very high				X
	High			X	
3. Transparency	No opinion				
	Low	X	X		
	Very low				

3.2.2. Lack of Clear Rules

Different respondents, depending on their direct experience on public forums, reported that another obstacle during the forums' development is the lack of clear rules of the functioning of the process. In some, it was not clear who would have to make the final decision, how to solve internal conflicts in the group, how to use a voting system, etc. This is a clear lack of transparency if transparency means that throughout the entire process "established channels for continuous dialogue and information sharing" exist and "timely response to information requests" [54] is provided. Without these kinds of rules, predominant participants were those most active, more able to talk and discuss what is introduced in the system. However, vagueness and ambiguity of rules may also follow an inverse phenomenon, i.e., they restrict oligarchy [55]. One reason for vagueness of rules functioning in favor of predominant actors in our case studies can be that predominant actors are often the public administrations, which organize participatory forums by *de facto* maintaining control on decision-making and thus unconsciously guiding the process itself. However, this issue remains unclear and should be further explored in future research. The level of transparency assessed through respondents' interviews, as expected, varies a lot with the forum. However, as shown in Table 3, higher transparency levels were reported for local-level processes (forums 3 and 4) and lower for those at national level (forums 1 and 2). Transparency often being mentioned as key element for the effectiveness of participatory processes [56], it might be argued that the forums act faster and better at local level—leading to faster dynamics of changes—because the "rules of the game" are more clearly stated at this more local level, rather than at a higher level. Indeed, this is consistent with the empirical findings of Maier et al. (2014) [31] who stated that "more effective (and less polarized) participation processes will likely take place at the local level, as opposed to the state policy level", and Hogl et al. (2012, p. 301) [34] who found implementation at national level, compared to sub-national levels, more difficult due to "hierarchical steering, sectoral isolationism and expert-centered decision-making".

3.2.3. Lack of Representativeness of Interests

Another commonly mentioned factor is the lack of representativeness of interests. Again, greater concerns are related to national-level processes, while those at local level seem to be able to have a higher capacity of involving stakeholders representative of different interests. Lack of representation of interests was evidenced especially for forest owners, and this is probably due to the fact that there is no nationally recognized association of forest owners. This is a very well known problem in Italy, and reported as a key issue by most of the interviewees. Being aware of this problem, a challenge during the participatory processes is to give all groups of stakeholders the chance to be heard and to represent different interests appropriately [5].

Concerning processes accessibility (see Table 3), interviewees reported that in both national forums the participatory process is based on draft documents circulated among participants, while changes are discussed during the meetings. Most of the outputs are non-binding documents and guidelines that might be used to address forest-management practices. Accessibility of the participatory forums, in terms of provision of adequate tools and resources [57], is not satisfactory because no advanced approaches or tools suitable to facilitate and support the participatory process have been used to increase the level of participation.

3.3. Lessons Learned (or Not) for Orchestrating Forest Policy in Italy

When analyzing the content of the interviews in depth, interesting aspects emerged concerning participatory forums and process development. One is related to the level of stakeholder satisfaction. A generally good level of satisfaction about participatory forums was highlighted by policy makers, while both scientists and other stakeholders were less positive in their evaluations even if they were aware that efforts made up to now have to be appreciated. For example, interviewees' perceptions include statements such as "Significant advancements have been made", "Public administrations made unexpected efforts", "There are some positive signals that they will finally listen to us", and "We cannot do better with the present conditions". However, there is a common perception that, even if we are on the right track and there are some good signs for the future, policy makers are still lacking experience in organizing effective participation forums. The different perception of the quality of participatory forums, higher for policy makers (public administrations) than for the other stakeholders involved can be interpreted in two ways. It can be a mirror of the remember principle of panarchy, being resistant to changes [35]; or a sign of limited awareness of the importance of having professional arrangements to be able to adopt more effective processes. In both cases, the result is that participatory forums are mainly naïvely organized in Italy.

Reasons for difficulties are assumed to include elements such as low inter-sectoral knowledge exchange (isolated sectors, not taking advantage of others' experience), and lack of empirical evidence of more effective results of participatory-based decisions with respect to hierarchical-based. It is hard to overcome the fragmented, obsolete and rigid legal-institutional context that is typical of the hierarchy: Some pre-existing conditions, especially at national level, are not easily changed. In panarchy, adaptive cycles can be dominated by higher or lower levels, not necessarily by higher levels as in a hierarchy: In other words, conditions can arise that trigger bottom-up change in the system [25]. However, positive experiences of these dynamics in some specific fields of interest (e.g., forest certification forums) are not easily replicable, and for other types of forest activities (e.g., protection of public goods or provision of ecosystem services) there is not a clear economic value combined with ecological and social interests. Underlying causes are the following: (i) it takes a long time to change the core policy beliefs of Public Authorities, which often do not recognize the value of participation needs and follow the Remember principle of panarchy, i.e., they maintain a conservative approach; (ii) no investments have been made in improving management of participatory processes—such as contracts for professionally qualified facilitators, training for the staff of organizations in charge of arranging and managing participation—thus no innovation to support the revolt principle of panarchy has been introduced; (iii) hidden lobbies (including those of different groups in the academic world) have greater influence than formally involved representatives, and this is one reason why stakeholders become demotivated.

Another lesson is that "the public" is not involved at all. Only organized stakeholders (even if not well represented or with problems of representativeness) are typically invited to forest-related forums. There is increasing potential for citizens to be involved, especially thanks to the various technological instruments that would allow even non-experts to participate in forestry issues. However, despite an increasing number of initiatives oriented towards the active involvement and cooperation of citizens with scientists in collecting, delivering, validating and sharing information to support decision making (for example, used in monitoring biodiversity), citizens' science initiatives are still marginally implemented in Italy. This situation does not create appropriate conditions for innovation derived from lower levels (bottom-up) to influence higher levels of decision making, thus further reducing the possibility for the revolt principle of panarchy to happen and induce changes.

Moreover, participatory processes are showing more clearly the problems connected with the lack of coordination at national level, lack of representativeness of key stakeholders and lack of general guiding strategies for the forest sector, with still confused and overlapping tasks among institutions at various levels, duplication of negotiation and discussion efforts, unclear and unique positions and visions, etc. (Figure 3). In the country forest policy is treated as a marginal part of the agricultural sector or more in general rural development, and as a marginal part of the environmental sector. This is demonstrated by the fact that the Ministry of Environment also has responsibility for biodiversity-related issues in forests and protected areas. All forest-related tasks are delegated to regions, the regional forest policies of which are not coordinated with each other. Thus, even in the case that forest actors would be able to aggregate themselves around common interests and lobbies at national level, they would not have a clear institution to refer to, when advocating for policy reforms. They simply do not know exactly which officials and offices they should contact to ask for changes, as tasks and responsibilities for forests remain highly fragmented.

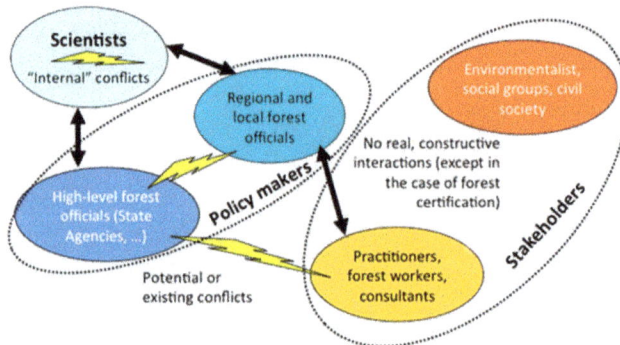

Figure 3. Relations between scientists, policy makers and stakeholders in Italian forest policy.

A final lesson emerged from some interviewees belonging to the group of scientists, is that the main conflicts among stakeholders are perceived not to be those, as normally expected, between forest owners and environmental organizations [58]. Rather, the main conflicts are those between the central state authority and the regional ones. But not all interviewees agreed on this. Indeed, some of them—belonging to the group of policy makers—reported that there is a high level of collaboration and coordination among the different administrative levels, thanks to the continuous and constructive activity of the State-Regions Conference. One interviewee even declared that in his opinion the participation process of the national Table on Forest-Wood Chain was a process of co-decision. The reason for these different perceptions, as well as other aspects (e.g., the reason why predominant participants take advantage of vagueness of rules instead of restricting oligarchy [55]), remain unclear and need further investigations. Moreover, our considerations about the state-of-the-art of forest-related participatory processes in Italy might become more robust if other participatory forums are analyzed and other stakeholders interviewed (thus increasing the empirical evidence based on case studies).

4. Conclusions

In conclusion, we can affirm that most of the experiences of public participation in forestry in Italy so far can be categorized as naïve experiences, not managed in a professional way, with a lack of knowledge and skills on how to arrange effective participation and so unable to keep stakeholders motivated and satisfied. There is still a clearly predominant role of the public administration (representatives of the policy makers), while scientists and stakeholders have a marginal role. This is reflected by the different levels of satisfaction with participation in forest-related forums: higher for public administrations, lower for scientists and other stakeholders. Hierarchy is still predominant, rather than panarchy—intended as a nested set of adaptive cycles that can be driven by bottom-up approaches to policy making. This reduces the options for an orchestration to be adopted among forest-related sectors and actors for introducing changes and policy reforms in forestry in the country. Because of the long-lasting but mostly ineffective implementation of forest participation in Italy, most of the analyzed participatory processes are in the intermediate phases of panarchy dynamics, i.e., in the conservation or release phases, with no corrective actions undertaken to change the situation. Moreover, they are significantly affected by the Remember principle, which tends to maintain the status quo with the attitude of forest public administrations to adopt a conservative approach. However, regional and local forest-related participatory forums seem to act faster and better than national ones, thus providing options for improvement to be transferred from lower to higher levels. What is probably still missing in the country are appropriate and wide mechanisms of connectivity between lower (regional) and higher (national) levels. In some way, participatory processes in forest-related issues in Italy are showing more clearly the problems connected with a lack of coordination at national level, a lack of representativeness and guiding strategies for the forest sector, with still confused and overlapping tasks among institutions at various levels, duplication of negotiation and discussion efforts, no clear and unique position and vision, etc. Having been unable to learn from past failures and introduce significant changes in decision-making processes, above all at national level, orchestrating forest policy in Italy remains a "mission impossible". It is likely that, before discussing orchestrating scientists, policy makers and stakeholders, a new common vision of what forests are should be built. From this common vision could start a new dialogue among institutions, actors and public opinion, moving towards a reorganization phase and, finally, predominance of the Revolt principle of panarchy.

Author Contributions: Conceptualization, L.S., A.P. and I.D.M.; Data curation, L.S., A.P., M.M. and F.C.; Investigation, L.S., R.R. and I.D.M.; Methodology, L.S., A.P. and F.C.; Supervision, L.S. and D.P.; Writing—original draft, A.P. and I.D.M.; Writing—review & editing, L.S.

Funding: This research received no external funding.

Acknowledgments: This study is related to the COST Action FP1207 "Orchestrating forest related policy analysis in Europe" (ORCHESTRA). The authors wish to acknowledge the participation and collaboration of the stakeholders who provided useful information and suggestions.

Conflicts of Interest: The authors declare no conflict of interest.

References

1. Lasserre, B.; Chirici, G.; Chiavetta, U.; Grafi, V.; Tognetti, R.; Drigo, R.; Di Martino, P.; Marchetti, M. Assessment of potential bioenergy from coppice forests through the integration of remote sensing and field surveys. *Biomass Bioenerg.* **2011**, *35*, 716–724. [CrossRef]
2. Paletto, A.; Giacovelli, G.; Pastorella, F. Stakeholders' opinions and expectations for the forest-based sector: A regional case study in Italy. *Int. For. Rev.* **2017**, *19*, 68–78. [CrossRef]
3. Secco, L.; Pettenella, D.; Gatto, P. Forestry governance and collective learning process in Italy: Likelihood or utopia? *For. Policy Econ.* **2011**, *13*, 104–112. [CrossRef]
4. Cantiani, M.G. Forest planning and public participation: A possible methodological approach. *iForest* **2012**, *5*, 72–82. [CrossRef]
5. Paletto, A.; Cantiani, M.G.; De Meo, I. Public participation in Forest Landscape Management Planning (FLMP) in Italy. *J. Sustain. For.* **2015**, *34*, 465–483. [CrossRef]
6. Kleinschmit, D.; Pülz, H.; Secco, L.; Sergent, A.; Wallin, I. Orchestration in political processes: Involvement of experts, citizens, and participatory professionals in forest policy making. *For. Policy Econ.* **2018**, *89*, 4–15. [CrossRef]
7. Allen, C.R.; Angeler, D.G.; Garmestani, A.S.; Gunderson, L.H.; Holling, C.S. Panarchy: Theory and application. *Ecosystems* **2014**, *17*, 578–589. [CrossRef]
8. Shannon, M. Design principles for science-stakeholder deliberation: A typology and tool box. In Proceedings of the COST Action ORCHESTRA Conference "Orchestrating Forest Policy MAKING: Involvement of Scientists and Stakeholders in Political Processes", Bordeaux, France, 23–25 September 2015.
9. FAO-ECE-ILO. *Public Participation in Forestry in Europe and North America*; Report of the FAO/ECE/ILO Joint Committee Team of Specialists on Participation in Forestry, Working Paper 163; Sectorial Activities Department, International Labour Office: Geneva, Switzerland, 2000.
10. World Bank. *The World Bank and Participation*; Operations Policy Department: Washington, DC, USA, 1994.
11. Intergovernmental Panel on Forests. Report of the Ad Hoc intergovernmental panel on forests on its fourth session. In Proceedings of the Commission on Sustainable Development, Fifth session (UN DPCSD E/CN. 17/1997/12), New York, NY, USA, 7–25 April 1997.
12. Pülzl, H.; Rametsteiner, E. Grounding international modes of governance into National Forest Programmes. *For. Policy Econ.* **2002**, *4*, 259–268. [CrossRef]
13. Sarvašová, Z.; Dobšinská, Z.; Šálka, J. Public participation in sustainable forestry: The case of forest planning in Slovakia. *iForest* **2014**, *7*, 414–422. [CrossRef]
14. Inglehart, R. *The Silent Revolution: Changing Values and Political Styles among Western Publics*; Princeton University Press: Princeton, NJ, USA, 1977.
15. Vilkka, L. *The Intrinsic Value of Nature*; Value Inquiry Book Series; Rodopi: Helsinki, Finland, 1997; ISSN 0929-8436.
16. Buttoud, G. How can policy take into consideration the "full value" of forests? *Land Use Policy* **2000**, *17*, 169–175. [CrossRef]
17. Sandström, C.; Lindkvist, A.; Öhman, K.; Nordström, E.M. Governing competing demands for forest resources in Sweden. *Forests* **2011**, *2*, 218–242. [CrossRef]
18. Benz, A. Multi-level governance. In *Formulation and Implementation of National Forest Programmes, Theoretical Aspects*; Glück, P., Oesten, G., Schanz, H., Volz, K.-R., Eds.; EFI Proceedings 30; European Forest Institute: Joensuu, Finland, 1999; Volume 1, pp. 73–84.
19. Daniels, S.E.; Walker, G.B. *Working Through Environmental Conflict: The Collaborative Learning Approach*; Praeger: Westport, CT, USA, 2001.
20. Appelstrand, M. Participation and societal values: The challenge for lawmakers and policy practitioners. *For. Policy Econ.* **2002**, *4*, 281–290. [CrossRef]
21. Kangas, A.; Saarinen, N.; Saarikoski, H.; Leskinen, L.A.; Hujala, T.; Tikkanen, J. Stakeholder perspectives about proper participation for Regional Forest Programmes in Finland. *For. Policy Econ.* **2010**, *12*, 213–222. [CrossRef]
22. Kouplevatskaya-Buttoud, I.; Buttoud, G. Assessment of an iterative process: The double spiral of re-designing participation. *For. Policy Econ.* **2006**, *8*, 529–541. [CrossRef]

23. Brunckhorst, D.J. Institutions to sustain ecological and social systems. *Ecol. Manag. Restor.* **2002**, *3*, 108–116. [CrossRef]

24. Garmestani, A.S.; Allen, C.R.; Mittelstaedt, J.D.; Stow, C.A.; Ward, W.A. Firm size diversity, functional richness and resilience. *Environ. Dev. Econ.* **2006**, *11*, 533–551. [CrossRef]

25. Garmestani, A.S.; Allen, C.R.; Cabezas, H. Panarchy, Adaptive management and governance: Policy options for building resilience. *Neb. Law Rev.* **2008**, *87*, 1036–1054.

26. Garmestani, A.S.; Benson, M.H. A framework for resilience-based governance of social-ecological systems. *Ecol. Soc.* **2013**, *18*, 9. [CrossRef]

27. Ruppert-Winkel, C.; Winkel, G. Hidden in the woods? Meaning, determining, and practicing of 'common welfare' in the case of the German public forests. *Eur. J. For. Res.* **2001**, *130*, 421–434. [CrossRef]

28. Faehnle, M.; Tyrväinen, L. A framework for evaluating and designing collaborative planning. *Land Use Policy* **2013**, *34*, 332–341. [CrossRef]

29. Mårald, E.; Sandström, C.; Rist, L.; Rosvall, O.; Samuelsson, L.; Idenfors, A. Exploring the use of a dialogue process to tackle a complex and controversial issue in forest management. *Scand. J. For. Res.* **2015**, *30*, 749–756. [CrossRef]

30. Aguilar, S.; Montiel, C. The challenge of applying governance and sustainable development to wildland fire management in Southern Europe. *J. For. Res.* **2011**, *22*, 627–639. [CrossRef]

31. Maier, C.; Lindner, T.; Winkel, G. Stakeholders' perceptions of participation in forest policy: A case study from Baden-Württemberg. *Land Use Policy* **2014**, *39*, 166–176. [CrossRef]

32. Secco, L.; Da Re, R.; Pettenella, D.M.; Gatto, P. Why and how to measure forest governance at local level: A set of indicators. *For. Policy Econ.* **2014**, *49*, 57–71. [CrossRef]

33. Gupta, J. Glocal forest and REDD+ governance: Win-win or lose-lose? *Curr. Opin. Environ. Sustain.* **2012**, *4*, 620–627. [CrossRef]

34. Hogl, K.; Kvarda, E.; Nordbeck, R.; Pregernig, M. (Eds.) Effectiveness and legitimacy of environmental governance—Synopsis of key insights. In *Environmental Governance: The Challenge of Legitimacy and Effectiveness*; Edward Elgar Publishing Ltd.: Cheltenham, UK; Northampton, UK, 2012; pp. 280–304.

35. Kumar, S.; Kant, S.; Amburgey, T.L. Public agencies and collaborative management approaches. Examining resistance among administrative professionals. *Adm. Soc.* **2007**, *39*, 569–610. [CrossRef]

36. Rametsteiner, E. The role of governments in forest certification—A normative analysis based on new institutional economic theories. *For. Policy Econ.* **2002**, *4*, 163–173. [CrossRef]

37. Gibson, C.C.; Ostrom, E.; Ahn, T.K. The concept of scale and the human dimensions of global change: A survey. *Ecol. Econ.* **2000**, *32*, 217–239. [CrossRef]

38. De Meo, I.; Ferretti, F.; Hujala, T.; Kangas, A. The usefulness of Decision Support Systems in participatory forest planning: A comparison between Finland and Italy. *For. Syst.* **2013**, *22*, 304–319. [CrossRef]

39. Yin, K.R. *Case Study Research: Design and Methods*; Applied Social Research Methods Series; Sage: Thousand Oaks, CA, USA, 2009.

40. Secco, L.; Pettenella, D. Participatory processes in forest management: The Italian experience in defining and implementing forest certification schemes. *Swiss For. J.* **2006**, *157*, 445–452. [CrossRef]

41. Carbone, F.; Savelli, S. Forestry programmes and the contribution of the forestry research community to the Italy experience. *For. Policy Econ.* **2009**, *11*, 508–515. [CrossRef]

42. Cesaro, L.; Romano, R. *Politiche Forestali e Sviluppo Rurale: Situazione, Prospettive e Buone Prassi, Quaderno n. 1*; Osservatorio Foreste INEA: Roma, Italy, 2008.

43. De Meo, I.; Ferretti, F.; Frattegiani, M.; Lora, C.; Paletto, A. Public participation GIS to support a bottom-up approach in forest landscape planning. *iForest* **2013**, *6*, 347–352. [CrossRef]

44. Cesaro, L.; Romano, R.; Zumpano, C. *Foreste e Politiche di Sviluppo Rurale: Stato Dell'arte, Opportunità Mancate e Prospettive Strategiche*; INEA: Roma, Italy, 2013.

45. Romano, R.; Marandola, M. Le Politiche Forestali in Italia: Tema di Nicchia Oppure Reale Occasione di Sviluppo Integrato Per Il Paese? Criticità, Opportunità E Strumenti Alle Soglie Della Programmazione 2014–2020. In Proceedings of the Second International Congress of Silviculture, Florence, Italy, 26–29 November 2014; pp. 775–779.

46. De Meo, I.; Ferretti, F.; Paletto, A.; Cantiani, M.G. An approach to public involvement in forest landscape planning in Italy: A case study and its evaluation. *Ann. Silvic. Res.* **2017**, *41*, 54–66.

47. Blackstock, K.L.; Kelly, G.J.; Horsey, B.L. Developing and applying a framework to evaluate participatory research for sustainability. *Ecol. Econ.* **2007**, *60*, 726–742. [CrossRef]

48. Menzel, S.; Nordstrom, E.M.; Buchecker, M.; Marques, A.; Saarikoski, H.; Kangas, A. Decision support systems in forest management: Requirements from a participatory planning perspective. *Eur. J. For. Res.* **2012**, *131*, 1367–1379. [CrossRef]

49. Reed, M.S.; Graves, A.; Dandy, N.; Posthumus, H.; Hubacek, K.; Morris, J.; Prell, C.; Quinn, C.H.; Stringer, L.C. Who's in and why? A typology of stakeholder analysis methods for natural resource management. *J. Environ. Manag.* **2009**, *90*, 1933–1949. [CrossRef] [PubMed]

50. Grilli, G.; Garegnani, G.; Poljanec, A.; Ficko, A.; Vettorato, D.; De Meo, I.; Paletto, A. Stakeholder analysis in the biomass energy development based on the experts' opinions: The example of Triglav National Park in Slovenia. *Folia For. Pol. Ser. A* **2015**, *57*, 173–186. [CrossRef]

51. Lockwood, M. Good governance for terrestrial protected areas: A framework, principles and performance outcomes. *J. Environ. Manag.* **2010**, *91*, 754–766. [CrossRef] [PubMed]

52. Saarikoski, H.; Tikkanen, J.; Leskinen, L.A. Public participation in practice—Assessing public participation in the preparation of regional forest programs in Northern Finland. *For. Policy Econ.* **2010**, *12*, 349–356. [CrossRef]

53. Tuler, S.; Webler, T. Voices from the forest: What participants expect of a public participation process. *Soc. Nat. Resour.* **1999**, *12*, 437–453.

54. Brinkerhoff, J.M. Assessing and improving partnership relationships and outcomes: A proposed framework. *Eval. Prog. Plan.* **2002**, *25*, 215–231. [CrossRef]

55. Hasanagas, N.D. Network analysis functionality in environmental policy: Combining abstract software engineering with field empiricism. *Int. J. Comput. Commun. Control* **2011**, *6*, 622–634. [CrossRef]

56. Ananda, J. Implementing participatory decision making in forest planning. *Environ. Manag.* **2007**, *29*, 534. [CrossRef] [PubMed]

57. Asthana, S.; Richardson, S.; Halliday, J. Partnership working in public policy provision: A framework for evaluation. *Soc. Policy Adm.* **2002**, *36*, 780–795. [CrossRef]

58. Lindstad, B.H. 'What's in it for me?'—Contrasting environmental organisations and forest owner participation as policies evolve. *For. Policy Econ.* **2018**, *89*, 80–86. [CrossRef]

forests

MDPI

Article

Stakeholder Participation in Natura 2000 Management Program: Case Study of Slovenia

Tomislav Laktić [1] and Špela Pezdevšek Malovrh [2,*]

[1]   Water and Investments Directorate, Cohesion Policy Division, Ministry of the Environment and Spatial Planning, 1000 Ljubljana, Slovenia; tomislav.laktic@gov.si
[2]   Department of Forestry and Renewable Forest Resources, Biotechnical Faculty, University of Ljubljana, 1000 Ljubljana, Slovenia
*   Correspondence: spela.pezdevsekmalovrh@bf.uni-lj.si; Tel.: +386-1320-3522

Received: 23 August 2018; Accepted: 24 September 2018; Published: 26 September 2018

Abstract: The Natura 2000 network, which is one of the largest networks of protected areas in the world, is the core pillar in the European Union's biodiversity conservation policy. To achieve the national enforcement of Natura 2000 and to overcome implementation problems, effective policy measures are needed, and participation among different stakeholders is required. The aim of this paper was to evaluate the process of formulation of the Natura 2000 Management Program (2015–2020) in Slovenia based on a set of criteria and indicators for the evaluation of a participatory process, and to present the differences between the stakeholders' evaluations according to their role in the implementation process. For that purpose, stakeholders were divided into two groups—stakeholders from project partner's institution, and others. The research was done in two steps: First, in-depth semi-structured interviews with 17 representatives of the key stakeholders' institutions or organizations that were involved in the formulation of the Natura 2000 Management Program were carried out; in the second, a survey with 266 stakeholders involved in the formulation and decision-making process was conducted. Overall, the results show that the participatory process was well organized, independent and fair; however, not all of the requisite conditions for successful participation were fulfilled, as was shown with the sub-criteria of transparency. The only difference between the groups exists in the evaluations of the sub-criteria transparency of the process. The group of stakeholders from the project's partner institutions were more satisfied with the transparency of the process, as the average value was 3.36, compared to the others where the average value of satisfaction was 2.80. This indicates the need for an improved, novel, and innovative approach that leads to multi-level governance.

Keywords: participation; Natura 2000; management program; stakeholders; Slovenia

1. Introduction

The Natura 2000 network, which encompasses the Birds and Habitats Directive of the European Union (hereafter EU) (Directive 79/409/European Economic Community, amended as 2009/147/European Commission and Directive 92/43/European Economic Community, respectively), is the core pillar in the EU's biodiversity conservation policy [1,2], and is one of the largest networks of protected areas in the world [3,4]. These directives are legally binding; thus, individual member states had to transpose them into their national legislation [1]. Moreover, the EU member states were relatively free to choose the most suitable means to achieve the main goals of the directives, and to design the implementation process and the management of the protected sites in line with the overall objectives of national policies. Thus, countries have adopted different, often top-down and science driven approaches, with widely varying degrees of participatory processes [5].

The implementation of the objectives of the Birds and Habitats Directives in the national legislation leads to changes in national legislation. These changes are based on the changing relationship in decision making at the national level, with the inclusion of new stakeholders in the national system of nature protection (which also causes changes in the traditional roles of the stakeholders) [6,7]. Non-governmental organizations (hereafter NGOs) in the field of nature protection and private forest owners have been given the opportunity for equal participation as governmental stakeholders in decision making in the process of implementation of Natura 2000 [3,7–10].

Several authors have recently studied the implementation processes and issues that have arisen in the national transposition and establishment of Natura 2000. According to [11], the implementation of Natura 2000 in various EU member states is linked to multiple processes at different policy levels, and depends on case-specific interplay. Consequently, a large diversity of implementation approaches has been adopted, and can be seen as advantageous as it enables learning for improved future functioning [12]. To improve the practical implementation in most countries, the authors suggest the need for more communication and participation with stakeholders [13], the importance of inter-sectoral cooperation [8,14], the need for co-responsibility of all relevant stakeholders who are involved in the process [15], and an awareness of the socio-economic and cultural contexts in which conservation planning takes place [16]. Moreover, research also deals with challenges for the implementation of Natura 2000 in forests. According to [17], the main challenges are related to balancing biodiversity conservation and timber production, integrating nature conservation and local stakeholders' demands, the development of an effective and accepted funding scheme for the implementation of Natura 2000 in forests, and how to integrate nature conservation policies with forest and other land use sector policies. However, the implementation process also has obstacles connected to poor participation and negative perceptions of stakeholders, inflexible regulations, and limited concern to local context [8,18]. In addition, the implementation process of Natura 2000 very often became a complex and lengthy undertaking, causing a variety of policy and management conflicts [19–21]. These conflicts were related to contradictory stakeholder's interests, values, and perceptions, as well as to different and competing land use principles [22]. To prevent further conflicts and to strengthen the implementation process, the involvement of and cooperation with various stakeholders, as well as stakeholder participation and coordination between institutions and organizations, have been stressed as important tools to increase the acceptance of Natura 2000 [23,24].

In Slovenia, the designation of Natura 2000 sites was a requirement for accession to the EU, so the majority of sites were designated by 2004. Therefore, in 2002, the Government launched the project Natura 2000 to elaborate a communication strategy and to present the Natura 2000 contents to relevant stakeholders. Therefore, Slovenia has chosen to develop an overall national document "Operational program—Natura 2000 Management Program", hereafter PUN. The Government of the Republic of Slovenia adopted in 2007, the Operational programme—Natura 2000 Management Programme for the 2007–2013 (hereafter PUN 2007–2013) and the second Natura 2000 Management Programme for the 2015–2020 (hereafter PUN 2015–2020), which outlines management requirements for sites, and which need to be incorporated in other planning documents that already regulate the use of natural resources, such as forest management plans [25]. Since the request of the EU, the Natura 2000 network in Slovenia covers 37.9% of the national territory [26], of which almost 70% are forests, followed by 23% of agricultural land and slightly less than 13% of different types of permanent grassland [27]. The Ministry of Environment and Spatial Planning (hereafter MESP) and Ministry of Agriculture, Forestry and Food (hereafter MAFF) are responsible for ensuring that management measures are incorporated in the management plans, and are implemented in the field. The overall approach in Slovenia for management plans related to forestry, hunting, and fishing is that the responsible institutions draft a plan and provide an opportunity for consultation among the stakeholders.

Slovenia is among the most forested countries in Europe; 1,184,369 ha of forests cover more than a half of the country's territory (forestation amounts to 58.4%). Notably, 76% of forests in Slovenia are private property, while 24% are public (owned by the state or communities). Larger and rarely fragmented estates of state-owned forests enable good professional management. Private forest estates are small, with an average area of only 3 ha, and even these are further fragmented into several separate plots. For the great majority of these estates, forests are not of economic interest. There are already 313,000 (461,000 when including co-owners) forest owners in Slovenia. The major fragmentation of forest property and the number of forest owners and co-owners present a serious obstacle to professional work in private forests, to optimal timber production, and to the utilization of forest potential. In Slovenia 58.4% of forest owners have a forest property smaller than 1 ha, while 41% of forest owners have a forest property between 1 and 30 ha. Less than 1% of forest owners have a forest property bigger than 30 ha [28–30].

Stakeholders can participate in the formulation and implementation of the Natura 2000 in different ways, e.g., from written consultation through steering committees and advisory boards, to discussion forums and workshops [31,32], but there are no specific recommendations about participation. Consequently, the degree of participation varies among countries according to national characteristics and peculiarities [33]. Moreover, the countries also differ greatly with regard to the levels of stakeholder involvement, including at what stage stakeholders were informed or involved (before or after site designation), the level of involvement (only information or negotiation), as well as which interest groups were consulted the most (nature conservation and/or landowners and users) [17].

However, the overall impression is that in the beginning of the domestic implementation process, public authorities in most EU countries did not involve landowners, whereas conservation experts were involved to provide their substantive technical knowledge and information about habitats and species which administrations were partially lacking. This can be largely explained by the fact that the designation of Natura 2000 sites had to be based solely on scientific criteria rooted in conservation biology, as stipulated in the EU's Habitats and Birds Directives. Site designation was mostly done without any consultation with, or informing of, landowners. With the progression from site designation to the current phase of management planning and practical measures, there is growing evidence of more dialogue and cooperation between nature conservation authorities and environmental groups on the one side, and agricultural and forestry authorities and land users' groups on the other [17].

For example, in the process of implementation of the Natura 2000 network in Slovakia, public stakeholders were involved through co-decision, while the other categories of stakeholders were involved at different levels (collaboration or information), demonstrating a willingness to follow a participative approach [34].

Based on the evaluation criteria for an effective participatory process [35], the objectives of the paper are to present how stakeholders' involvement and participation have evolved in the process of formulation of the Natura 2000 Management Program (2015–2020) in Slovenia, and to present the differences between stakeholders' evaluations of the participatory process according to their role (stakeholders from project partner's institution and other stakeholders) in the implementation process. The usefulness of this research is to improve the participatory process in future PUN and other nature conservation and forest-related programs.

2. Conceptual Framework of the Study

2.1. Natura 2000 Network and Participatory Process in Slovenia

In Slovenia, it is possible to identify three main separate stages in the implementation of the Natura 2000 network and the participation of stakeholders:

1. The first one corresponding to the designation of Natura 2000 network starting at the end of 2002 and ending in May 2004, with the entry of Slovenia into the EU. In 2002, the Government launched the project Natura 2000 and established two working groups: The first for communication and the second as an expert-technical group. Expert-technical working group was responsible for the expert preparation of the designation process, autonomous in its co-operation with external expert partners and was responsible for commissioning expert evaluation from NGOs and other

sources of expert knowledge. In this phase, the efforts of the communication working group (Three institutions were partners in communication group, namely the Institute of the Republic of Slovenia for Nature Conservation (hereafter IRSNC), Slovenian Forest Service (hereafter SFS) and the Chamber of Agriculture and Forestry of Slovenia (hereafter CAFS).) were focused on the elaboration of a communication strategy with different key stakeholders (local authorities, land owners, the broad public etc.), but because of a lack of time, this communication was more focused on informing them about the possibilities, expected problems, and restrictions than on actually including them in the process. Consequently, broader interests were not expressed, and contra interests were not clearly articulated in the stage of preparing a list of potential protected areas. Hence, conflicts have been very limited in this phase [25,36–39]. Therefore, the process of implementation of the Natura 2000 (designating the sites) was highly centralized and top-down orientated [36].

2. The second one, starting in 2005 and ending in 2007, coincided with the elaboration of the Operational programme—Natura 2000 Management Programme for the 2007–2013. The PUN 2007–2013 designated protection objectives and measures at Natura 2000 sites, as well as defining the competent sectors and responsible implementers of these protection measures [40]. A further goal in this respect was to enable horizontal links with strategic plans and development programs. These activities took place within the LIFE pilot project "Natura 2000 in Slovenia—Management Models and Information System" at five different Natura 2000 sites, by laying the theoretical foundations for the elaboration of PUN. The coordinator of the project was the IRSNC. Numerous workshops in cooperation with representatives from the hunting, forestry, agriculture, fisheries, water management sectors, and local community representatives were held to provide systemic solutions for efficient and sustainable management of Natura 2000 sites in Slovenia. The first draft of PUN 2007–2013 was presented to research and education institutions, nature conservation institutions, and managers of protected areas for modification of objectives and measures. The final draft of PUN 2007–2013 was then presented during two workshops to local authorities, NGOs, and public services, before the official adoption by the Government in the end of 2007 [40].

It can be concluded that the process of elaborating the PUN 2007–2013 was highly centralized and top-down oriented, with the beginnings of inclusiveness of stakeholders from public services who are responsible for the implementation of Natura 2000 measures.

3. The third one starting at 2012 and ending at 2015 (Figure 1) corresponded to the elaboration of the PUN 2015–2020, where the main focus was on preparation of detailed conservation objectives for Natura 2000 sites, measures for achieving conservation objectives, and the institutions responsible for their implementation. The activities necessary for the adoption of this operational program were supported by the LIFE [+] project, whose coordinator was the MESP and whose associated partners were the IRSNC, the SFS, Fisheries Research Institute of the Republic of Slovenia (hereafter FRIS), Institute for Water of the Republic of Slovenia (hereafter IWRS), and the CAFS as a representative of agriculture and forest land owners. As the initial top-down implementation in former phases provoked severe resistance from different stakeholders, mostly land and forest owners, changes towards a participatory governance mode occurred in this phase, which enabled participatory elements. Therefore, the communication plan included stakeholder analysis, different ways of involving stakeholders from different sectors in each phase with the aim of giving information, as well as consultation and participatory decision-making regarding the management of Natura 2000 sites in the future. The stakeholders could have influenced the content of the management program of Natura 2000. However the aim of PUN was not to change the boundaries of Natura 2000 in Slovenia, and in addition, stakeholders did not have influence over the defined boundaries of Natura 2000. The institutions responsible for the content of PUN 2015–2020 were IRSNC and MESP. They made decisions regarding which comments were to be accepted, and which ones rejected. The individual detailed protection objectives and measures were the main part of stakeholders' comments, for which the criteria for the acceptance of comments or arguments were based on scientific and expert knowledge.

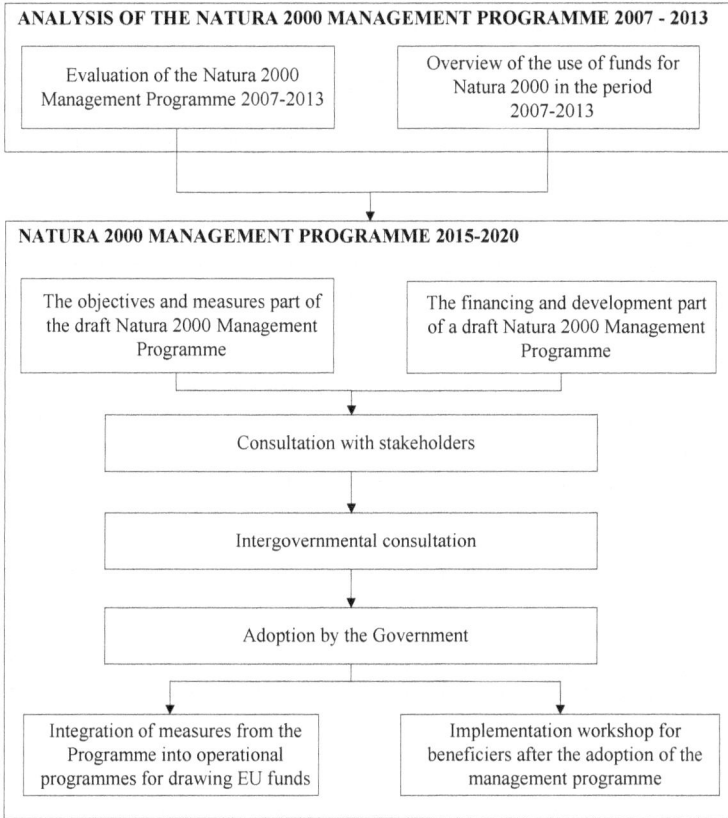

Figure 1. The formulation process of the PUN 2015–2020 in Slovenia.

The first draft of PUN 2015–2020 was prepared by IRSNC in cooperation with project partners, researchers, and experts on plant and animal species, during 77 workshops. The draft was discussed in 6 targeted roundtables (public meetings) in different parts of Slovenia. The amended draft of PUN 2015–2020 was a subject of intergovernmental consultation, which consisted of consultation meetings with all ministries and their public bodies, as well as the Chamber of Commerce. The PUN 2015–2020 was adopted in April 2015. After the adoption, eight workshops were organized for beneficiaries—including workshops with presentations of beneficiaries working on the elaboration of management programme for a wider audience of other colleagues (public servants) from public services, and also the private sector, especially to those responsible for future implementation of measures from the programme—and managers of nature parks to spread awareness about PUN. In the elaboration of the new management program, despite the aforementioned adopted approach, there was still a top-down approach; there was a switch from more communication with stakeholders to consultation with key stakeholders [37].

When Natura 2000 was implemented, the Government of the Republic of Slovenia asserted that no additional costs for its implementation, maintenance, and management would be needed. The funding needed for Natura 2000 was predicted to come only from EU. Consequently, no financial compensation for restrictions to landowner's activities under Natura 2000 were predicted in PUN. However, a budgetary fund, the so called Forest Fund, was established (Article 33) based on the changes in the organizational structure of the state forest management and the adoption of the new Act in 2016, the "Management of State Forest Act". The Forest Fund is financed by the income of the state forest management. The assets of forest funds are intended to also cover measures in the area of Natura 2000 in private forests in accordance with the PUN 2015–2020 and program investments in forests, established on the basis of the National Forest Program prepared by SFS in accordance with the Forest Act (Management of State Forests Act, Official gazette of Republic of Slovenia number 9/2016) [41].

2.2. Theoretical Context

Claims for participation are often made in terms of democracy. The basic principles that characterize democracy, and which are recognized in most democratic constitutions, are "popular control" and "political equality". This view, often referred to as the liberal constitutionalism of political participation [42], is based on the idea that a liberal constitution should facilitate pluralistic politics, i.e., politics that acknowledge various interests in society [33]. Participatory approach in this context would imply equal participation of different interest groups, and is often called "stakeholders' participation". From a democratic point of view, the essential question with regard to this kind of stakeholders' participation approach is whether it manages to include all the different interests in a fair and legitimate way; according to [43] "public participation" encompasses a group of procedures designed to consult, involve and inform the public to allow those affected by a decision to have an input into that decision.

The extent to which participatory processes fulfill the expectations is nevertheless highly dependent on how stakeholders are involved in the process. Instructions for a successful participatory process do not exist, because the form and structure of the process can vary widely, depending on e.g., the policy in consideration, the institutional context, and the operational environment. Researchers have, however, suggested some general criteria for an effective participatory process, based either on theoretical analysis (see e.g., [35,44,45]) or empirical studies in a particular context (see e.g., [46–48]).

In our study, the set of criteria outlined by [35] was selected as the basis for the evaluation of the participatory process (Table 1). The authors divided the evaluation criteria into *acceptance* criteria (including representativeness, independence, influence and transparency), which are related to the effective construction and implementation of a procedure, and *process* criteria (including resource accessibility, cost-effectiveness and task definition), which are related to the potential acceptance of a procedure.

The authors agree that an effective participation process should be inclusive and representative, meaning that participation should comprise, from the very beginning, a broadly relevant and representative sample of the population of the affected public, and that no stakeholders who are willing to participate be deliberately excluded from the process [45,47].

Another key characteristic for genuine participation is that the participation process should be conducted in an independent, unbiased way, which means that management and facilitation of the participation process should be completely independent from all affected interests, and should be fair. On a general level, fairness stipulates that all stakeholders are equally able to express opinions, defend them, and request evidence and justification from other stakeholders. One of the main complaints about the participation processes is that they have often been perceived as ineffectual, simply being used to legitimize decisions or give an appearance of consultation without there being any intent of acting on recommendations [35]. One approach that might lead to fulfilling these criteria is to ensure that stakeholders have an influence on the final decision, so that the output of the process has a genuine impact on policy. The key element defining stakeholder participation is the will of the stakeholders to influence the policy. The point of a participatory process is also to obtain the opinions and perceptions of stakeholders. Sometimes, according to [45], capacity building can be a more enduring and widespread outcome of the participatory process than an actual agreement. The influence of stakeholders depends on their degree of involvement. According to [49], it is possible to divide the degree of involvement into five degrees of participation, with different possibilities for stakeholders to influence the outcome: (a) Information: Explanation of the project to the stakeholders; (b) Consultation:

Presentation of the project to stakeholders, collection of their suggestions, and then decision making with or without taking stakeholders' input into account; (c) Collaboration: Presentation of the project to stakeholders, collection of their suggestions, and then decision making, taking stakeholders' input into account; (d) Co-decision: Cooperation with stakeholders towards an agreement for solution and implementation; (e) Empowerment: Delegation of decision-making over project development and implementation to the stakeholders. Finding a jointly acceptable strategy and course of action is important, if the goal is to commit stakeholders to the implementation of the policy/program.

Table 1. Evaluation Criteria and Indicators for an Effective Participatory Process.

Criteria	Sub-Criteria	Related Indicators
Acceptance criteria	Representativeness	Early involvement
		Relevance of the involved stakeholders
		A broad range of stakeholders involved
		No stakeholder who is willing to participate is deliberately excluded
		Exclusion of stakeholders from the process
	Independence	Management and facilitation of the process
		Equality of stakeholders in the process
	Influence	Stakeholders views and opinions are heard
		Stakeholders responses and proposals are respected
		Stakeholders have the power to influence the process and its outcomes
		Capacity building
Process criteria	Transparency	Transparency of the process
	Accessibility	Information resources
		Material resources
		Time resources
	Cost-effectiveness	Financial resources
	Task definition	Nature and scope of a participation process

A central element of effective collaboration is also transparency. The process should be transparent so that the stakeholders can see what is going on and how decisions are being made. Stakeholders should be informed, e.g., about what the purpose of their participation and involvement is, what they can influence, and who can participate and how. A genuine attempt to share information means that organizers actively ensure that all stakeholders are aware of, and understand, the relevant information [47].

It is clear that an effective decision-making process requires access to appropriate and relevant information, which means that land owners should have free access to this information to enable them to successfully fulfill their briefing. Accessibility includes: (a) information resources and access to them (summaries of the pertinent facts); (b) material resources (e.g., overhead projectors, whiteboards), and (c) time resources (stakeholders should have sufficient time to make a decision). Possible difficulties in a participatory process also revolve around the issues of finance. The participatory process should be cost-effective, which means that appropriate finances are available, and that the key concern for those organizing a participatory process is a value for money. Prior to conducting a participatory process, it is clearly sensible to take into account the potential costs of the methods, in both time and money, and to consider the extent to which they fulfill the objectives [34]. Each participatory process should clearly define the nature and scope of the participation tasks. It is important to ensure that there is as little confusion and dispute as possible regarding the scope of a participation process, its expected output, and the mechanisms of the procedure. All of these aspects should be clearly defined at the outset [35].

According to this conceptual framework, the 266 surveys and 17 interviews were analyzed with focus on the participatory process for the PUN (2015–2020).

3. Materials and Methods

The methodological approach in this analysis can be characterized as a theory-driven study of the latest PUN process in Slovenia. For insight into the formulation of process PUN 2015–2020, we reviewed and analyzed the project reports of the Operational Program for Managing Natura 2000 sites in Slovenia. The analysis provided us with information about the activities that have taken place, their order, and their duration. Furthermore, through the review of the reports, we obtained a list of the stakeholders, who participated in the process of formulating the PUN 2015–2020.

In order to analyze how stakeholders' involvement and participation have evolved in the process of formulation of PUN 2015–2020 in Slovenia, the methodological approach in the present study is based on a triangulation of methods—qualitative and quantitative. Therefore, in the first phase, in-depth semi-structured interviews with representatives of the key stakeholders' institutions or organizations that were responsible for, or involved in the formulation of, the PUN were carried out. In the second phase, a survey of all stakeholders involved in the formulation and decision-making of PUN 2015–2020 was conducted.

3.1. Interviews with Stakeholders

Detailed interviews were administrated to 17 stakeholders representing 11 institutions or organizations, which accounted for 11.7% of the total institutions that were involved in the formulation of the PUN (Table 2). When identifying and selecting stakeholders, a purposive sampling was used, where information-rich stakeholders are chosen, because they have particular features or characteristics which will enable detailed exploration and understanding of the central theme and puzzles which the research wish to study [50]. This approach aims to ensure that all key stakeholders are taken into account, and that some diversity is included.

Table 2. Stakeholders Involved in the Interview.

Name of Institution/Organization/Association	N° Respondents
Ministry of Environment and Spatial Planning	3
Institute of the Republic of Slovenia for Nature Conservation	1
Slovenia Forest Service	3
Institute for Water of the Republic of Slovenia	1
Triglav National Park	2
Farmland and Forest Fund of the Republic of Slovenia	1
University of Ljubljana—Biotechnical Faculty	1
Agricultural Institute of Slovenia	1
Centre for Cartography of Fauna and Flora	1
Chamber of Agriculture and Forestry of Slovenia	2
Fisheries Research Institute of Slovenia	1
Total	17

N°—number.

The first version of the semi-structured interview was pre-tested in May 2015, and interviews conducted between December 2015 and January 2016.

The framework of the interview was prepared with the aim to analyze five main aspects of the participatory process: (1) the role of the institution and the personal role of the interviewee in the process of formulating the PUN (2015–2020); (2) the level of participation and stakeholders involvement; (3) the evaluation of participation in the process; (4) the transparency of the process and (5) the management and organization of the process. Table A1 in the Appendix A presents the distribution of interviewees according to the evaluation Sub-Criteria for an Effective Participatory Process.

All interviews were voice recorded and later transcribed. The length of interviews ranged between 20 and 35 min, with an average of 22 min. The data collected with interviews was analyzed through a content analysis using codes and their sub-codes (e.g., role of institution, participation, conflicts, agreement) with the support of MAXQDA 10 software (VERBI GmbH: Berlin, Germany).

3.2. Survey of the Stakeholders

In the first step of the research, a preliminary list of stakeholders involved in PUN was created based on list of participants in the workshops and reports of the LIFE [+] project. Eight hundred and fifteen stakeholders from 94 different institutions and organizations (i.e., ministry, public forest administration, NGOs, university and research institutions, private forest owners and farm associations) were identified as a study population. In the second phase, a sample size of 261 stakeholders was calculated with the help of Sample Size Calculator [51], (confidence level 95%). In order to maximize the response rate and to reduce survey error, Dillman's Tailored Design Methods (TDM) [52] was adopted. Therefore, the questionnaire was sent by post to 767 identified stakeholders using the 1KA web survey program (https://www.1ka.si). The response rate was 34.8%, with 266 filled questionnaires from representatives of 39 different institutions, which is 41.2% of all involved institutions.

Stakeholders were categorized into seven groups of interest: Governmental organizations (e.g., ministries), public administrations (e.g., SFS), local communities (e.g., municipalities), environmental NGOs, Chambers and Associations (e.g., CAFS, Association of Forest Owners), universities and research institutes (e.g., Biotechnical Faculty) and others (private companies). The majority of respondents involved in our survey were from public administration (42.9%), followed by respondents from Chambers and Associations (21.1%), governmental organizations (9.7%), universities and research institutes (4.9%), others (4.1%), environmental NGOs (2.3%) and local communities (0.7%) (Table 3).

Table 3. Stakeholders Involved in the Survey.

Group of Stakeholders	N° Sample	N° Respondents
Governmental Organizations	98 (12.8%)	26 (9.7%)
Public Administrations	244 (31.9%)	114 (42.9%)
Local Communities	25 (3.2%)	2 (0.7%)
Environmental NGOs	19 (2.5%)	6 (2.3%)
Chambers and Associations	298 (38.9%)	56 (21.1%)
Universities and Research Institutes	42 (5.4%)	13 (4.9%)
Others	41 (5.3%)	11 (4.1%)
Total	767	266 (Because the survey was anonymous, 38 (14.3%) respondents choose not to reveal their institution.)

The questionnaire—comprising both open-ended and closed-ended questions—consisted of five sections seeking information on: (1) nature protection policies; (2) the participatory process and the relations between stakeholders; (3) the influence of stakeholders on the process; (4) conflicts and (5) the socio-demographic characteristics of stakeholders. In Sections 2 and 3, several questions were asked with the aim of gathering information on stakeholders' involvement and participation in PUN 2015–2020. The questions in the survey were developed based on the criteria for the evaluation of a participatory process [35].

For each of the indicators for the evaluation of the participatory process, the stakeholders were asked to evaluate their satisfaction on a five-point Likert scale (1—fully unsatisfied to 5—very satisfied). Based on these evaluations, average values were calculated for each indicator. The questionnaire was pre-tested in June 2016, and the survey data was collected in September and October 2016.

Collected data was coded and inserted in MS Excel and processed in IBM's SPSS Statistics 21.0 software (International Business Machines Corporation: New York, United States of America).

In order to check the quality of the data and to detect errors, outliers, and missing values, all data was first checked with frequencies. In order to determine the normality of the data, normality tests (Kolmogorov-Smirnov and Shapiro-Wilk) were used [53]. In this case, data were not normally distributed; therefore, non-parametric tests were used [53]. As one of the objectives of this study was to present the differences among the stakeholders' evaluation of the participatory process according to their role (stakeholders from project partner's institution and other stakeholders) in the implementation process, a comparison was made between those stakeholders' groups. Hence, in order to determine the differences between these two groups, the non-parametric Mann-Whitney U test was used. The differences between groups were statistically tested using the Mann-Whitney U test ($p \leq 0.005$). The statistical results of Mann-Whitney U test are presented in Table A2 in Appendix A.

4. Results

4.1. Stakeholders' Evaluation of the Acceptance Criteria

4.1.1. Representativeness of the Participatory Process

The participatory process applied in the formulation of PUN 2015–2020 in Slovenia involved different stakeholders from the early beginning of the process. First, the project coordinator (MESP), together with project partners, made a detailed analysis of the stakeholders, based on sector involvement and type of institution, and prepared influence/interest matrix for potential stakeholders' prioritization and a precise plan of their involvement in different stages of the process: "The selection criterion for stakeholders to participate in the process was that they are institutions with an interest and competency in the planning and implementation of nature protection policies and measures" [MESP 1]. Consequently, the formulation of PUN 2015–2020 brought together a broad range of stakeholders from different sectors, which is confirmed by the results of the interviews, in which 70.7% of interviewed stakeholders agreed with the statement that the process was representative, 17.6% were indifferent, and 11.7% disagreed. On the other hand, the results from the survey also confirmed these findings, where the indicators for the evaluation of participatory process, i.e., "A broad range of stakeholders involved", and "Relevance of the involved stakeholders", were evaluated respectively as satisfied (average value 3.24 ± 0.983 and 3.18 ± 0.937). Moreover, the Mann-Whitney U test did not show statistically significant differences between stakeholders from the project's partner institutions and other stakeholders for both indicators.

On the other hand, the stakeholders also felt that some groups were excluded from the process (landowners and people living in the Natura 2000 areas i.e., general public etc.): "Landowners under Natura 2000, especially those who are economically dependent on agriculture and forestry were not directly involved" [CAFS 2].

4.1.2. Independence of the Participatory Process

The management and facilitation of the PUN 2015–2020 was in the hands of project coordinator and partners. The draft version of PUN 2015–2020 was discussed in 6 workshops with a wider group of key stakeholders (219 representatives). The workshops were prepared separately for different sectors (forestry, agricultural, fisheries, and water) and local communities (workshops for the eastern and western part of Slovenia). The participation was based on consultation, collection of propositions/comments, and the elaboration of a final draft. The final draft was sent for intergovernmental consultation following the adoption of the PUN 2015–2020. After the adoption of PUN 2015–2020, 8 workshops were held for project partners and other target audiences (623 stakeholders) to spread information about PUN 2015–2020.

The results of the survey showed that the indicator "Workshop organization" was evaluated on average as 3.80 ± 0.848, and "Management and facilitation of the process" as 3.71 ± 0.831, which means that stakeholders were satisfied with the organization of the participatory process. The Mann-Whitney U test did not show statistically-significant differences between stakeholders from project partner's institutions and other stakeholders for both indicators. The survey results are also supported by the results of interviews, where 82.3% of interviewed stakeholders agreed with the statement that the process was independent, 11.8% were indifferent, and 5.9% disagreed, which is seen in the respondents' opinions: "The workshops were conducted according to established rules, useful and successful. We were absolutely satisfied with the workshops, because there was always the possibility to express our opinion" [CAFS 1]; SFS 1: "In the draft process, when we coordinated with the content, we received a record where it was clear what was changed. We updated the content regularly on the screen during the workshops and got a record that we could put comments on and comments were taken into account".

On the other hand, a few stakeholders were not satisfied with management and facilitation of the process and fairness of workshops and feedback from organizer of process: "We were invited to the workshop for presentation the draft, and at that time we were not given the full picture of the content creation process, how the process is proceeding, and whether we are expected to have any other proposal, cooperation" [representative of RDA Green Karst 1].

The role of facilitator is important for an effective participatory process. The facilitators were nominated by project partners who identified internal persons from IRSNC. The stakeholders' highlighted that the role of facilitator was crucial in the participatory process for the formulation of PUN 2015–2020. Particularly, the facilitator encourages trusting relationships and equality between the participants, so that mostly all stakeholders' views were heard and respected. The results of the survey evaluated the indicator for the evaluation of the participatory process "Stakeholders views and opinions are heard" on average as 3.43 ± 1.016, and "Equality of stakeholders in the process" as 3.35 ± 1.021, which means that stakeholders were fairly satisfied. The Mann-Whitney U test did not show statistically significant differences between stakeholders from project partner's institutions and other stakeholders.

The survey results are also supported by the interviews, where respondents stated: "I think that everybody has the possibility to express their opinion, if they have an interest" [IFRS 1].

4.1.3. Influence in the Participatory Process

The stakeholders who were present at the workshops contributed to the co-creation of PUN 2015–2020 through their active participation, as highlighted also by a respondent: "The stakeholders' contribution to the PUN was great; they made it more practical, so it has more chance for implementation. It can be said that, from the draft to the proposal, a great improvement has been made" [IRSNC 1] and "At workshops where we prepared the draft, we self-coordinated about the substance and then got feedback from the IRSNC, so we simultaneously repaired the draft and presented it on big screen. Most of the notes were included in the draft later on" [SFS 1].

The level of participation is strictly linked to the power of stakeholders to influence the process and outcomes in the participatory process. Yet, the stakeholders' experiences of having an actual influence on the formulation process varied. Stakeholders from partner institutions felt that their interests and proposals were adequately represented in the final version of PUN, which is also confirmed by the results of the interviews, where 70.7% of interviewed stakeholders agreed with the statement that they had influence on the process, 17.6% were indifferent, and 11.7% disagreed. The results from the survey also confirmed this finding, where stakeholders were least satisfied with the indicator "Stakeholders' response and proposals were respected" (average value 2.99 ± 0.976) and "Power of stakeholders to influence the process and its outcomes" (average value 2.82 ± 0.996). Moreover, the Mann-Whitney U test did not show statistically significant differences between stakeholders from project partner's institutions and other stakeholders. The results also show a slightly higher influence of the institutions than the influence of individuals, as the average estimate of the influence of institutions was 2.97.

However, not all interviews supported the results of the survey. Some of stakeholders felt that they had little, if any, influence on the final version of PUN 2015–2020. According to the representative of NGOs, they had an impact on the draft version of PUN, but not all their proposals were considered at the end. "The ultimate decision-making power was possessed by IRSNC, who took the final decision concerning the draft and some of our proposals were ignored. We also made comments on the public meeting about some measures, but we were not heard and did not get an answer" [Bird Life NGO 1].

Also, partner institutions noticed that other stakeholders had limited influence, arguing that: "Some content stayed uncoordinated. We were approaching the finalized management program through the entire process, but we were not completely harmonized with other stakeholders about some of the content" [SFS 1] or "IRSNC did not take comments into account when they were contrary to the opinions of Bird Life NGO" [SFS 3]. However, some representatives from partner institutions claimed that they had limited power to influence the process and its outcome. For example, the representative of IWRS 1: "We had to accept some of the objectives and measures and try to understand them as best as possible" and representative of SFS 3: "In the reconciliation of the PUN, we could only make our comments by problematizing the issue of the positive opinion of the institution in intergovernmental consultation".

Capacity building is clearly the most concrete outcome of the formulation of the PUN 2015–2020 process that has taken place during the period 2012–2015. It was a valuable forum to receive and exchange information, and to some extent, to build shared understanding of nature conservation problems. In addition, the results of the survey show that "Capacity building" was evaluated with an average value 3.04 ± 0.842. The Mann-Whitney U test did not show statistically significant differences between stakeholders from project partner's institutions and other stakeholders.

The interviews supported the results of survey, which is seen in a statement of the representative of IWRS 1: "I think that the team from the ministry was great, probably could not be better. I know, because I participated in other projects with other ministries where we had problems. Here everything went smoothly. When we presented our program on workshops I always had a feeling of security and I believed in the program". A further statement which supports the results is from a representative of MESP 1: "There is an important benefit of the partnership, because each of the five partners is a responsible authority in their field of work, and project results require joint understanding and joint solutions. As an example, analyses were prepared e.g., for Natura 2000 on agricultural land by the IRSNC, which has knowledge and experience on the nature conservation aspect of achieving objectives and the measures implemented, and by CAFS, which has knowledge and experience on the agricultural aspect of implemented measures, and stakeholders experience. Without both institutions involved there would be only one analyzed perspective. The same goes for forests and waters and fisheries".

Capacity building has influenced the sharing of information as representative of CAFS 1 explained: "For our needs within the framework of some project activities, we needed data that we did not have. The MESP has easily provided us with this information" and IWRS 1: "Any information or data we needed concerning the project from any project partner was not a problem to get it".

4.1.4. Transparency of the Participatory Process

The transparency of the participatory process was evaluated generally as satisfactory, as 45.6% of the respondents of the survey answered that the process was transparent or very transparent (average value of transparency: 3.26 ± 0.917). The results were also supported by the results of the interviews, where 76.6% of interviewed stakeholders agreed with the level of transparency of the process, 11.7% were indifferent, and 11.7% disagreed. Among the answers of the interviewees: "For me, the PUN process was transparent, because I was part of the team and got an inside look on how the system works" [SFS 1] and "Yes, I believe that the process was transparent, because the starting point was adjusted to ensure the transparency of the process and it worked" [CAFS 1].

Those respondents from the survey and interviews who were not satisfied with the transparency of the participatory process (21.3%) mentioned that the main reason for inadequate transparency was the lack of feedback on their comments at the workshops, the lack of information on the content editing (what can be changed and what cannot), which phase the process of creating the PUN is in, the state of future cooperation methods, who has already been involved in the process of creating the PUN, and who else will be.

This is also evident from the survey answers and statements of interviewed stakeholders: "There was a selective invitation for participation at the process and the workshops were not recorded … participants did not receive any feedback, the final version of PUN was changed beyond the knowledge of the participants" [SFS 1], "There was a lack of stakeholders' awareness and their insufficient involvement. Those of us, who were involved, did not receive any information about the process itself" [NIB 1], and "The process is not transparent, because there is no feedback or information on what stage the PUN is in" [SFS 3]. The final version of the PUN 2015–2020 was adopted by MESP and IRSNC, and did not take all the recommendations of stakeholders into consideration, as pointed out by a respondent in the survey: "The final decisions were not drawn up at the workshops, but later, after the reconciliations between the ministries, where the other stakeholders were no longer included. Final decisions should be taken with the consensus of all stakeholders" [Biotechnical Faculty 1], and "There were no clear instructions, participants were selectively invited, workshops did not keep a record, there was no feedback, the final document was changed without the participants' knowledge, different information was provided at the time of preparation on the meaning and usability of PUN, as it turned out later upon confirmation" [NGO 1].

Moreover, the Mann-Whitney U test showed statistically significant differences between stakeholders from project partner's institutions and other stakeholders (U = 1260.500, z = −3.018, p = 0.03). Stakeholders from project partner's institutions evaluated sub-criteria with an average value of 3.60, while other stakeholders evaluated it with an average value 2.80, which means that stakeholders from project partner's institutions were more satisfied with the transparency of the process than other stakeholders.

4.2. Stakeholders Evaluation of the Process Criteria

4.2.1. Accessibility of the Participatory Process

Access and flow of information in the process of formulating the PUN 2015–2020 was satisfactory, as different means of communication were used, depending on which stakeholders they wanted to reach i.e., mass media, project website, formal invitations on workshops. The main source of information was the project's website, where all of the project's steps were published.

Stakeholders were generally satisfied with the accessibility of the participatory process, which is confirmed by the results of the interviews, where 82.4% of interviewed stakeholders agreed with the accessibility of the information in the process, 5.9% were indifferent, and 11.7% disagreed.

Stakeholders were satisfied with the material resources as the answers of the survey, where the indicator "Intelligibility of the material and information" was evaluated with an average value of 3.47 ± 0.861, and indicator "Notification of certain activities" with 3.11 ± 0.951. The stakeholders' satisfaction with material resources was pointed out by the interviewed representative CAFS 1: "We got access to the data that the ministry had for our needs, so we cooperated well". Respondents also evaluated the indicator "Free access to all information" as satisfactory (average value 3.41 ± 0.881).

Stakeholders had sufficient time for reflection, to make a decision and to express their opinions, comments and thoughts about the content of PUN, as stated by representative: " … there was almost one year of these public announcements, so whoever had an interest had the opportunity to suggest or comments the PUN" [IFRS 1].

4.2.2. Cost-Effectiveness

Analysis of the costs of PUN is crucial in order to assess cost-effectiveness. The project's budget was 1,706,914.00€, where the financial contribution of European Union was 50.0% [54]. The largest share of the expenses represents personnel costs as 62 staff worked on the project altogether. The largest expenses in this budget line represents the implementation of the workshop for preparing the first draft, followed by project management and analysis of measures implemented and objectives achieved from PUN 2007–2013. The planned financial resources were sufficient through internal redeployment of resources, according to representative IRSNC 1: "The planned financial resources for workshops for the first draft were underestimated at the time of application, but we provided enough money with internal redeployment of funds".

Stakeholders from project partner's institutions who were directly involved in the project were satisfied with the cost-effectiveness, which is confirmed by the results of the interviews, where 47.1% of interviewed stakeholders agreed with the cost-effectiveness of the project, and 52.9% of stakeholders are indifferent because they were not directly involved in the project and could not respond to the question.

4.2.3. Task Definition

The nature and scope of the participation task were defined between the partners of the PUN 2015–2020 project in a document called "Process and Communication Plan". The document represents a detailed plan of the implementation of this project, defining the project phases, actions of each phase, its activities, and the tasks of each partner. Additionally, the time schedule and correlations between tasks were defined as well.

The stakeholders who were involved as partners knew from the very beginning what their role and tasks were. "I think that, even within the institution, we have explained how things are going, so everyone knew where to get the information and to whom to send comments … so the tasks were clearly defined" [SFS 1].

However, not all stakeholders knew that; this was especially the case among stakeholders who were included only in the phase of the draft presentation. This is seen from the survey answer of the of rural development agency Green Karst 1: "We were invited to the workshop where the draft was presented and at that point, we were not given the whole picture of the creation process. We also did not receive any feedback after the workshop regarding which proposal was taken into account, how the process was proceeding, and whether we are expected to offer any other proposals or cooperation".

Additionally, 47.1% of interviewed stakeholders agreed with the task definition and the document "Process and Communication Plan". The rest of the stakeholders were indifferent because they were not involved in the project (LIFE $^+$), so they were not familiar with the document.

5. Discussion

The success of the nature conservation policy depends on stakeholders' co-operation and participation [19]. The analysis of stakeholders' involvement in the decision-making process of PUN 2015–2020 in Slovenia is a prerequisite for successful participation in order to highlight a wide range of interests and point-of-views between different groups of stakeholders.

This study shows that the set of acceptance and process criteria and indicators can be used as a basis for the evaluation of the participatory process in nature conservation and forest policy. The information retrieved from the survey and interviews of the stakeholders indicates that the participatory process was designed by considering the representativeness, independence, influence, transparency, resource accessibility, cost-effectiveness, and task definition criteria. In some EU countries, the participatory process was organized in the same way [31,32,34]. This shows a positive trend from a command and control approach towards a participatory approach not only in PUN 2015–2020, but also in other EU countries [1,55,56], which means that various groups of stakeholders could influence policy outcomes in several ways [8]. Overall, however, not all of the requisite conditions for successful participation were created by the evaluated participatory process. Studies [31,32,34,57] show that some conditions for successful participation were not achieved. This indicates the possibility for an improved, novel and innovative approach that leads to multi-level governance and a better participatory process.

For example, with regard to the stakeholders' involvement in the process, the main limit was that some groups of stakeholders felt excluded from the process or were only included in the final phase (i.e., landowners and general public), or that stakeholders did not perceive that other stakeholder groups were also involved in process of the formulation of PUN 2015–2020. A reason for that could be the organization of separate workshops for different sectors, in which only a part of the stakeholders were present, or the fact that only the interests of privileged stakeholder groups were included in the decision-making process, while other marginalized stakeholders were only informed about the results of the process in the final phase. During the consultation meetings on intergovernmental consultation, MAFF did not give his consent to MESP for the adoption of PUN 2015–2020, until a consensus between SFS and IRSNC about the measures related to forestry and nature conservation under Natura 2000 was reached; e.g., one of the topics was the amount of dead wood that should stay in the forest. The aim of the meeting was to reach a consensus on an expert level. Therefore, more participation of marginalized stakeholder groups at the early and final stages is needed, and should be based on consensus decision making and involvement which would improve the participation process.

Moreover, to improve the management and facilitation of the process, it is important to know the role that stakeholders play in the process. Therefore, more attention should be given by project partners to the presentation of the Process and Communication Plan. In addition, the facilitators of the PUN 2015–2020 were very professional and capable of facilitating an open-discussion, and equally, of supporting stakeholders in expressing their opinions, needs, and interests. However, the choice of the facilitator should be reconsidered and assigned to an external person who is not affected by the topic of discussion in order to improve the independence of the participatory process [31,35]. The role of facilitator is essential to ensure equality among stakeholders in the process [31,58].

The adopted participatory approach in PUN 2015–2020 was less centralized than the previous one, but it was still a top-down oriented approach with a switch from a more information-based communication with stakeholders, as in the previous process, to a consultation with key stakeholders. This finding is in accordance with studies evidencing that the initial top-down approach practiced in most countries was changed to more socially inclusive and participatory bottom-up approaches [34,59–61]. Still, stakeholders felt that all their responses and proposals were not taken into consideration. To improve the perceived influence of stakeholders, project partners need to connect participation with co-responsibility [15].

The level of participation is strictly linked to the power of stakeholders to influence the final decision in the participatory process [62]. According to the results, stakeholders had conflicting opinions on their real ability to influence the policy process and the final decisions. This is not surprising, as project partners were the organizers of the process, the actors who selected the stakeholders, the participants in process, and facilitators at the same time. Accordingly, the stakeholders who were involved in the process participated with different levels of involvement and power to influence the process. This finding confirms that using participation techniques does not guarantee that the stakeholders will feel empowered, but they are very useful tools to get the stakeholders' opinions [57]. According to the results, the ultimate decision-making power was held by the MESP and IRSNC who made the final decision. This is not problematic per se, according to [63] (p. 329): "As with all public policy issues, the government should have the last word" (p. 329). Other studies have shown similar problems regarding the influence and the power of stakeholders that emerged as a consequence of the adopted top-down approach [61,64,65]. It seems that, in future participatory processes, it would be beneficial if project partners would conduct another round of workshops to consult a broad range of stakeholders. These workshops can be considered to incorporate the stakeholders' ideas and comments.

This would improve the transparency of the process and obtain new opportunities for stakeholders to exert their influence.

In general, the transparency of the decision-making process in PUN 2015–2020 was guaranteed, but 21.3% of stakeholders were still not entirely satisfied with it. That was also confirmed by the Mann-Whitney U test. The project partners were more satisfied with the transparency of the participatory process than other stakeholders. Therefore, a more transparent process would represent a significant asset, including adequate information for included stakeholders and the land owners (via website), proper planning and preparation of the consultation process, respecting process results, and an improved communication gap between policymakers and other relevant stakeholders, which means keeping them constantly informed about the process and any further steps. With regard to the stakeholders involved in the process, the limit was the low level of participation of land owners. In order to increase the future legitimacy and transparency of the PUN formulation process, land owners should be involved more actively in the decision-making. This was similarly deduced in the formulation of the National Forest Program in the Czech Republic [31]. In this respect, a good example is represented by the Austrian Forest Forum, where land owners had the chance to participate in the dialogue process via an Internet platform or through written statements. Moreover, in the Austrian Forest Forum, land owners were kept informed through a Forest Dialogue Newsletter [66].

The important resources in a participatory process are information, and material and time resources. The availability of these resources is progressing in PUN, but it needs more capacity and resources in terms of better communication with stakeholder groups. According to the organizer of the process, financial resources were sufficient. Restrictions on financial resources are liable to have an impact on the quality of the participation process. It is essential to ensure that appropriate finances are available.

Concerning the task definition criteria, tasks between stakeholders who were involved as project partners were very precisely defined. However, stakeholders from other institutions or stakeholders from the same institutions who were involved in a later phase of the process (presentation of draft) did not know that. To improve those criteria, what the purpose of stakeholders' participation is/their tasks and what is expected from them should be clearly stated at the beginning of each workshop.

When interpreting the results of this study, one should note that the approach has some limitations. The analysis only considers the opinions regarding the evaluation of the participatory process of formulation of PUN 2015–2020 of stakeholders who participated in the study. Relying on the answers of respondents (266), this study does not reveal the opinions of the silent majority on the evaluation of the participatory process of the formulation of PUN 2015–2020, nor how they will act in the future. Moreover, the present results could be biased, because only representatives of private landowners were involved in the process and consequently in our study, but they are especially relevant for the management of Natura 2000. Their absence from the process was recognized by the respondents, and is indicated in the results section. The real reason for their absence is unknown. The most likely reason could be a lack of interest in participation in the process; less likely reasons could also be a boycott, which can result in serious opposition in the future or in political radicalization. Those two reasons are less likely, because representatives of the Association of Forest Owners were invited to round tables (public meetings) where they could have given input on a draft version of PUN, but they did not attend. In addition, they were present in last phase of workshops when the PUN 2015–2020 was already adopted by the government. The reason for the lack interest could be a lack of knowledge on how to argument their opinions in order to influence the draft version, or because the aim of PUN 2015–2020 was not changing the borders but to establish measures to achieve conservation objectives of species and Natura 2000. However, the number of respondents in the survey was large enough to present the full picture of the process.

6. Conclusions

There is increasing contention that public participation in policy-making is necessary to reflect and acknowledge democratic ideals and to enhance trust in regulators and transparency in regulatory systems [35]. Consequently, the participation of stakeholders in nature conservation policies, such as PUN, and especially in the management of protected areas, such as Nature 2000, has gained importance in the last decades.

The detailed analysis described in the previous sections gives us a general overview of the participatory process and stakeholders' involvement in the process of formulation of PUN 2015–2020 in Slovenia. The process was assessed based on evaluation criteria for an effective participatory process, taking into account the acceptance criteria (representatives, independence, influence and transparency) and process criteria (accessibility, cost-effectiveness and task definition). It is evident from research that these criteria do not function independently, and it is therefore important to pay attention to all of them. For example, a transparency criterion relates to influence criteria and vice versa.

Therefore, evaluating the participatory processes is essential to increase the quality of future processes. Such an evaluation can be useful for the organizations responsible for planning and management, as well as stakeholders and policy makers. In order to obtain good results, the managers need to have guidelines to help them in designing and carrying out a successful participatory planning process.

While a number of different participation techniques have been developed, their relative usefulness is difficult to ascertain, because there are either no systematic comparisons between them or they are rare. Our evaluation was done ex post, based on quantitative and qualitative analysis, and carried out by a researcher related to the participatory processes as an observer. We are aware of the fact that the chosen criteria can have multiple meanings, and that stakeholders might define the criteria in a different way.

The results of this study can be applied for stakeholder involvement in the participatory processes in other natural resource management situations or similar processes of policy formulation activities. The usefulness of this research is to improve participatory processes in future PUN and other nature conservation and forest-related programs.

Finally, every process of formulating the policy is different, and the participation techniques used should be considered on a case by case basis. However, lessons can be learned from this case and used to develop participation techniques and more customer-oriented planning.

Author Contributions: The authors have contributed to this study in equal parts. The results and article are part of emerging Ph.D. at Biotechnical Faculty, University of Ljubljana of T.L. and his supervisor Š.P.M., with the title: *"The characteristics of the Social Networks and Participation of Stakeholders in the Management of Natura 2000 Sites"*.

Funding: This research was funded by Pahernik foundation. Authors wish to thank to the foundation for supporting the publishing of results.

Acknowledgments: First: we wish to thank to all the interviewees and respondents who took part in this research and made it possible by sharing their experiences. We also want to thank to the anonymous reviewers for their contributions and special issue guest editor of section "Forest Economics and Human Dimensions" Alessandro Paletto. Finally we would like to thank to Saša Petrovič and Ana Marija Laktić for proofreading.

Conflicts of Interest: The authors declare no conflict of interest. Both authors declare that they are not NGO members and that they do not own forest land.

Appendix A

Table A1. Distribution of Interviewees According to the Evaluation Sub-Criteria for an Effective Participatory Process.

Sub-Criteria	Agreed	Undefined	Disagreed
Representativeness	SFS 1 MESP 1, MESP 2, MESP 3, CAFS 1, FRIS 1, IRSNC 1, IWRS 1, TNP 1, AIS 1, BF 1, FFFRS 1	CCFF 1, SFS 3, TNP 2	CAFS 2 SFS 2
Independence	MESP 1, MESP 2, MESP 3, IRSNC 1, SFS 1, SFS 3, SFS 2, IWRS 1, AIS 1, BF 1, CCFF 1, CAFS 1, FRIS 1, FFFRS 1	TNP 2, CAFS 2	TNP 1
Influence	MESP 1, MESP 2, MESP 3, IRSNC 1, SFS 1, SFS 2, AIS 1, BF 1, CCFF 1, CAFS 1, FRIS 1, FFFRS 1		SFS 3, IWRS 1, TNP 1, TNP 2, CAFS 2
Transparency	MESP 1, MESP 2, MESP 3, IRSNC 1, SFS 1, SFS 2, IWRS 1, TNP 1, TNP 2, AIS 1, CAFS 1, FRIS 1, FFFRS 1	CCFF 1, CAFS 2	SFS 3, BF 1
Accessibility	MESP 1, MESP 2, MESP 3, IRSNC 1, SFS 1, SFS 2, IWRS 1, TNP 2, AIS 1, BF 1, CAFS 1, CAFS 2, FRIS 1, FFFRS 1	CCFF 1	SFS 3, TNP 1
Cost-effectiveness	MESP 1, MESP 2, MESP 3, IRSNC 1, SFS 1, IWRS 1, CAFS 1, FRIS 1	SFS 3, SFS 2, TNP 1, CCFF 1, TNP 2, AIS 1, BF 1, CAFS 2, FFFRS 1	
Task definition	MESP 1, MESP 2, MESP 3, IRSNC 1, SFS 1, IWRS 1, CAFS 1, FRIS 1	SFS 3, SFS 2, TNP 1, TNP 2, AIS 1, BF 1, CCFF 1, CAFS 2, FFFRS 1	

Table A2. Statistic results of Indicators for an Effective Participatory process.

Indicators	Mann-Whitney U	Z	p	Mean	Std. Deviation
Management and facilitation of the process	1926.500	-0.586	0.558	3.71	0.831
Workshop organization	2036.00	-0.165	0.869	3.80	0.848
A broad range of stakeholder involved	1728.500	-0.638	0.524	3.24	0.983
Relevance of the involved stakeholders	1889.000	-0.383	0.701	3.18	0.937
Access to information	1887.500	-0.869	0.385	3.41	0.881
Intelligibility of the material and information	2122.000	-0.299	0.765	3.47	0.861
Notifications of certain activities	1931.500	-0.904	0.366	3.11	0.951
Stakeholders views are heard and respected	1851.500	-0.819	0.413	3.43	1.016
Stakeholders responses and proposals are respected	1732.000	-1.043	0.297	2.99	0.976
Power of stakeholders to influence process and its outcomes	1641.000	-1.465	0.143	2.82	0.996
Capacity building	1391.000	-0.375	0.708	3.04	0.842
Transparency	1260.500	-3.018	0.003	3.26	0.917

References

1. Keulartz, J. European nature conservation and restoration policy—Problems and perspectives. *Restor. Ecol.* **2009**, *17*, 446–450. [CrossRef]
2. Jones-Walters, L.; Çil, A. Biodiversity and stakeholder participation. *J. Nat. Conserv.* **2011**, *19*, 327–329. [CrossRef]
3. Weber, N.; Christophersen, T. The influence of non-governmental organisations on the creation of Natura 2000 during the European Policy process. *For. Policy Econ.* **2002**, *4*, 1–12. [CrossRef]
4. Kati, V.; Hovardas, T.; Dieterich, M.; Ibisch, P.L.; Mihok, B.; Selva, N. The challenge of implementing the European network of protected areas Natura 2000. *Conserv. Biol.* **2015**, *29*, 260–270. [CrossRef] [PubMed]

5. Rauschmayer, F.; Van den Hove, S.; Koetz, T. Participation in EU biodiversity governance: How far beyond rhetoric? *Environ. Plan. Gov. Policy* **2009**, *27*, 42–58. [CrossRef]

6. Beunen, R.; De Vries, J.R. The Governance of Natura 2000 Sites: The Importance of Initial Choices in the Organisation of Planning Processes. *J. Environ. Plan. Manag.* **2011**, *54*, 1041–1059. [CrossRef]

7. Šobot, A.; Lukšič, A. The Impact of Europeanisation on the Nature Protection System of Croatia: Example of the Establishment of Multi-Level Governance System of Protected Areas NATURA 2000. *Socijalna Ekologija Časopis za Ekološku Misao i Sociologijska Istraživanja Okoline* **2017**, *25*, 235–270. [CrossRef]

8. Ferranti, F.; Beunen, R.; Speranza, M. Natura 2000 network: A comparison of the Italian and Dutch implementation experiences. *J. Environ. Policy Plan.* **2010**, *12*, 293–314. [CrossRef]

9. Cent, J.; Mertens, C.; Niedziałkowski, K. Roles and Impacts of Non Governmental Organizations in Natura 2000 Implementation in Hungary and Poland. *Environ. Conserv.* **2013**, *40*, 119–128. [CrossRef]

10. Stringer, L.C.; Paavola, J. Participation in environmental conservation and protected area management in Romania: A review of three case studies. *Environ. Conserv.* **2013**, *40*, 138–146. [CrossRef]

11. Borrass, L.; Sotirov, M.; Winkel, G. Policy change and Europeanization: Implementing the European Union's Habitats Directive in Germany and the United Kingdom. *Environ. Politics* **2015**, *24*, 788–809. [CrossRef]

12. Winkel, G.; Blondet, M.; Borrass, L.; Frei, T.; Geitzenauer, M.; Gruppe, A.; Winter, S. The implementation of Natura 2000 in Forests: A trans-and interdisciplinary assessment of challenges and choices. *Environ. Sci. Policy* **2015**, *52*, 23–32. [CrossRef]

13. Van Apeldoorn, R.C.; Kruk, R.W.; Bouwma, I.M.; Ferranti, F.; De Blust, G.; Sier, A.R.J. *Information and Communication on the Designation and Management of Natura 2000 Sites*; The Designation in 27 EU Member States; Main Report 1; EC Publications: Brussel, Belgium, 2009.

14. Sarvašová, Z.; Šálka, J.; Dobšinská, Z. Mechanism of cross-sectoral coordination between nature protection and forestry in the Natura 2000 formulation process in Slovakia. *J. Environ. Manag.* **2013**, *127*, 65–72. [CrossRef] [PubMed]

15. Gil, A.; Calado, H.; Cost, L.T.; Bentz, J.; Fonseca, C.; Lobos, A.; Vergilio, M.; Benedicto, J. A Methodological Proposal for the Development of Natura 2000 Sites Management Plans. *J. Coast. Res.* **2011**, *64*, 1326–1330.

16. Margules, C.; Sarkar, S. *Systematic Conservation Planning*; Cambridge University Press: Cambridge, UK, 2007.

17. Weiss, G.; Sotirov, M.; Sarvašova, Z. Implementation of Natura 2000 in forests. In *European Forest Institute—What Scienec Can Tell Us*; Sotirov, M., Ed.; European Forst Institute: Joensuu, Finland, 2017; pp. 39–58. Available online: http://www2.efi.int/files/attachments/publications/wsctu7_2017.pdf (accessed on 25 September 2018).

18. Blicharska, M.; Orlikowska, E.H.; Roberge, J.M.; Grodzinska-Jurczak, M. Contribution of social science to large scale biodiversity conservation: A review of research about the Natura 2000 network. *Biol. Conserv.* **2016**, *199*, 110–122. [CrossRef]

19. Alphandéry, P.; Fortier, A. Can a territorial policy be based on science alone? The system for creating the Natura 2000 Network in France. *Sociol. Rurals* **2001**, *41*, 311–328. [CrossRef]

20. Hiedanpää, J. European-wide conservation versus local well-being: The reception of the Natura 2000 Reserve Network in Karvia, SW—Finland. *Landsc. Urban Plan.* **2002**, *61*, 113–123. [CrossRef]

21. Milligan, J.; O'Riordan, T.; Nicholson-Cole, S.A.; Watkinson, A.R. Nature conservation for future sustainable shorelines: Lessons from seeking to involve the public. *Land Use Policy* **2009**, *26*, 203–213. [CrossRef]

22. Blondet, M.; Koning, J.; Borrass, L.; Ferranti, F.; Geitzenauer, M.; Weiss, G.; Turnhout, E.; Winkel, G. Participation in the implementation of Natura 2000: A comparative study of six EU member states. *Land Use Policy* **2017**, *66*, 346–355. [CrossRef]

23. Reed, M.S. Stakeholder participation for environmental management: A literature review. *Biol. Conserv.* **2008**, *141*, 2417–2431. [CrossRef]

24. Saarikoski, H.; Tikkanen, J.; Leskinen, L.A. Public participation in practice—Assessing public participation in the preparation of regional forest programs in Northern Finland. *For. Policy Econ.* **2010**, *12*, 349–356. [CrossRef]

25. Ferlin, F.; Golob, A.; Habič, Š. Some principles for successful forest conservation management and forestry experiences in establishing the Natura 2000 network. In *Legal Aspects of European Forest Sustainable Development, Proceedings of the 7th International Symposium, Zlatibor Mountain, Serbia, 11–15 May 2005*; Schmithüsen, F., Herbst, P., Nonic, D., Jovic, D., Stanisic, M., Eds.; ETH Zürich: Zürich, Switzerland, 2005; pp. 1–11.

26. Commission Staff Working Document. The EU Environmental Implementation Review Country Report—SLOVENIA. 2017. Available online: http://ec.europa.eu/environment/eir/pdf/report_si_en.pdf (accessed on 25 September 2018).

27. Petkovšek, M. *Slovenian Natura 2000 Network in Numbers*; Varstvo Narave: Ljubljana, Slovenia, 2017; Volume 30, pp. 99–126.

28. Pezdevšek Malovrh, Š.; Leban, V.; Krč, J.; Zadnik Stirn, L. *Slovenia: Country Report*; COOL—Competing Uses of Forest Land: Ljubljana, Slovenia, 2012.

29. Medved, M.; Matijašić, D.; Pisek, R. Private property conditions of Slovenian forests in 2010. In Small scale forestry in a Changing World. Opportunities and challenges and the Role of extension and technology transfer. In Proceedings of the IUFRO Conference, Bled, Slovenia, 22–23 December 2010.

30. Pezdevšek Malovrh, Š. Influence of Institutions and Forms of Cooperation of Private Forest Owners on Private Forest Management. Ph.D. Thesis, Biotechnical Faculty, Department of Forestry and Renewable Forest Resources, Ljubljana, Slovenia, 2010.

31. Balest, J.; Hrib, M.; Dobšinská, Z.; Paletto, A. The formulation of the National Forest Programme in the Czech Republic: A qualitative survey. *For. Policy Econ.* **2018**, *89*, 16–21. [CrossRef]

32. Kovács, E.; Kelemen, E.; Kiss, G.; Kaloczkai, A.; Fabok, V.; Mihok, B.; Bela, G. Evaluation of participatory planning: Lessons from Hungarian Natura 2000 management planning processes. *J. Environ. Manag.* **2017**, *204*, 540–550. [CrossRef] [PubMed]

33. Primmer, E.; Kyllönen, S. Goals for public participation implied by sustainable development, and the preparatory process of the Finnish National Forest Programme. *For. Policy Econ.* **2006**, *8*, 838–853. [CrossRef]

34. Brescancin, F.; Dobšinská, Z.; De Meo, I.; Šálka, J.; Paletto, A. Analysis of stakeholders' involvement in the implementation of the Natura 2000 network in Slovakia. *For. Policy Econ.* **2018**, *89*, 22–30. [CrossRef]

35. Rowe, G.; Frewer, L.J. Public participation methods: A framework for evaluation. *Sci. Technol. Hum. Values* **2000**, *25*, 3–29. [CrossRef]

36. Boh, T. *Shielding Implementation from Politicisation? Implementation of the Habitats Directive in Slovenia*; OEUE Phase II, Occasional Paper; Dublin European Institute: Dublin, Ireland, 2004.

37. Gallo, M.; Malovrh, Š.P.; Laktić, T.; De Meo, I.; Paletto, A. Collaboration and conflicts between stakeholders in drafting the Natura 2000 Management Programme (2015–2020) in Slovenia. *J. Nat. Conserv.* **2018**, *42*, 36–44. [CrossRef]

38. Hlad, B.; Kline, M. *Natura 2000. Končno Poročilo o Izvajanju Komunikacijske Strategije. Priloga 10: Priporočila za Upravljanje in Komuniciranje Natura 2000*; Ministry of the Environment and Spatial Planning: Ljubljana, Slovenia, 2004.

39. Nastran, M.; Pirnat, J. Stakeholder participation in planning of the protected natural areas: Slovenia. *Sociologija i Prostor Časopis za Istraživanje Prostornoga i Sociokulturnog Razvoja* **2012**, *50*, 141–164. [CrossRef]

40. Bibič, A.; Ogorelec, B.; Podobnik, J.; Mikuletič, J.; Vaupotič, M.; Bedjanič, M.; Midžić, Z.; Trebar, B. *Natura 2000 Site Management Programme: 2007–2013 Operational Programme*; Ministry of the Environment and Spatial Planning: Ljubljana, Slovenia, 2007; Volume 61.

41. Management of State Forests Act. In *Official Gazette of the Republic of Slovenia*; No. 9/16 of 12 February 2016; Ljubljana City Library: Ljubljana, Slovenia, 2016.

42. Dryzek, J.S. *Deliberative Democracy and Beyond. Liberals, Critics, Contestations*; Oxford University Press: Oxford, UK, 2000; p. 195. ISBN 0198295073, 9780198295075.

43. Smith, L.G. *Impact Assessment and Sustainable Resource Management*; Longman: Harlow, UK, 1993.

44. Healey, P. *Collaborative Planning: Shaping Places in Fragmented Societies*; Macmillan: London, UK, 1997; ISBN 1403949204, 9781403949202.

45. Innes, J. Consensus building: Clarification for the critics. *Plan. Theory* **2004**, *3*, 5–20. [CrossRef]

46. Webler, T.; Tuler, S.; Krueger, R. What is good participation process? Five perspectives from the public. *Environ. Manag.* **2001**, *27*, 435–450. [CrossRef]

47. Mascarenhas, M.; Scarce, R. The Intention Was Good: Legitimacy, Consensus Based Decision Making, and the Case of Forest Planning in British Columbia, Canada. *Soc. Nat. Resour.* **2004**, *17*, 17–38.

48. McGurk, B.; Sinclair, A.; Diduck, A. An assessment of stakeholder advisory committees in forest management: Case studies from Manitoba, Canada. *Soc. Nat. Resour.* **2006**, *19*, 809–826. [CrossRef]

49. Luyet, V.; Schlaepfer, R.; Parlange, M.B.; Buttler, A. A framework to implement Stakeholder participation in environmental projects. *J. Environ. Manag.* **2012**, *111*, 213–219. [CrossRef] [PubMed]

50. Ritchie, J.; Lewis, J. (Eds.) *Qualitative Research Practice: A Guide for Social Science Students and Researchers*; SAGE Publications Ltd.: London, UK, 2003; ISBN 0 7619 7109 2.
51. Creative Research Systems: Sample Size Calculator. Available online: https://www.surveysystem.com/sscalc.htm (accessed on 11 September 2018).
52. Dillman, D.A. *Mail and Internet Survey: The Tailored Design Method*, 2nd ed.; John Wiley & Sons: Hoboken, NJ, USA, 2007; ISBN 978-0471323549.
53. Field, A. *Discovering Statistics Using SPSS*, 3rd ed.; Sage Publications Ltd.: London, UK, 2009; p. 857.
54. PUN 2000. Life + Management. Available online: http://www.natura2000.si/en/life-management/ (accessed on 18 August 2018).
55. Sotirov, M.; Lovric, M.; Winkel, G. Symbolic transformation of environmental governance: Implementation of EU biodiversity policy in Bulgaria and Croatia between Europeanization and domestic politics. *Environ. Plan. C Gov. Policy* **2015**, *33*, 986–1004. [CrossRef]
56. Ferranti, F.; Turnhout, E.; Beunen, R.; Behagel, J.H. Shifting nature conservation approaches in Natura 2000 and the implications for the roles of stakeholders. *J. Environ. Plan. Manag.* **2014**, *57*, 1642–1657. [CrossRef]
57. Kangas, A.; Heikkilä, J.; Malmivaara-Lämsä, M.; Löfström, I.; Case, P. Evaluation of a participatory urban forest planning process. *For. Policy Econ.* **2014**, *45*, 13–23. [CrossRef]
58. Domínguez, G.; Tena, J. Monitoring and evaluating participation in national forest programmes. The Catalan case. *Schweizerische Zeitschrift fur Forstwesen* **2006**, *157*, 438–444. [CrossRef]
59. Turnhout, E.; Behagel, J.; Ferranti, F.; Beunen, R. The construction of legitimacy in European nature policy: Expertise and participation in the service of cost-effectiveness. *Environ. Politics* **2015**, *24*, 461–480. [CrossRef]
60. Eben, M. Public participation during site selections for Natura 2000 in Germany: The Bavarian case. In *Stakeholder Dialogues in Natural Resources Management*; Stollkleemann, S., Welp, M., Eds.; Springer: Berlin/Heidelberg, Germany, 2006; pp. 261–278, ISBN 978-3-540-36916-5.
61. Kluvánková-Oravská, T.; Chobotová, V.; Banaszak, I.; Slavikova, L.; Trifunovova, S. From government to governance for biodiversity: The perspective of central and Eastern European transition countries. *Environ. Policy Gov.* **2009**, *19*, 186–196. [CrossRef]
62. Lecomte, N.; Martineau-Delisle, C.; Nadeau, S. Participatory requirements in forest management planning in Eastern Canada: A temporal and interprovincial perspective. *For. Chron.* **2005**, *81*, 398–402. [CrossRef]
63. Gane, M. *Forest Strategy: Strategic Management and Sustainable Development for the Forest Sector*; Springer: Godalming, UK, 2007; ISBN 9781402059643.
64. Geitzenauer, M.; Hogl, K.; Weiss, G. The implementation of Natura 2000 in Austria—A European policy in a federal system. *Land Use Policy* **2016**, *52*, 120–135. [CrossRef]
65. Weber, N. Participation or involvement? Development of forest strategies on national and sub-national level in Germany. *For. Policy Econ.* **2017**, *89*, 98–106. [CrossRef]
66. Hogl, K.; Kvarda, E. *The Austrian Forest Dialogue: Introducing a New Mode of Governance Process to a Well Entrenched Sectoral Domain*; InFER, University of Natural Resources and Applied Life Sciences: Vienna, Austria, 2008.

forests

MDPI

Article

Approaching Local Perceptions of Forest Governance and Livelihood Challenges with Companion Modeling from a Case Study around Zahamena National Park, Madagascar

Nathalie Bodonirina [1], Lena M. Reibelt [2], Natasha Stoudmann [3], Juliette Chamagne [3], Trevor G. Jones [4], Annick Ravaka [1], Hoby V. F. Ranjaharivelo [1], Tantelinirina Ravonimanantsoa [1], Gabrielle Moser [3], Arnaud De Grave [5], Claude Garcia [3,6], Bruno S. Ramamonjisoa [1], Lucienne Wilmé [7,8] and Patrick O. Waeber [3,*]

[1] ESSA Forêts, Ankatso, BP 175, Antananarivo 101, Madagascar; nathaliebodonirina@gmail.com (N.B.); ravakaannick@gmail.com (A.R.); ranjarivelo@gmail.com (H.V.F.R.); tantely.buk@gmail.com (T.R.); bruno.ramamonjisoa@gmail.com (B.S.R.)

[2] Madagascar Wildlife Conservation, Lot 17420 bis, Avaradrova Sud, Ambatondrazaka 503, Madagascar; reibelt.lena@gmail.com

[3] Forest Managament and Development Group, Ecosystems Management, ETH Zurich, Universitätsstrasse 16, 8098 Zurich, Switzerland; n.stoudmann@hotmail.com (N.S.); juliettechamagne@gmail.com (J.C.); gabi.moser1@gmail.com (G.M.); claude.garcia@usys.ethz.ch (C.G.)

[4] Department of Forest Resources Management, Forest Sciences Centre, 2045–2424 Main Mall, University of British Columbia, Vancouver, BC V6T 1Z4, Canada; trevor.jones@ubc.ca

[5] Association Bricolages Ondulatoires & Particulaires, 37 lot. Champ du Bourg, 38570 Goncelin, France; arnaud.de.grave@gmail.com

[6] UR Forêts et Société (F&S), Département Environnement et Sociétés du CIRAD, Campus International de Baillarguet, 34398 Montpellier CEDEX 5, France

[7] Missouri Botanical Garden, Madagascar Research & Conservation Program, BP 3391, Antananarivo 101, Madagascar; lucienne.wilme@mobot-mg.org

[8] World Resources Institute, Madagascar Program, BP 3884, Antananarivo 101, Madagascar

* Correspondence: patrick.waeber@usys.ethz.ch; Tel.: +41-44-632-49-92

Received: 24 August 2018; Accepted: 3 October 2018; Published: 10 October 2018

Abstract: Community-based natural resource management (CBNRM) is a widely used approach aimed at involving those utilizing resources in their management. In Madagascar, where forest decentralization has been implemented since the 1990s to spur local resource users' involvement in management processes, impacts remain unclear. This study aimed to investigate farmers' perceptions and practices regarding forest use under various forest governance systems, using a participatory gaming approach implemented in the Zahamena region of Madagascar. We report on (i) the conceptual models of the Zahamena socio-ecological system; (ii) the actual research tool in the form of a tabletop role-playing game; and (iii) main outcomes of the gaming workshops and accompanying research. The results allow the linking of game reality with real-world perceptions based on game debriefing discussions and game workshop follow-up surveys, as well as interviews and focus group research with other natural resource users from the study area. Results show that the Zahamena protected area plays the role of buffer zone by slowing down deforestation and degradation. However, this fragile barrier and CBNRM are not long-term solutions in the face of occurring changes. Rather, the solution lies in one of the main causes of the problem: agriculture. Further use of tools such as participatory gaming is recommended to enhance knowledge exchange and the development of common visions for the future of natural resource management to foster resilience of forest governance.

Forests **2018**, *9*, 624

Keywords: participatory modeling; role-playing games; stakeholder engagement; transdisciplinary research; slash-and-burn agriculture; deforestation; forest degradation; community-based forest management

1. Introduction

Agriculture is one of the main drivers of global deforestation, and shifting agriculture is responsible for 92% of tree cover loss in Africa [1]. Land conversion for agricultural production affects water, soil, air, forests, and many ecosystems [2]. It caused the clearing or conversion of 70% of grasslands, 50% of open woodlands, 45% of temperate dry forests, and 27% of tropical forests, with some five to ten million hectares of forest annually being lost to agriculture [3] or 39 million ha in Africa for the period of 2001–2015 [1]. Former studies showed that the combination of population growth and shifting cultivation is the main cause of deforestation in pantropical regions [4]. A recent study showed that tree cover increased globally between 1982 and 2016 due to a net loss of forest cover in the tropics and a net gain in the extra-tropics [1].

In Madagascar, less than 18% of terrestrial surface was covered by forests in early 2000 [5]. These ecosystems contain a high degree of biodiversity, being home to more than 12,000 endemic plant species [6] and over 700 endemic species of vertebrates [7,8]. The tree cover is continuously decreasing, with a steep increase since 2013 and a maximum in 2017 [9,10]. Deforestation at the national scale remains unchecked with some 100,000 hectares lost annually. Average deforestation rates have been on a progressive rise since 2005, where annual nationwide rates translated to around 1.5% [11]. In certain areas of Eastern Madagascar, the rates of humid dense forest loss surpassed four percent in 2017 [9]. Forest ecosystems and their biodiversity are, therefore, highly threatened due to ongoing deforestation and degradation despite concerted international conservation efforts over the past 30 years [12,13]. A mix of proximate and ultimate factors were identified linking forest loss to human activities, such as agricultural and pastoral expansion [8,14,15], cash-cropping [14,16], changing international commodity prices [16], increased access through new roads to formerly remote areas [17], or large-scale agri-business [14]. Farmers' slash-and-burn practices (Malagasy *tavy*) heavily participate toward the depletion of forest biodiversity [18–22]. These practices are, however, necessary for the great majority of the population, as they often lack alternative options in securing their livelihoods [23–25].

The positive outcomes of community-based natural resource management (CBNRM) still need to be proven [26]. According to Mesham and Lumbasi [27], decentralization is a foreign idea in many countries, and had the opposite impact to the intended empowerment of communities, thus hampering partnerships at the cost of conservation. Pollini and Lassoie [28] put similar critiques forward in the case of Madagascar, where the World Bank, together with international backers and non-governmental organizations (NGOs) began pushing the Malagasy government to start establishing and implementing forest decentralization in the late 1990s in the hope of slowing deforestation and degradation rates while conserving biodiversity, in the form of VOIs (*vondron'olona ifotony*, an association of rural stakeholders that entered into management contract in the GELOSE process) [13,29,30]. The GELOSE (*Gestion Locale Sécurisée*, secure local management) policy, which regulates the transfer of property rights to local communities, is integrated into the national forestry policy (Law 97-107 and Decree 97-1200), and is applicable to forests, pastures, water, and wildlife. However, because implementing the GELOSE was deemed cumbersome (e.g., Reference [31]), the Contractual Forest Management (GCF, *Gestion Contractualisée des Forêts*) was issued as an enabling legislation with focus on forests (Order No. 2001-122). The GCF aimed to reduce the contract signatories from a tri-partite agreement to the State (Ministry of Environment, Water, and Forests), and to the community residents (VOI) [32–34]. However, its implementation was widely criticized due to its top-down nature. The interests of the actual community management associations (local communities) were being trumped by external environmental mediators (international NGOs) driving the process (e.g., References [28,35]), despite the fundamental necessity of CBNRM to be a partnership, in this case between local communities and the Forest Administration. The large-scale implementation of community-based natural resource management agreements—to date, some 450 agreements were signed across Madagascar [36]—still leaves many households with little incentive to protect their natural resources [22,37].

The current form of community-based natural resource management (CBNRM) in Madagascar is a type of co-management where responsibilities are shared among various stakeholders, all of whom are committed to ensuring that the local communities are actively involved in the management of the natural resources. The *cantonnement forestier* is part of the Forest Administration and is the institution responsible for the management of the forests under its jurisdiction. The forest administration formalizes the contracts of management transfer to the local communities, or VOIs. It supervises and advises the VOI and assists the judicial police role. Through the transfer of management, some forest management responsibilities are delegated to the local communities, including those pertaining to user rights of the forests. The VOIs set surveillance patrols, apply the *dina* (local rules) when possible, or report to the forestry administration otherwise. In the case of protected areas managed by the Madagascar National Parks (MNP), a management unit can invite members of a local community; they are grouped in a *Comité Local du Parc* or CLP to participate in the protection of natural resources by joining surveillance patrols or zoning tasks. The MNP can also initiate a transfer of an area of forests to local communities to strengthen the protection of a park, but final approval and the official contract linked to the transfer remain the responsibility of the Forest Administration (cf. [38] for more details on management structures in relation to protected area management). It remains unclear whether community-based natural resource management contributed to a slowing down of deforestation and degradation at the local scale [39,40].

There are claims of transdisciplinary and participatory approaches to natural resource management that overcome sectorial divides in the context of decentralized governance [41]. Many of these concepts, however, lack operability [42]. More experience is required to help practitioners implement such approaches, and particularly to understand how information and research can feed decision-making at individual, community, and political levels, and ultimately be incorporated into the policy process [43–45]. Potential barriers are lacking guidelines and technical expertise. How to go about breaching the science–policy–practice divide is still debated [11,46], and most are simply "muddling through" (cf. Reference [47]). One of the conditions that international donors imposed on the government of Madagascar is decentralization [12]; however, in fact, the decision-makers themselves lack guidelines on how to implement decentralized territorial governance spatially and temporally [13]. Since the devolution of forest governance in the 1990s in an act to give rural people more responsibilities over their resources (cf. Reference [48]), the number and interactions of stakeholders involved at various levels of forestry increased in complexity. Forest governance now has all the ingredients of a wicked problem (sensu Reference [49]), where a mélange of value and knowledge systems meet, i.e., where international, national, and local interests are competing over the same resources, over access (or lack thereof), user rights, and over decision power. Reasons for, or drivers of, deforestation are found in a complex interrelationship of socio-economic and political factors [4]. To better understand forest and other land use by rural resource users, we investigate here farmers' practices and perceptions under different forest governance systems.

This research is part of the AlaReLa project (Alaotra Resilience Landscape, 2013–2017) based on a transdisciplinary approach where a multitude of research methods were applied to facilitate participation and involvement of local stakeholders and regional decision-makers in the Maningory watershed. In this paper, we focus on participatory gaming based on companion modeling (ComMod), a participatory modeling approach aimed to tackle natural resource management issues [50] around Zahamena National Park. Given the complex nature of this natural resource management (NRM) problem (forest governance regimes), and its interactions with various stakeholders, our research approach is exploratory in nature; through our gaming approach, we first try to understand the system under scrutiny (which is Objective 1 of ComMod, cf. [50] to see whether any patterns or surprises may emerge. Secondly, this spurs discussions revolving around these outcomes (governance/livelihoods). The engagement with various stakeholders (as done in this research), especially with groups that usually go unheard (e.g., subsistence farmers), combined with the sharing of results (ongoing), has the potential to better involve and inform the actual decision-makers and their policymaking at various levels (Objective 2 of ComMod, cf. [50]).

This contribution centers on a forest governance game developed in and for the Zahamena region, eastern Madagascar. Our overarching research objectives were to explore farmers' perceptions of forest use and governance in the light of CBNRM and its impacts in the Zahamena region, which currently remain unclear. Our two working questions were as follows: (i) (In what way) does the management system impact farmers' (gaming) decisions and behavior? (ii) (In what way) does the management system have an impact on (virtual and real-life) forest status/deforestation? In this publication, we report on the actual research tool, a role-playing game (RPG). The results focus on (i) the co-constructed conceptual models of the Zahamena socio-ecological system (interactions of actors and resources, land-type dynamics), which informed and guided the collaborative development of the RPG; (ii) main outcomes of the gaming workshops (virtual world), including game debriefing and discussions addressing game reality and decisions (with a focus on stakeholder perspectives on the governance challenges accounting for deforestation); (iii) land-cover mapping to gain insights into forest cover change in the past 25 years; (iv) real-world perceptions and narratives of resource users, linking the game experience with the real world. Further interviews, focus groups, and "ethnophotography in environmental science"—an approach using ethnography-inspired photography to explore landscape and land-use perceptions of local resource users [51]—were used in complement.

2. Materials and Methods

2.1. Study Area

Our research focuses on the Zahamena forest (Figure 1), which, in most parts, is under a protected area management regime. Zahamena is located in the east of the country, at the nexus of the Alaotra-Mangoro and Analanjirofo regions. The region of Lake Alaotra represents an agricultural hub as a center of rice production, and the Analanjirofo as its vernacular name implies (*jirofo* means clove) is an important region in cash crop production. Zahamena was the third Strict Nature Reserve of Madagascar established 31 December 1927. A central enclave was part of the reserve in 1927, and the continued presence of several villages over time led to the area of the enclave being degazetted in 1966. These villages kept growing and a further area was degazetted when Zahamena was partly turned into a National Park in 1997 (International Union for Conservation of Nature (IUCN) category II; the idea was to open the Strict Nature Reserve and turn it into a National Park, thus allowing visits not only from tourists but also—and especially—communities). In 2007, Zahamena was amongst six rainforest sites listed as a World Heritage Site, known as the Rainforests of Atsinanana. In 2015, the total Zahamena area became a National Park, and today, it covers a total area of 64,935 hectares under the management of Madagascar National Parks (MNP, formerly ANGAP). MNP agents ensure the ecological monitoring and conduct forest patrols. The local forest patrols (CLP, *Comité Local du Parc*, usually villagers from around Zahamena) support the MNP in forest monitoring. The VOI is responsible for the forest and *dina* (local rules) monitoring in the community-based forest areas. The association reports directly to the Forest Administration (*chef de cantonnement* or *chef du cantonnement forestier* as a subdivision of Forest Administration at the district level) and deals with usage requests from the community. There are currently 14 VOIs around Zahamena, six of which are in the commune of Antanandava. The two communes of Antanandava and Vavatenina mostly depend on agricultural production for both subsistence and markets, with a majority of villagers engaging in self-subsistence activities. Many do have more than just one activity, which is typical for the region (e.g., Reference [52]).

Zahamena National Park has a zoning plan; this *plan de gestion* is composed of the core zone for conservation (*noyau dur*) and the buffer zone (*zone tampon* with ZOC—*zone à utilisation contrôlée*, ZUD—*zone à utilisation durable*, and ZS—*zone de service*). The local population is officially not granted access to the conservation area, but they can extract wood for personal use and cultivate in the buffer zone. The VOIs are supposed to manage these activities, but due to them lacking capacity and staffing, VOIs and the MNP are not able to ensure control of the geographic boundaries of the different zones, resulting in illegal logging, slash-and-burn cultivation, and lemur and other bushmeat hunting. The management and conservation of in situ biodiversity and ecosystems is anchored in the protected area legislation *Code des Aires Protégées* (N. 848-05/N. 2001-05). It was legalized in 2003 and revised in 2015 under the *Refonte du Code des Aires Protégées* (N. 2015-005). This was supplemented with an updated protected area code, the Updated Malagasy Environment Charter (*Charte de l'Environnement Malagasy actualisé* N. 2015-003).

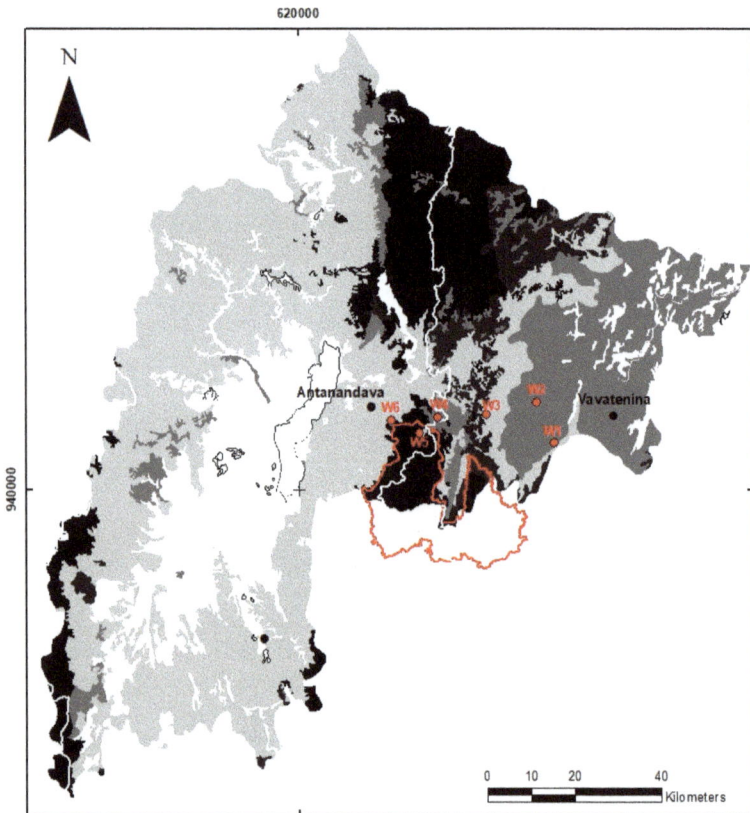

Figure 1. Study area within the Maningory watershed, with locations of the six game workshops (W1–6). Humid dense forest (black), degraded forest (dark grey), agricultural matrix (grey), and grasslands, including wooded grasslands (light grey), other land types such as water bodies, wetlands, and irrigated rice fields (white). Contours in red show National Park boundaries of Zahamena, and grey delimits Lake Alaotra, where other interviews to this research were conducted. The white north–south dividing line is the political boundary between the two districts, Analanjirofo in the east and Alaotra-Mangoro in the west. Land types are based on Landsat imagery classification using Labord projection (LC8158072_0722014236; LC8158073_0732014236; LC8159072_0722014259 Courtesy of the United States (US) Geological Survey).

2.2. Companion Modeling and Role-Playing Games

To explore farmers' decisions and perceptions of forest use and governance in the light of CBNRM, we co-developed a role-playing game following the companion modeling (ComMod) approach. ComMod is a participatory modeling approach [50] used to develop shared visions (mental models) of socio-ecological systems (e.g., Figure 2 in the Section 3). This means that all relevant stakeholders share their own reality, i.e., their perception of the system at hand, and how it works. The resulting model contains more information than single stakeholders' initial mental models. The co-development of these shared representations of reality already initiate a learning process, and can then serve for research and stakeholder engagement processes in resource management planning, outreach, negotiations, and policy decisions. The approach was proven to be a suitable and helpful negotiation and planning tool in various forest and agriculture settings (e.g., [53–57]. Thus, ComMod is more than a research approach; it enables mutual learning, dialogue, promotes collective decision-making, and can guide co-development of (alternative) management plans.

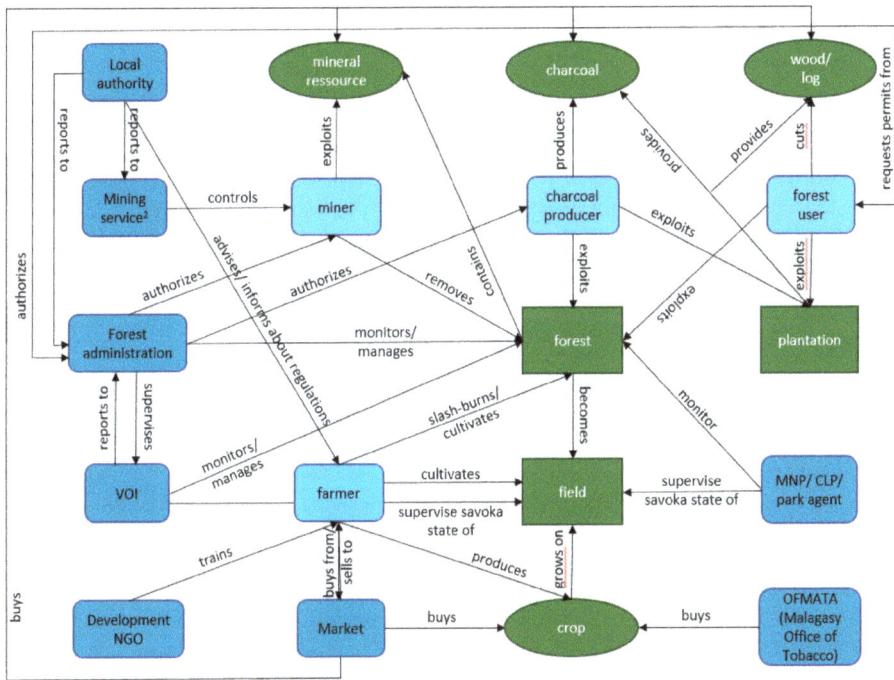

Figure 2. Conceptual model of the Zahamena socio-ecological system showing the interactions of actors and resources. Actors represent either individuals (light blue) or institutions (dark blue); resources are either primary (square boxes) or commodities (ellipsoids). The verbs on the arrows describe the main interactions between two entities.

The first step in this complex participatory process (ComMod) is the definition of a problem together with concerned stakeholders. Following this, its related actors, resources, dynamics, and interactions (called ARDI; cf. Reference [58]) are identified collectively. The emerging model contains social, ecological, and economic dimensions, and can, in a next step, be translated into a tabletop role-playing game (RPG;). This RPG then serves as a tool to explore and discuss complex interactions and feedback loops in the system at hand, and can involve the exploration of potential future scenarios. As such, the RPGs are suitable for stakeholder involvement, exchanging information, promoting negotiation processes, and serving science and policy as a valuable tool for understanding stakeholder decisions and preferences in natural resource management realms.

2.2.1. Development of an RPG

The development of a game consists of two main phases: a diagnostic phase and a gamification phase, which is the transformation of mental models into a game with a game board, rules, and players. The diagnostic phase served to identify actors, resources, dynamics, and interactions [37] of forest decline and land conversion. This diagnostic phase entailed a multitude of approaches including focus groups and interviews with different stakeholders, and ethno-photography of sciences (details on content and participant selection are specified in Appendix A.1). One of the photography methods used to collect these narratives was created for the project [51]. It was aimed at obtaining a corpus of visual narratives shared between the photographer (necessarily biased) and the people (or landscape) portrayed. During "iteration workshops", a set of pictures taken during previous visits to the area were analyzed by the participants, and missing aspects identified by them were added afterward. The interactions between the participants and the photographer gave a complement of information by addressing the same issues as the scientific methods used, but from a collaborative and artistic perspective.

The obtained information was used to develop a mental model of the Zahamena socio-ecological system (SES) and to identify the dynamics in the system such as regrowth and fallow stages after burning and cultivation. These conceptual models served to develop the components and rules of the RPG (role-playing game; see a detailed description of the RPG used in this research in Appendix A.2): Actors were turned into players, resources into tokens, and dynamics into rules. After an extensive testing phase for calibration and parameterization based on stakeholder engagements (Appendix A.1), a total of six full-day workshops with 30 participants took place, where two scenarios were played with respective debriefings. Follow-up surveys were performed the day after. The research team facilitating a game workshop comprised four researchers sharing facilitation, note-taking, and game-updating tasks.

During a game, which consists of five rounds, five players make decisions on land-use activities to pursue their livelihoods. They can do a variety of activities such as agriculture (rice, maize, beans, tobacco, cassava, and clove) and use small or big amounts of fertilizer to improve outputs. Players can further invest in charcoal production, wood exploitation, mining, install firebreaks to prevent propagation of fires, and do reforestation. The game board (Figure A2) represents the typical forest-dominated Zahamena landscape. The goal for the players is to make a living, i.e., to be able to pay their household's livelihood costs at the end of every round. How they achieve this depends on their personal preferences and decisions. The game has two scenarios, each consisting of five turns, representing five years each. The two different scenarios were developed to be able to observe and discuss different strategies and decisions based on different forest management regimes. In Scenario 1, all 35 forest plots of the "gamescape" (the game board representing the landscape) are state-owned. In Scenario 2, different governance systems underlie the forest plots: Three cells remain state forests, eight plots are under a community-based regime (VOI), and 24 cells represent the protected area (Zahamena National Park). We developed the two scenarios to explore and understand the impacts these different land management regimes may or may not have on the behavior, land use, and/or livelihood strategies of local people. A detailed description of the game components and functioning is provided in Appendix A.2.

2.2.2. Game Workshop Structure

One gaming workshop lasted a whole day: the game sessions for both scenarios were divided between morning and afternoon. After a brief introduction of the research team and the gaming approach, participant information was collected (gender, age, schooling, and livelihood activities). After explaining the course of the game to the players, scenario 1 was played, followed by a small debriefing discussion that addressed players' feelings about the game and about results during the first scenario (what happened on the game board), reasons for the activities pursued or not (mostly in the forest), and the problems or constraints they encountered during the game. After a lunch break, the facilitator explained the landscape changes and new rules before playing scenario 2. The second debriefing contained additional aspects as compared to the first debriefing, including comparisons between scenarios 1 and 2, the adopted strategies for the different scenarios, the changes observable on the game board, and, most importantly, the transition to reality from what they see on the game board or what they encountered during the game. At the end of the workshop, follow-up surveys were scheduled for the next day, with the team visiting the individual participants at their homes to address additional questions surrounding daily life activities, forest management, landscape changes, and governance.

2.2.3. Game Workshop Participants

A total of 30 resource users participated in the six game workshops, conducted along the Vavatenina–Antanandava transect (Figure 1). Workshops 1–3 were conducted far away from Zahamena (over 1–2 days walking distance) in the district of Vavatenina, where there is almost no dense forest remaining. The forests available to these villages are only vestiges of forest-covered landscapes from long ago. Workshops 4–6 took place in villages close to the Zahamena forest block, in the commune of Antanandava, where people still have daily access to significant forest resources (from less than an hour to two hours walking distance from Zahamena National Park). The selection criteria for participants were (i) local residence, (ii) agriculture as main livelihood activity, and (iii) aged at least 18 years. Additionally, participants were not to be from the same households, and particular attention was paid to having a balanced gender representation.

At sites 1–4, participants were from the ethno-linguistic group of the Betsimisaraka, and, at sites 5 and 6, all players were a mix of Betsimisaraka and Sihanaka. Primary activities in sites 1–3 were cash crop cultivation (clove, coffee, and vanilla) and shifting rice cultivation, while the landscape was dominated by clove fields and *savoka* (seral stage of regeneration after slash-and-burning). There was only a small surface area suitable for irrigated rice. In sites 4–6, the primary occupation was farming of irrigated and rainfed rice, as well as bean and cassava production. The landscape was characterized by dense forest, *savoka*, and the presence of grasslands and wooded grasslands.

The age of participants ranged from 18 to 80 years (split in the following age categories: ≤20 (1); 21–30 (6); 31–40 (5); 41–50 (7); 51–60 (7); 61–70 (2); 71–80 (2)). Of the 21 men and nine women, the educational levels ranged from not identified (2), illiterate (2), primary level (14), secondary level (8), high school (3), to university (1).

2.3. Land-Cover Mapping and Dynamics Assessment

For this article, we used Landsat imagery to identify real-world trends in forest cover change, and to allow for a comparison with landscape development in the game setting (one scenario representing 25 years). For this, we acquired two Landsat images from the United States Geological Survey (USGS) Earth Resources Observation and Science Center (Sioux Falls, SD, USA). The images provide coverage over the study area for 19 June 1990 and 24 August 2014 and were acquired atmospherically corrected to Level-2 surface reflectance [59]. Images were selected approximately 25 years apart, as constrained by cloud cover. While the 1990 image was cloud free, the best available image circa 2012–2016 contained some cloud-cover. Landsat data were selected as they are freely available, provide a moderate (i.e., 30-m) spatial resolution, and have a proven capacity for detecting forest cover disturbance [60]. An unsupervised iterative self-organizing classification algorithm (i.e., ISOCLUST; Clark Labs: Worcester, MA, USA) was used on the 2014 image to group pixels into dominant cover types based on their shared spectral properties and to target areas dominated by shadow and/or cloud. Areas dominated by cloud and the boundaries of the study area were applied as a mask to both dates of imagery.

Land-cover types were defined based on an aggregation and augmentation of existing categories (i.e., dense forest, degraded forest, wooded grassland, grassland, agriculture mosaic, rice agriculture, and water vegetated swamp) into spectrally discernible classes (i.e., dense forest, sparse/degraded forest, grass/shrub-dominant, soil-dominant, water, and vegetated swamp/wet agriculture (e.g., rice)). Examples of target land-cover categories were initially located with reference to ground data collected by the AlaReLa project in October 2015. Additional examples of target classes were located with reference to an existing national-level land-cover map (i.e., Reference [61]), fine-spatial-resolution satellite imagery viewable in Google Earth Pro, and the appearance/spectral values of the Landsat data. For all classes, training and testing sites were derived to calibrate and validate an image classification algorithm. The maximum-likelihood classification algorithm was selected given a proven track record for classifying different types of forest cover [62]. Resulting map accuracy was quantified using a confusion matrix to cross-tabulate with independent testing (i.e., validation) data. The Kappa index of agreement (KIA) assessed how better than random the map was [63]. Due to a lack of historical reference data, the 1990 classification was not validated. The mapping results were used to establish historic (i.e., 1990) and contemporary (i.e., 2014) forest extent and quantify dynamics (i.e., loss, persistence, and gain) from 1990 to 2014 using the Idrisi Land Change Modeler (Clark Labs, Worcester, MA, USA).

2.4. Analysis of Gaming and Debriefing Data

Before, during, and after the gaming sessions, various qualitative and quantitative data were collected. Qualitative data were participant statements, noted and audio-recorded during discussions and debriefings (as described in Section 2.2.2). We screened the obtained qualitative data for exemplary statements, i.e., to represent the range of different perceptions, strategies, and realities.

Quantitative data were collected with data sheets during both scenarios, and captured participant behavior and decisions such as type, number, and location of activity (cultivation of rice, maize, beans, tobacco, cassava, clove; use of small or big amounts of fertilizer; charcoal production, wood exploitation, mining, firebreak installation, and reforestation), as well as the consequences (production, i.e., income/livelihood and forest cover; see also Table A4 for detailed lists of variables measured during gaming).

For statistical analysis, livelihood and forest cover over time showed normally distributed errors. Therefore, we used two-sample *t*-tests to analyze the effect of the scenarios. To investigate the interactive effects of the scenario and workshop on these, we used linear models. All variables related to players' strategies showed non-normal error distributions. Non-parametric Wilcoxon rank sum tests were, therefore, used to check for differences between scenarios. All analyses were done using the R software version 3.0.3. [64].

3. Results

This section is based on outcomes of the transdisciplinary approach described in the methodology section. In the first part of this section, we disentangle the interactions between stakeholders and their resources, and land-type dynamics, i.e., what are the responses of land-cover types to the interactions described. These results stem from the diagnostic phase, including interviews, surveys, and focus groups. In a second part, we juxtapose game outcomes (virtual world) with Landsat-based land-type dynamics and with outcomes from discussions (stemming either from debriefings, interviews, or focus group meetings), i.e., real-world perceptions of the stakeholders. When quotes are used, a code indicates their origin (FG = focus group, I = interview, W = (game) workshop, and P = player).

3.1. Understanding the Socio-Ecological System

3.1.1. Interactions of Actors and Resources

The main local actors in the socio-ecological system of the Zahamena region are farmers, charcoal producers, and miners, as well as institutional actors such as the Forest Administration, mining service, traditional authorities, and NGOs. Different forms of land use, such as wood exploitation or the establishment of farmland, transform dense forest into other land-use forms and provide cash crop and food crop to the local population. Authorities and institutions mainly hold indirect roles in supervising resource use and the application of laws (Figure 2).

3.1.2. Land-Type Dynamics

Dense forests turn into *ala lany* (exploited forest) when precious woods and construction timber are harvested. When farmers establish fields for farming, forest is burned, mostly by practicing *tavy* (slash-and-burn agriculture). Amongst Betsimisaraka, *tavy* is also the name of the resulting field that is used for rice cultivation or a mixed culture consisting of rice (the main crop) and beans or corn. In the second year, farmers have the choice to either cultivate rice or mixed culture once more, or another annual culture such as maize or cassava, or a perennial culture such as coffee, clove, or litchi. Betsimisaraka use the word *tsabo* for rain-fed crops (annual or perennial, except for rice) and for the terrain these grow on. *Tsabo* does not require burning to prepare the ground for cultivation; a simple exposure of the superficial part of the soil (*kapakapa*) or a light ploughing similar to the system of cultivation under vegetal cover is enough. In contrast, a burn always precedes the *tavy*, and as the farmers usually do not use fertilizer, rice is cultivated in the first year when soil fertility is highest (Figure 3).

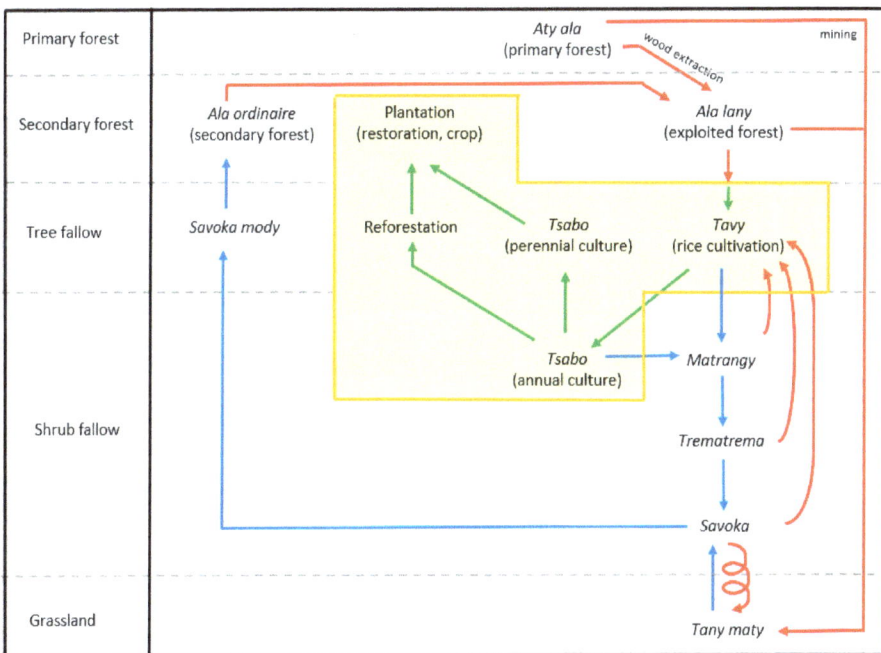

Figure 3. Conceptual model of land-use change and (post-) cultural dynamics in the Zahamena socio-ecological system. Natural succession (left panel) is compared with agricultural practices (yellow box); processes: *tavy* and/or cutting (red arrows), cultivating (green arrows), and resting periods/fallows (blue arrows). Local Betsimisaraka terms are in italics.

After three or four years of cultivation, farmers leave the plot fallow for at least three years and then redo the *tavy* to cultivate it again. Generally, farmers only plant perennial crops such as cloves once the quality of the soil diminishes and no longer gives a satisfactory annual crop production, particularly if they do not have much land. When fields are no longer cultivated, different fallow stages follow, such as *matrangy* in the first year of regeneration (1–2 years), *trematrema* with higher vegetation (2–15 years), and *savoka* with high and dense vegetation, including woody vegetation (3–15 years). After 10–20 years, *savoka mody* develops, which turns into secondary forest after (20–)50–60 years (time data from field work and Reference [21]). This secondary forest can then form the starting point for a new cycle (Figure 3). Restoration is also a possibility. While this leads to the development of plantations, annual cultures enter the fallow cycle, unless they are cultivated again. If burning cycles move closer together, i.e., fallows are no longer allowed or are burned at an earlier stage than *savoka*, the risk that the terrain becomes uncultivable (*tany maty*) increases. This also happens after mining, when the nutritious layers of the soil are removed or dislocated.

3.2. Forest Change, Livelihood Challenges, and Governance

3.2.1. Virtual World—Gaming

Though the game represents an artificial world, it still has meaning to the players. Impacts of actions and decisions on how to use (or not use) a land unit are directly visualized on the game board (Figure 4).

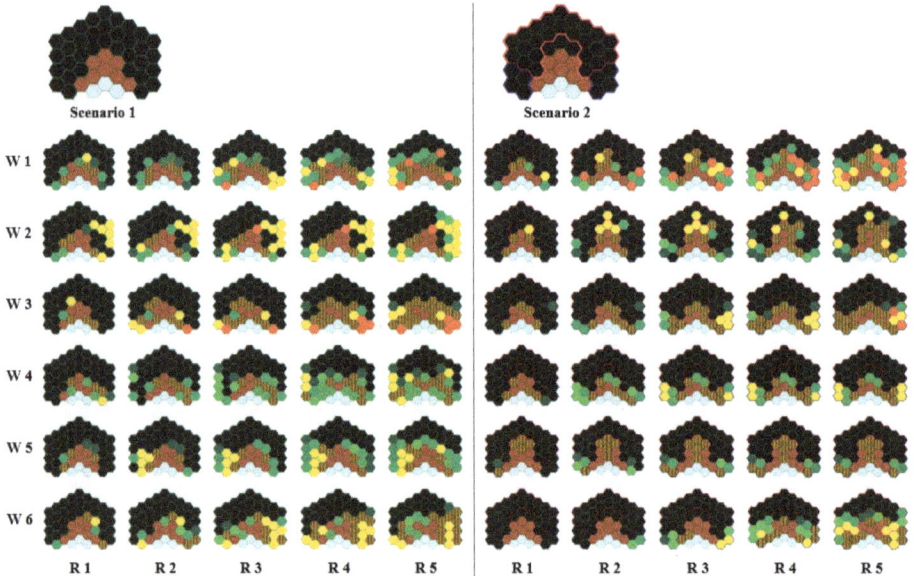

Figure 4. Game landscape (gamescape) change over the course of 25 years, comparing scenario 1 (left) vs. scenario 2 (right); W1–6 represent the six game workshops. Over the period of five rounds—five years per round (R)—land types change according to players' decisions and activities. Original landscape for both scenarios (top) is composed of 35 dense forest cells (dark green), eight *tanety* cells (brown), and three irrigated rice field cells (blue). Scenario 2 has, in addition, a National Park (red contours), and community-based forests (blue contours). The other greens (darker to bright) represent forest degradation (low to high), red represents a completely deforested cell; yellow represents a fallow (*savoka*), green hashed represents reforestation, and brown hashed represents a clove plantation. Details of the actual role-playing game (RPG) are provided in Figure A2.

Players' gaming behavior and its impacts were statistically analyzed. Testing for differences in livelihood and the state of the forest revealed several results. Between scenario 1 and scenario 2, there is a statistically significant difference in the amount of degraded natural forest (more in Scenario 1; Figure 5, Table A4), comprising all forest degradation levels summed together, but not in livelihood. Testing for differences in game activities showed a statistically significant difference between scenario 1 and 2 only in six out of 40 variables (Table A5). However, when testing for differences between workshops 1–3 (situated far from dense forest) and 4–6 (close to dense forest), many variables showed a statistical difference; there were significant differences in livelihood (much higher in W4–6), moderately degraded forest (more in W1–3), very degraded dense forest (more in W1–3), and completely deforested plots (more in W4–6; Table A4). Of the 40 game activity variables, 26 were tested to be significantly different between forest-close and forest-distant workshops (Table A6). When testing for the effect of interactions between the scenario and workshop on land, there was, on average, a significant difference between scenarios 1 and 2, and a slight but non-significant difference between workshops 1–3 and 4–6. There was more dense forest, and less degraded forest in scenario 2 than in scenario 1, and this difference intensified slightly for workshops 4–6. Indeed, in workshops 4–6, there was a bit more forest and a bit less degraded forest in scenario 2 than in workshops 1–3 (Figure 5).

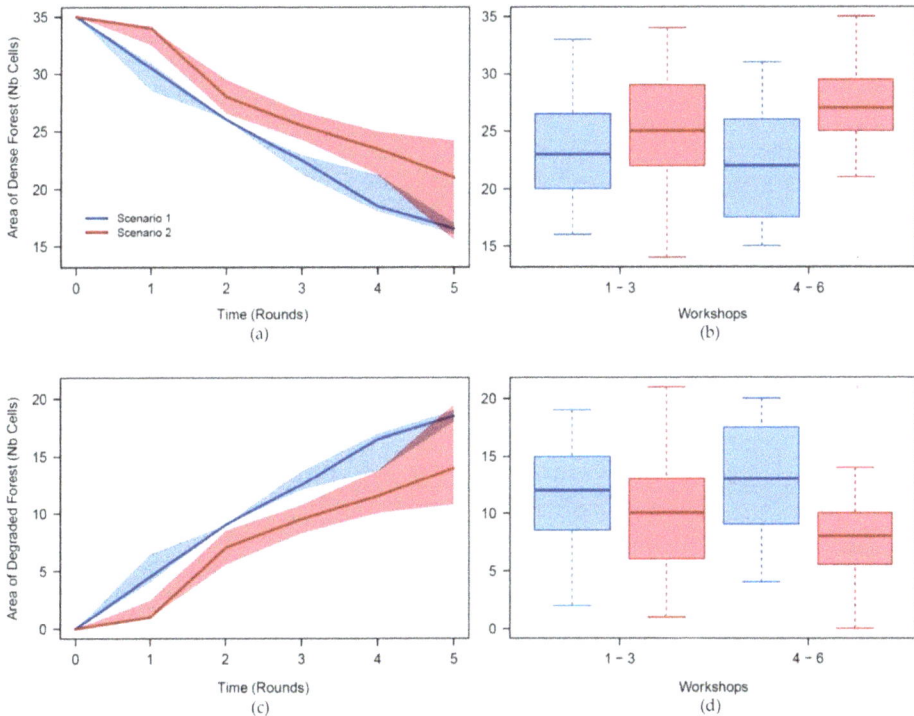

Figure 5. Deforestation and degradation happening in the gamescape. Scenario comparison for the number of plots with dense forest (**a**) and the area of degraded forest (**c**); interactive effects between scenario and workshop for dense forest (boxplots, (**b**)) and for degraded forest (**d**).

Participants of the game workshops stated that they based their gaming behavior mainly on daily life (67%), gaming rules (17%), and on what others did in the game (10%). The rest did not give a reply (6%). Some players added remarks, for example, that the use of fertilizer is not yet common and is rarely done, that activities like cloves are not yet practiced in the sites of the commune of Antanandava (W4–6), and that the exploitation of forest resources for commercial purposes—selective cutting, charcoal production, and mining—is not very frequent. During the game, players have to establish strategies and make decisions regarding which activities to pursue, including the adherence or non-compliance to management rules and laws. Below, discussion outcomes from the debriefing sessions are presented in relation to forest exploitation activities during the game sessions.

The most common reasons for not undertaking charcoal production were the difficulty in obtaining a permit and the fear of being punished.

"I did not produce charcoal for fear of exploiting the forest." (game workshop participant W1P5)

"Even when following (the law), it is difficult to obtain a permit." (W6P5)

Other players did not use the forest [selective cutting] because they lacked permits and/or feared law enforcement.

"This activity requires authorization but I did not have enough money to apply for a permit and was stopped by forest protection." (W4P5)

"I exploited the forest but stopped after being fined by the Forest Administration." (W2P3)

"I thought about it [mining], and then I thought maybe I need authorization from the *chef de cantonnement* and the mining department, but I did not have enough money to start the process." (W4P5)

Reasons for exploiting the forest were mainly income and the development of arable land.

"I cut wood to earn a lot of money." (W2P2)

"To extend cultivable land." (W6P4)

"I have a cutting permit but I have also made illegal cuts." (W4P1)

"Mining can earn you a lot of money if you are lucky." (W2P2)

"I waited until the plot was completely deforested before mining." (W1P2)

The establishment of the protected area in scenario 2 shifted the focus of cultivation to the VOI forests; however, these could not fully cover the needs of the players in the long run, and, once these were exhausted, the neighboring forest cells were increasingly exploited and transformed into farmland.

"The land is not enough, the community forest has been exhausted, the parcels of forest in the park are close to the lands already developed . . . so I still used the parcels of the park." (W3P3)

"The land is not enough; we had to expand farmland. Also, the number of dependents [members of household] continues to increase." (this participant still used the parcels of the park; W3P5)

"When the VOI forest was completely destroyed, it was necessary to enter the protected area. The degradation is done continuously starting from the parcels closest to those already used." (W5P4)

"We must be cautious about the use of reserve [protected areas] resources because even if we follow the law, we do not obtain authorization." (W6P5)

3.2.2. Land-Cover Mapping and Dynamics Assessment

Mapping and dynamics results indicate a general and widespread decrease in dense forest and an increase in sparse/degraded forest (Figure 6). Within the study area, from 1990–2014, approximately 96,607 ha of dense forest was lost, 82,153 ha remained unchanged (i.e., persistence), and 4077 ha were gained. Dense forest primarily transitioned to sparse/degraded forest but also to grass/shrub-dominant and agricultural land-use areas. Dense forest loss occurred almost entirely outside the boundaries of the Zahamena National Park. According to the confusion matrix (Table A7), the 2014 classification resulted in a highly accurate map, with an overall accuracy of 99.6% and a KIA of 0.99.

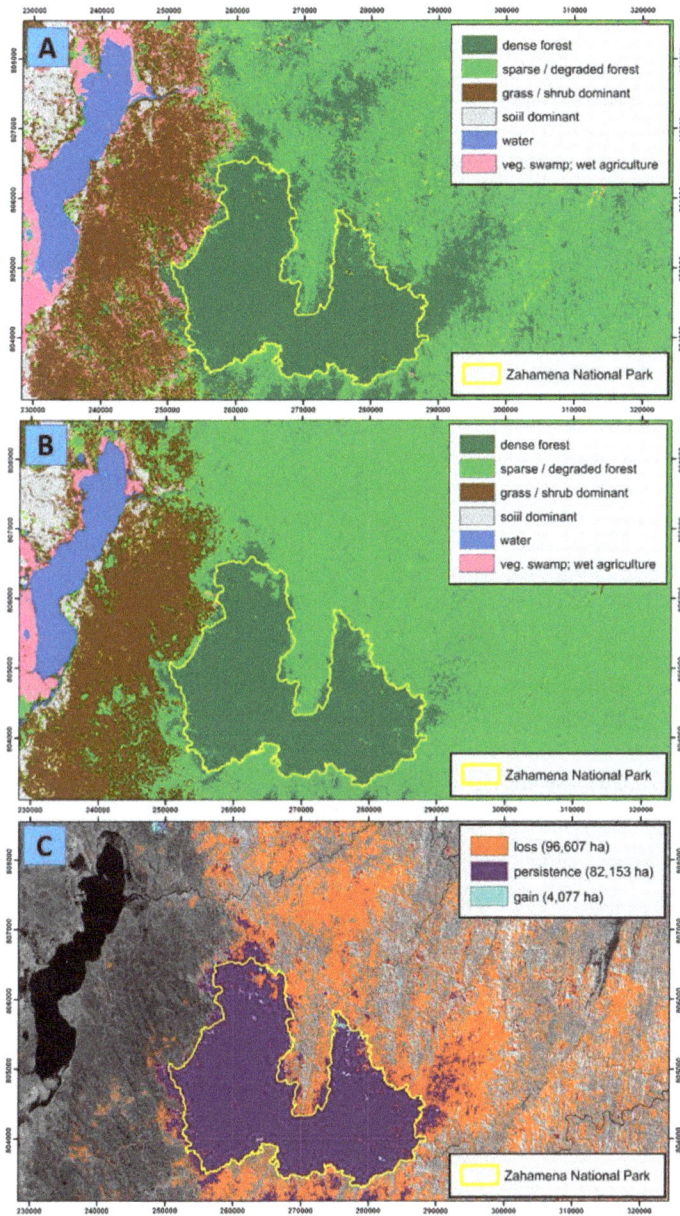

Figure 6. Supervised maximum-likelihood classification results of Landsat imagery showing the distribution of targeted land-cover classes in the study area for June 1990 (Panel A) and August 2014 (Panel B). Panel C shows loss, persistence, and gain in dense forest from 1990–2014 in and around the Zahamena National Park. The background image is the near-infrared (NIR) band of Landsat imagery from August 2014. The boundary for the Zahamena National Park appears in yellow in all panels. The coordinates for all panels are Universal Transverse Mercator (UTM) Zone 39 South.

3.2.3. Real World—Land-Use Narratives and Perceptions

Land-use characterizations allow for a better understanding of local realities, and thus, an improved understanding of gaming results. This section provides detailed insights into individual realities and general livelihood strategies, based on discussions during interviews, game workshop debriefings, follow-up surveys to the game workshops, and focus groups.

The majority of people do not produce their own charcoal, but collect firewood. People closer to the Zahamena National Park realize that finding firewood requires more and more walking due to ongoing deforestation when compared to 20 to 30 years ago; even the degraded forests are not producing enough firewood anymore. People blame the growing population for this situation. For example, the woman selling charcoal (Figure 7) walks about three hours every day from her home to the charcoal production site, then three hours with the charcoal bag on top of her head to the market in Andreba-Gare where she stays until she has sold all the charcoal. One stack of charcoal costs about 100 Ariary (Ar; 0.03 €). The walk back home is also about three hours. Families produce charcoal for the market, where a bag of charcoal costs around 7000 Ar (1.83 €). Firewood collection in dense and degraded forests was free; in recent times, people started selling firewood produced on their plantations. Clove production is very common on the eastern side of Zahamena (district of Vavatenina), where many households possess clove trees. These were introduced by the French colonial powers together with coffee, vanilla, and litchi. It takes seven years for the first production of cloves, and mature trees produce cloves every 2–3 years. The trees are also appreciated for their leaves, which are used for essential oil production during lean periods (Figure 7).

"Growing vanilla and cloves is a good alternative [to rice production or mining], as these two crops can bring in lots of money." (Interviewee I6, Antanandava)

"Before, we grew nothing but rice, but then the foreign foremen taught us how to grow cloves, coffee, and vanilla, and we continued." (W3P1)

"Clove is allotted for export, it generates money even with a small production." (W4P1)

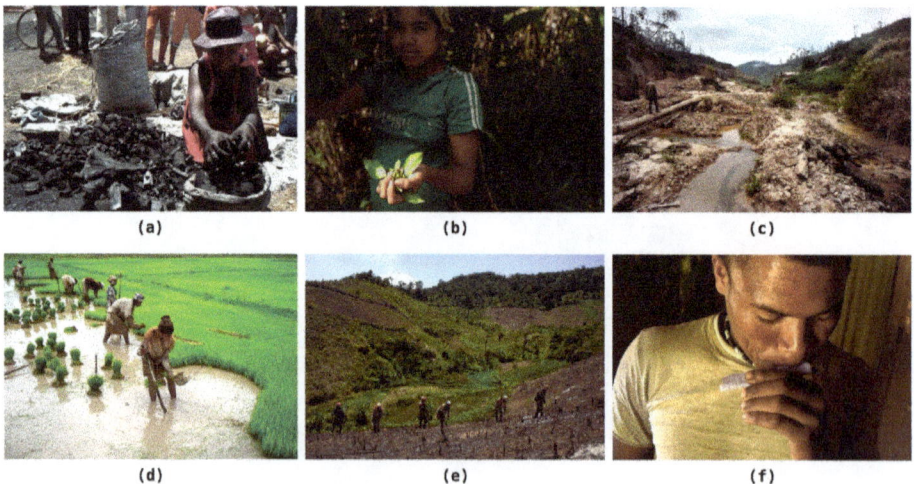

(a) (b) (c)

(d) (e) (f)

Figure 7. Characteristic land use in the Zahamena socio-ecological system. The ethno-photography in environmental science method (cf. Reference [58]) allows for a shared vision of the landscape uses and proposes a visual characterization of user-defined important livelihood activities surrounding the remaining dense forests: charcoal (**a**), clove cultivation (**b**), artisanal small-scale mining (**c**), rice in flooded paddies (**d**), *tavy* (**e**), and tobacco (**f**).

Dense forests are not only disturbed by the actual mining, which involves heavy use of water and digging (Figure 7), but also by deforestation linked to miners' subsistence (e.g., *tavy* for crop and clearing of land for housing; see statements in Appendix A.3). While the RPG offered the opportunity for researchers and locals to explore different management scenarios, the follow-up survey after the game workshops served as a tie to the real world, allowing the elicitation of perceptions of forest governance and enforcement bodies (see also Appendixs A.4 and A.5). The statements stem from the survey questionnaires, and they all relate to the real world (not the game). When players were asked about their opinion regarding the state of forests if there was no protected area, the majority indicated that it would be worse without such a governance system, and that the existence of VOIs prevented people from exploiting forest in the protected area, but only as long as these provided sufficient output. The presence of enforcement bodies was also judged influential. The consensus amongst the game workshop participants was that local management would still be best suited to serve the local population, who has livelihood security as the highest priority.

Perceptions of forest users on forest governance in real life:

"All forest may be destroyed but the park would be used as last recourse." (W1P3)

"As long as there is no manager, people always take advantage of it to clear the forest in secret. Also, the *chef de cantonnement* only comes very rarely." (W2P3)

"We are so afraid to enter and use the forest, whatever its status, but we still dare because a lack of options." (W4P1)

"Communities first use the VOI forest and once it is fully depleted, communities will be forced into the protected area as they need to use the forest and its resources." (W4P2)

"If there are only state forests and the park, the state forests will be degraded first. On the other hand, the presence of forests managed by communities prevents them from entering the reserve." (W5P4)

"If there is nobody in charge at the local level, the forest will be destroyed, even in the protected area, starting from the outside." (W6P4)

The principal problems players observe in forest management in the real world are mostly related to (mis)communication, the fuzziness of responsibilities, and the lack of power.

"ANGAP (new MNP) does not allow the local population to enter the park, but the staff enter the park to harvest the wood and sell it. Only ANGAP staff benefit from the Park's presence and there are no tangible benefits to the communities. If there are people from the riparian community who want to harvest or cut, they do not get authorization, yet the big exploiters do not have that problem. In sum, the Zahamena forest is protected but there is no consideration of the subsistence of the population." (W4P5)

"For the VOI forest: It is the people outside the local communities and the leaders who actually manage and exploit the forest. As far as the Park is concerned, it is an NGO that manages it with foreigners, which is why the forest is protected. If we let the Malagasy manage, the degradation could not be prevented." (W4P1)

"There is not enough collaboration and support from ANGAP and the *chef de cantonnement*. For example, the construction of a dam for agriculture, so that communities are obliged to progressively expand their fields for cultivation on the edges of forests." (W5P5)

"There is not enough communication with local people from forest management." (W6P4)

"The institution of the VOI is nothing but a name; it has no management power. If people with an authorization from the *chef de cantonnement* come to exploit, the VOI remains powerless. If people change their land to cultivate, the reason is that the soil of the plots they used until then are no longer productive. Exploitable lowlands for irrigated rice cultivation exist but without a dam, farmers prefer to exploit forests." (W6P5)

4. Discussion

In Madagascar, the impacts of policies to involve local resources users in forest management remain unclear. This study explored farmers' perceptions and practices regarding forest use under various forest governance systems with the help of a role-playing game and accompanying research.

The land-cover mapping results indicated that during the past 25 to 30 years—since the inception of the National Environmental Action Plan (NEAP)—deforestation of dense forests or their transformation into degraded or other land-cover types led to a loss of almost 100,000 hectares in the Zahamena area. While the general trend of transition from dense to sparse/degraded forest is readily apparent and easily explained, the less dominant transitions and interplay between other land-cover categories requires further efforts for more robust, thorough, and tangible context. For example, there is an increase in sparse/degraded forest observable in the western portion of the study area beyond the borders of the Zahamena National Park. This transition from grass/shrub-dominant areas and those dominated by wet agriculture to sparsely forested areas, while possible, requires further investigation, confirmation, and explanation. Despite the accuracy of the mapping results, the maps resulting from this study and their associated dynamics should be considered preliminary. Additional ground reference is required to further refine class definitions, provide additional examples of all targeted categories, and ultimately help contextualize all observed transitions. Bearing in mind the uncertainty associated with less common and more subtle changes, it is clear that, overall, throughout the study area, there was a substantial shift from dense to sparse/degraded forest, and that the vast majority of this change occurred beyond the boundaries of the Zahamena National Park. In addition to overall loss, mapping and dynamics results illustrate that areas of continuous dense forest already had wedges of sparse/degraded forest in 1990, which, in many areas, were since enlarged and completely cleared of dense forest. This trend supports the findings of Reference [11], which showed that almost half of Madagascar's forests are now located less than 100 km from a forest edge. A problem that should not remain ignored is the loss of forest quality (cf. Reference [65] linking fragmentation and biodiversity loss). Globally, focus has been to date on forest lost, but studies like Reference [66] highlight that loss of forest quality is under-researched in the Brazilian Amazon, contributing to much larger-scaled losses having so far gone undetected.

Our study showed that deforestation is taking place in both scenarios, while forest quality loss could be slowed down in the second scenario through different layers of governance, with some players mentioning being hesitant to enter the protected area. Though there was no difference between the two scenarios for most variables, it was interesting to see that there were more significant differences between the workshops far from the forests versus the ones close to Zahamena. This is not a coincidence, as distance to resources was shown to play an important role, as found for example by Reference [67] in the context of lemurs and the protected area at Lake Alaotra, a wetland complex just west of Zahamena. Similarly, Reference [24] depicted that people were more attached to forests in Manompana, some 200 km to the north of Zahamena, when living closer to forest edges. In our game, people closer to the Zahamena protected area did deforest and degrade forest less than those from the Vavatenina (eastern Zahamena) side. One reason could be that people living close to forests "care" more since they see the obvious benefits of having forests as a livelihood resource (e.g., References [25,68]). Or, people living further away may have already "forgotten" the forests, as most of the bigger forest vestiges were cleared off more than a generation ago.

While a number of studies documented that forest governance is not working due to a combination of lack of financial, human, and technical resources [37,69,70], it remains unclear what exactly is not working. In addition to these resource shortages, there are also those assigning fault to an incompatibility and a misalignment between village managers and the Forest Administration—an issue highlighted during the interviews conducted in this study, where local participants expressed their frustration about the power dynamics at play. Pascual and colleagues [71] put forward the crucial importance of equity when developing and implementing payments for ecosystem services. Power imbalances that are not taken into account during the design phase of instruments, and a lack of cultural and social consideration were shown to influence ecological outcomes. Here, we gained deeper insight from farmers' (rural resource users') perspectives. One main aspect related to governance emerged during the game workshops: the fear of crossing boundaries, i.e., the fear of fines from the state authorities. In the game, the Forest Administration was represented by a member of the workshop organizers positioned "far away" from the main game table. During a full game session, he/she was seldom visited, since most players did not believe they would receive proper support or the right permits. However, the presence of the Forest Administration, or the knowledge of its existence, was reason enough to slow down the crossing of the park boundaries in the game. Only when the population pressure grew bigger and livelihood needs were more difficult to satisfy did people start entering the protected area. At the end of each round, every household grew by one family member, which translated into a doubling of the household, and therefore, of the total population. This mechanism was implemented intentionally to reflect the reality of the Zahamena area.

The biggest concern for all game workshop participants was livelihood security. Land represents the biggest investment, and forests are seen by many as a reservoir for future agricultural production, since agricultural production—annual crops or cash crop production—is the sole way of ensuring subsistence and/or creating additional income. During the game, participants first tried to meet their livelihood needs outside forests, and supplement this with activities within forests. The first occupied forest parcels were those managed by the local communities (VOIs). Reasons for this were that the parcels were seen as belonging to them; therefore, they feared no consequences despite the risk of breaking the local laws (*dina*), or that possible fines issued by VOI members would be less severe than in the case of parcels under the management of the Forest Administration.

As confirmed in this study, VOI representatives lack authoritative power, as social bonds are de facto suppressing efficient law enforcement [72,73]. A central problem revolving around community-based natural resource management (CBNRM) is the use of the term "community". As already stated by Berkes [74], communities are not homogeneous entities, but consist once again of a large number of different interest groups and power relations. This can hamper any community-based initiatives, be it in a governance or conservation context (e.g., Reference [75]). In Antanandava, between 4% and 82% of households adhere to the local VOIs (1–23% of the local population), and most VOIs are yet to receive their official contracts for management transfer, due to a lack of funds to undertake the necessary evaluation for contract renewal. This abeyance is causing further conflicts in forest management including mismanagement, illicit exploitation, threats against VOI members (as also known from other sites), and conflicts within the associations.

In the Zahamena area, VOI-managed forests, i.e., under a CBNRM regime, are the first to be depleted. A majority of the VOIs around the Zahamena currently do not have a proper status: Either they were signed in to the GELOSE but their contracts expired and are awaiting renewal, or they are awaiting signing since the GELOSE enactment. In both cases, there is de facto no functional VOI, and thus, the communities do not respect any leaders of these associations. There is a general mistrust in authorities, starting at the VOI level where community members blame those in charge of abusing their power and knowledge. This translates up the authoritative ladder to the Forest Administration and the MNP. Another issue is the quality of reporting (or lack thereof) by the VOI to the Forest Administration. Community members in charge are either not used to writing reports, or the Forest Administration agents do not understand what the locals are reporting, and thus, ignore their messages. Adding to this is the lack of monitoring from the Forest Administration side, which is supposed to ensure that the VOIs are functional. The current under-staffing of the Forest Administration and the MNP, however, does not allow ensuring the VOI's functionality.

In our research, participants of the game workshops found that real-life issues were clearly reflected. Nevertheless, games are "only" models, and thus, have their limitations (e.g., Reference [76]). They are not about forecasting or anticipating futures; rather, they are about providing discussion platforms. Gaming is an extremely powerful tool to do so. Role-playing games allow one to recognize reality, and to discuss, explore, and experience different potential management regimes, future scenarios, or climate changes and their respective implications—while being distant enough from reality to learn and exchange in a safe environment. RPGs are not just models reflecting more or less detailed versions of reality; more importantly, they are fun and allow participants to forget their day-to-day struggles. This enjoyment also means that engagement is high, and discussions are more meaningful because participants feel personally involved. The game workshops represent a "safe space" where people are able to freely discuss and voice issues, concerns, and critiques, when there is usually a fear of repercussion. Bridging the virtual and real world is facilitated by the gamescape (game board) and subsequent debriefing discussions [77]. These settings allow the participants to explore their own livelihood situation, and enable them to visualize what their individual decisions, e.g., the "household level" in the game, may have at the landscape level. In another game (the Alaotra wetland game) [78], several players mentioned that the context of the game workshop was the first time that they actually looked at the system as a whole. This gave them the opportunity to think not only within their respective "space of activity" (e.g., their piece of land in the marshes), but to think and experience the connectedness and linkages of all the zones in the landscape. For researchers, tracking the decision clues, such as type, frequency, and spacing of activities, increased the understanding of livelihood strategies of the main actors in the Alaotra wetland [78]. Projects like AlaReLa (Alaotra Resilience Landscapes, 2014–2017) bring together various stakeholders by participating in games such as the one presented in this study to explore possible future scenarios. Doing so can help strengthen these stakeholders' capacity to better cope with changes and future shocks (e.g., Reference [79]). In the long run, this might help regional decision-makers to better balance conservation and development goals, and it might ease local communities into becoming a more integrated part of the decision-making process. As stated by Reference [80] (p. 101), "Economic and financial policies have not provided much support . . . reflecting in part the very low political weight of the rural and farming population." The game approach is certainly not a panacea, but it allows engaging with various stakeholders that, structurally, would otherwise not have the chance to voice their experiences.

5. Conclusions

This study was interested in exploring farmers' perceptions and practices in the light of CBNRM at the local scale using a case study in the Zahamena region of Madagascar, as well as providing local perspectives on existing livelihood–forest dynamics. The National Park of Zahamena suffered from deforestation and degradation, mainly in the northeast, while the remaining park boundaries are mostly respected by surrounding communities. This shows that the protected area does have an effect—if only slowing down deforestation and degradation. In that light, even community-based and managed forests play a precise role of buffer forests, able to absorb the first larger wave of anthropogenic pressures. However, once these are depleted, no park boundaries will be able to halt people in search of subsistence. CBNRM may be a short-term solution, but it is not a sustainable stand-alone one (cf. Reference [81]). Given the complexity of current forest governance, the solution cannot be imposed top-down, as pushed by the international community in the 1990s decentralization effort. Boundaries will be ineffective as long as people's livelihood situation is not improved. The only option for the majority of rural resource users is to transform any land they can access. Without access to better education, agriculture is their only guarantee of survival. Thus, long-term solutions for forest conservation needs to come from outside of forests—from agriculture. Although currently a global driver of deforestation, improved and sustainable agriculture could allow land users to increase their per capita and land unit production (e.g., References [82–84]). This could come in the form of access to new tools, techniques, and land-use rights. However, this bares two important risks. Firstly, despite better production, people will deforest even faster and more efficiently, and/or, secondly, any new agricultural development program will not be accepted and embodied by the local population—as is the current case with VOIs—while further deforestation takes place. Thus, participatory gaming approaches have the potential to help the rural poor by opening up opportunities for knowledge

Forests 2018, 9, 624

exchange, developing shared understandings through the elicitation of mental models that juxtapose multiple worldviews, and by exploring different possible futures.

Author Contributions: Conceptualization, P.O.W. and L.M.R. Methodology, N.B., L.M.R., A.D.G., N.S., and P.O.W. Validation, N.B., A.R., H.V.F.R., and T.R. Formal analysis, J.C., T.G.J., and G.M. Investigation, N.B., H.V.F.R., T.R., A.R., N.S., and A.D.G. Data curation, L.M.R., G.M., and J.C. Writing—original draft preparation, P.O.W., L.M.R., N.B., and N.S. Writing—review and editing, N.S., C.G., B.S.R., and L.W. Visualization, L.M.R., J.C., N.B., and T.G.J. Supervision, P.O.W. Project administration, P.O.W. Funding acquisition, P.O.W. and C.G.

Funding: This research was funded by the Swiss Program for Research on Global Issues for Development under the research grant number IZ01Z0_146852.

Acknowledgments: The authors thank all authorities from the Alaotra Mangoro Region who supported and participated in our research. We further thank all farmers and further stakeholders for their trust and participation in this study. Our acknowledgements are extended to our research assistants on the ground and to the two anonymous reviewers and editor for their helpful comments on a former version of the manuscript. This research was funded by the Swiss Programme for Research on Global Issues for Development under research grant IZ01Z0_146852 as part of the AlaReLa Alaotra Resilience Landscape project.

Conflicts of Interest: The authors declare no conflicts of interest. The funders had no role in the design of the study; in the collection, analyses, or interpretation of data; in the writing of the manuscript, and in the decision to publish the results.

Appendix A.

Appendix A.1. Approaches Used during the Game Development Process

All research described in this publication followed the ethical code of conduct and recommendations by Wilmé and colleagues [85], i.e., all participants were ensured anonymity and confidentiality unless wished differently from the participants themselves.

Appendix A.1.1. Diagnosis Phase

The game development was informed by results from ten focus groups (42 local participants) and eight expert interviews with local and regional authorities from the Municipalities, Forest Administration, and MNP (Madagascar National Parks). Additionally, this phase was complemented with interviews with local resource users, mainly farmers ($n = 36$), miners ($n = 30$), and further authorities ($n = 17$), to triangulate and verify the output of the focus groups—including ethno-photography of sciences (for an explanation, see page 6 of the manuscript).

In a first step, three focus groups were conducted in the respective villages tackling different but complementary themes: (i) types of land use and land status; (ii) land-use rights and rules for each type of land use; (iii) VOI functioning and rules of land use and forest use (mainly those assigned to the management of VOI).

Participant selection for focus groups (FG) 1 and 2 was based on the following criteria: (i) main activity in agriculture; (ii) aged 20–60, (iii) origin (locals and migrants); (iv) gender balance; (v) familiarity with the practices and rules applied in the *fokontany* (village); and (vi) belonging to different households. For FG3, the presence of simple VOI members was intended as their views on the rules on the use of forest resources in practice and the roles actually played by VOIs, as well as their personal position to that, could be different from those of the bureau members. The criteria were communicated to the *chief fokontany* or village chief and he was in charge of finding the participants.

Table A1. Details on the focus groups (FG) of the diagnostic phase. VOI means vondron'olona ifotony, being an association of rural stakeholders involved in natural resource management.

Parameters	Focus Group 1	Focus Group 2	Focus Group 3
Topic	Identification of land-cover types: characteristics, the existing and used resources. Land tenure dynamics.	Land tenure rules for each land use: rules and rights in force (who can do what, where does this right come from, what are the problems/conflicts encountered, etc.). Land acquisition patterns and types of land transactions. Evolution of rules over time and space.	Role and function of VOI in forest tenure. Views of VOIs on the use of forests and community mgmt. VOI links with the forestry administration and the NPM.
Participants	4: 2 men and 2 women	6: 3 men and 3 women	4: 2 office members maximum
Age	Belonging to different age groups: young people (20 to 30 years) and adults		
Origin	Residents	Autochthonous and migrants	-
Main activity	Agriculture	Agriculture	-
Others	Knowing the reality in the *fokontany (village)*, able to discuss in group		

The interviews were mainly used to complement and cross-check the information obtained during the FGs and also to gather the views of key informants, i.e., public authorities at the commune level and forest managers at regional or district level.

Table A2. Details on the interviews with regional and local authorities.

Authority (*n* = 8)	Interview Topics
Fokontany (village) chiefs, or their assistants	— Knowledge and application of land and forest laws by the *fokontany* — Roles and implications of *fokontany* in land matters and use of different forest types
Sector Head of Zahamena National Park	— History of Zahamena National Park — The rules and laws in force — NPM interventions in park management and surrounding crop plots — Offenses and infringements encountered, problems — Involvement of local communities in management
Forest managers (Head of Forestry, Director DREEF Alaotra-Mangoro)	— Rules and laws in force concerning forests — Procedures to be followed for forest resource uses — Situation/problems of management transfer to VOI — The main offenses and infringements encountered
Land Office Agent Antanandava and Deputy Mayor	— Role of the land office — Mode of land acquisition and land security in the locality
Service domain	— Mode of accession to land — Land transfer procedures — Flaws in land management and problems faced by the service

Table A3. Details on stakeholders and approaches that further complemented the above-described ones; VOI means vondron'olona ifotony, being an association of rural stakeholders involved in natural resource management; NP means National Park.

Participants and Approach	Topics	Details
Authority interviews (*n* = 17)	Main sources of income of the population Main uses of the open landscape (in terms of agriculture) Problems related to *tavy* (slash and burn) practices Circuit flow of agricultural products and market problems *	Mayor and Deputy Mayors of the communes Ambohijanahary, Antanandava, and Vavatenina Agricultural Service Center of Vavatenina President of Water Users around Lake Alaotra Presidents of the VOI of Antanandava Chief sectors responsible for the conservation of the NP Zahamena-Antanadava and head of the conservation component of the NP at Fénérive Est
Miner interviews (*n* = 30)	Participants' perceived impacts of mining, on a personal and community level **	The village heads helped in the search for miner participants, in order to follow the cultural etiquette.
Farmer ARDI (*n* = 19)	Problem definition Main actors involved Resources needed for this use: land tenure, existing labor management systems Dynamic: ecological changes and dynamics on the evolution of forest cover, open and swampy landscapes	In three workshops; village heads helped in the search for suitable participants.
Ethno-photography of sciences	What is your landscape? Which aspects are missing in the presented pictures/should we include and why?	Random selection on village markets; criteria: natural resource users and village residents

* For a detailed account on this, see Ravaka and colleagues [86]. ** For a detailed presentation of farmers' perceptions on mining, see Stoudmann and colleagues [87].

Appendix A.1.2. Game Development and Calibration Phase

The RPG was then tested and adapted during 11 workshops with 60 participants (university students from the forestry department in Antananarivo, natural resource users from the study area, conservation NGOs from the Alaotra, and Forest Administration members). Another nine workshops with 45 participants (all natural resource users from the study area) served to calibrate the game. Calibration was based on implicit reality, where game components and activities were still meaningful to the players. A final test workshop with villagers living close to the Zahamena forest helped to structure and fine-tune the actual data collection workshops.

For field workshops during the development and calibration phase of the game, the main criteria for selecting participants were as follows: living in the village, subsistence activity based on agriculture and not having played the game before, and gender balance whenever possible.

Appendix A.2. Game Description of the Zahamena RPG

The course of the game was the same for both scenarios. At the beginning of each scenario, the general course and the rules of the game were explained with the help of a poster that was displayed in the room (Figure A1). The five players each chose a color and receive a colored badge with a corresponding number, as well as same colored pins (game tokens to mark their plots) so that the research team could easily follow their activities. Each player represented one household, and received a starting capital of 2500 game money. They then consulted together for about two minutes to define the location of their village and mark the chosen parcel with small houses that were given to them.

Players could then start investing their money at the market to receive seeds and compost, and to do activities on the game board; prices of the activities on the market were displayed (together with the corresponding images of the activity cards, to facilitate the "reading"). Players placed their colored pins on the parcels that they used on the game board. At the beginning of round 1 of each scenario, each player was to use at least three parcels. At the beginning of each new round, players could decide to mark new parcels with the help of their colored pins. It was possible for two or more players to mark the same parcel, but they had to agree amongst themselves on how to use the parcel (in turn, or at the same time and share costs and outputs), the activity they were going to do, and the proportion of cost sharing and production.

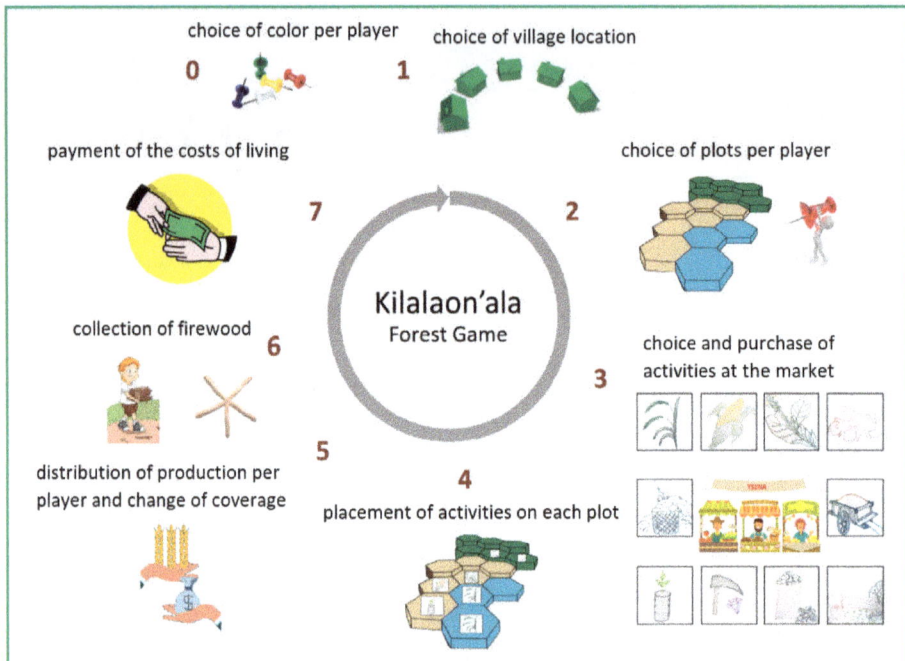

Figure A1. Course of the forest game. This poster illustration was used to explain the general course and rules of the game to the players.

At the market, players bought the activities (in the form of cards) that they did and then placed the cards on the parcels marked with their pins (Figure A2). Players could only put one activity card on a parcel, except for fertilizer and firebreak, as these were supplements to a main activity. In the rice fields (blue cover), only rice could be cultivated. Mining could only be done in the forest plots (green cover), while logging and charcoal could be done in forest plots and also in *tanety* plots if there was reforestation (brown cover; see details in Figure A2). All other activities could be done both in the forest and *tanety* plots. At the beginning of round 1, players had 10 minutes to shop at the market and place the activities on their marked parcels. For the remaining rounds, this time was reduced to five minutes.

The productions obtained by the players varied according to plot fertility and climate conditions. To account for the decline in fertility as plots were used, parcels were marked with colored points with each use. The facilitator used this coding system to determine the fertility, i.e., production output. To determine the climate during the round, the facilitator used cards: there were two "good climate" cards, two "insufficient rain" cards, and one "drought" card in the stack, each having a different impact on agricultural output. The order of the climate cards was predefined for data collection/comparability. However, this fact was unknown to the players to maintain the impression of random effects of the climate on the agricultural production and to see how players dealt with unpredictability. After announcing the climate to the players (good weather, drought, good weather, drought, insufficient rain, insufficient rain), the facilitator counted the production of each player's activities and gave them the sum in the form of money. This sum could be determined by looking at a pre-existing table summarizing the productions of different activities depending on the climate and making the necessary calculations.

At the end of each round, players had to pay livelihood costs corresponding to their yearly expenses, i.e., representing (non)consumables which they did not produce themselves. At the beginning, the cost of living was 2500 for five persons. After each round, a person, i.e., 500, was added to each household, so that the population doubled at the end of each scenario. The payment of the living costs in its entirety was obligatory (see details below, what happened if players could not afford it). Before the next round started, the game facilitator removed pins from marked but unused forest patches.

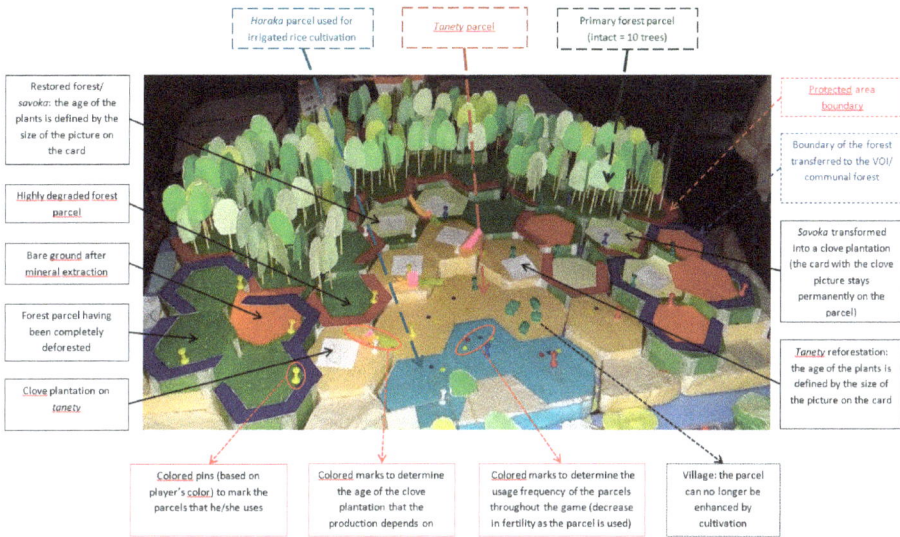

Figure A2. Original game board with tokens, pins, and the symbols indicating agricultural activities.

When a player placed an activity in the forest, the trees were removed and there could be changes in cover, depending on the chosen activity.

- Firewood. Each round, players had to collect firewood in the form of stacks of wood in the forest plots. The number of stacks of wood to collect was equal to the number of people in a household. Players could still take more stacks of wood than necessary and keep the surplus for the next rounds. If a player failed to obtain the required number of stacks of wood, he had to cut a tree if the number of stacks of wood missing was less than five and two trees if the missing number was greater than five. However, if the player had a reforestation plot (aged from T2 = 10 years), it was assumed that he collected his firewood in his reforestation plot and no longer needed to cut trees in the forest.
- Living costs. When a player did not have enough money to pay the obligatory costs of living at the end of a turn, he had to cut down a number of trees proportional to the amount he had to pay (one tree = 250). Each forest cell carried 10 trees and two piles of wood (corresponding to dead branches collected for firewood). The piles of wood were renewed at each turn in the following way: if the number of trees in the parcel was greater than five then there would still be two piles of wood; if the number of trees was less than five, there would be one pile of wood; if there were no trees on the plot, there would be no pile of wood.

Clearing cards. When there was a crop for the first time on a forest plot, it was assumed that there was clearing and burning. Since the behavior of the fire was unpredictable due to factors such as wind, burning method, etc., cards were drawn to find out what happened. In total, there were 10 cards in three categories: (i) two cards: no fire spread to neighboring plots; (ii) four cards: spreading of fire in four different directions; (iii) four cards: obligation to make a firewall. If the player made a firewall before the card was drawn, then the card was cancelled. If there was not yet a firewall, the player was obliged to pay a fine defined by the other players.

Forestry Administration. At the beginning of Scenario 1, we explained to the players that all the activities related to forests were the responsibility of the forest administration and that if they had any questions, they had to ask the manager, who stayed far from the play area (e.g., a neighboring room). The decision to go see the quartermaster was entirely up to the players. The time allowed for discussion with the quartermaster went hand in hand with the time to buy the activities at the market. The authorizations given by the quartermaster were only valid for one round and had to be renewed if necessary. At the end of round 2, the quartermaster made an appearance to raise awareness among the players. During round 3, he made another appearance and sanctioned players who did an activity in the forest during the round without authorization. The players had to go to the head quartermaster's office to pay the fines. This was to see if the players had the initiative to pay the fines. The quartermaster did not make any appearances during scenario 2.

A debriefing followed the first scenario, allowing the discussion of gaming behavior and strategies, to bridge game characteristics and reality, and to exchange on underlying mechanisms and relationships.

At the onset of scenario 2, the facilitator explained the observable changes on the game board to the players and elaborated on the forest management rules that followed. Players were invited to consult each other; one of them was chosen as a volunteer for the CLP (Local Committee of the Park); he or she oversaw the surveillance and the protection of the protected area. Another player was elected as VOI president. Rules for the use of the VOI forest/community forest were discussed amongst the players. The CLP was provided with monitoring sheets for infractions in the protected area, which he completed each turn. Sanctions against activities in the protected area depended on the report made by the CLP. Once all the roles were assigned, the game session with scenario 2 started and the sequence was the same as for scenario 1.

Table A4. Statistical Analysis of Players' Gaming Behavior and Land-Type Changes. Statistically significant differences between (i) Scenario 1 and 2 for very degraded forest; and (ii) between workshops 1–3 vs. 4–6 for the remaining variables. A Welch two-sample t-test was used with the R software version 3.0.3.

Variable	t	df	p	95% Confidence Intervals		Sample Estimates	
						Mean of x	Mean of y
Very degraded forest (S1–S2)	2.1038	51.764	0.04028	0.01791148	0.75986630	0.6388889	0.2500000
Livelihood	4.3275	202.23	2.37×10^{-5}	6507.807	17402.060	17975.667	6020.733
Moderately degraded forest	2.2297	62.928	0.02934	0.05763622	1.05347489	1.2222222	0.6666667
Very degraded forest	2.1038	51.764	0.04028	0.01791148	0.75986630	0.6388889	0.2500000
Completely deforested plots	−2.3073	47.361	0.02545	−1.14383016	−0.07839206	0.3055556	0.9166667

Table A5. Statistical Analysis of Players' Gaming Behavior and Land-Type Changes. All game variables that showed a statistically significant difference between Scenario 1 and 2 are listed below. A non-parametric Wilcoxon rank sum test with continuity correction was used with the R software version 3.0.3.

Variable	W	p	More in
Clove in *tanety* (number)	10,176.5	0.02893	Scenario 1
Reforestation in *tanety* (number)	11,930	0.02396	Scenario 1
Maize/bean in forest (number)	12,530.5	0.0145	Scenario 1
Activities in forest (total number)	13,631	0.0008079	Scenario 1
Loss of forest to maize/bean	12,223.5	0.002153	Scenario 1
Active loss	13,109	0.004556	Scenario 1

Table A6. Statistical Analysis of Players' Gaming Behavior and Land-Type Changes. All game variables that showed a statistically significant difference between workshops 1–3 vs. 4–6 are listed below. A non-parametric Wilcoxon rank sum test with continuity correction was used with the R software version 3.0.3.

Variable	W	p	More in
Charcoal in forest (nb)	12,000	0.01529	W1–3
Forest exploitation (nb)	13,275	6.608×10^{-7}	W1–3
Restoration exploitation (nb)	10,800	0.01361	W4–6
Fertilizer, big (nb)	7651.5	3.422×10^{-8}	W4–6
Fertilizer, small (nb)	7622.5	1.194×10^{-7}	W4–6
Reforestation in forest (nb)	10,579	0.03441	W4–6
Maize/bean in *tanety* (nb)	8170	8.176×10^{-7}	W4–6
Mines (nb)	12,675	6.965×10^{-6}	W1–3
Tobacco/cassava in forest (nb)	9841	0.0007592	W4–6
Tobacco/cassava in *tanety* (nb)	9615	0.0003511	W4–6
Firebreak (nb)	9660	0.004117	W4–6
Rice in forest (nb)	12,075	0.006133	W1–3
Activities (nb)	8199	4.127×10^{-5}	W4–6
Loss for livelihood (nb)	8092	2×10^{-8}	W4–6
Forced loss (nb)	7862	1.465×10^{-7}	W4–6
Production (nb)	13,796	0.0007046	W1–3
Reforestation *tanety* (nb)	10,434	0.006736	W4–6
Maize/bean in forest (nb)	12,601	0.009898	W1–3
Activities in forest (nb)	14,484	5.352×10^{-6}	W1–3
Active loss (nb)	14,602	3.137×10^{-7}	W1–3

Table A7. LandSat Accuracy Classification. Confusion Matrix Produced for Classification of August 2014 Landsat Data using Independent Validation Data.

	1	2	3	4	5	6	Total	Users (%)	Commission (%)
Dense forest (1)	45	3	0	0	0	0	45	100	0
Sparse/degraded forest (2)	0	45	0	0	0	0	45	100	0
Grass/shrub-dominant (3)	0	0	36	0	0	1	37	97.3	2.7
Soil-dominant (4)	0	0	0	36	0	0	36	100	0
Water (5)	0	0	0	0	36	0	36	100	0
Vegetation swamp; wet agriculture (6)	0	0	0	0	0	35	35	100	0
Total	45	45	36	36	36	36	234		
Producers (%)	100	100	100	100	100	97			
Amission (%)	0	0	0	0	0	3			
Overall	99.6								
Kappa	0.99								

Appendix A.3. Perceptions on Livelihoods and Change in the Zahamena Socio-Ecological System (Statements of Miners in Interviews, and Farmers in Game Workshops)

"Quarries and mines are not the same as they used to be—there are many more today. It is a good thing, as its consequences for myself are that I can buy more things as my standard of living increases." (I3, Vohimarina)

"Today Zahamena has also become an exploited area—there are many crystals and rubies there, with high prices. It is a bad thing because the protected area is getting destroyed and this leads to decreasing rain and volumes of water." (I6, Vohimarina)

"It (mining) leads to forest destruction and decreasing numbers of animals, as they have to cut down forests to exploit the mines. It has also led to an increasing number of bandits." (I18, Antanandava)

The Alaotra region is the rice granary of Madagascar. This crop is the staple food in Madagascar, and Malagasy tend to eat rice three times a day if they can afford it (cf. Figure 7).

"Rice farming is fundamental and it must be done because rice is irreplaceable." (W6P3)

"Even if one has a lot of money but not enough rice, then all the money will be spent on buying rice for consumption." (W6P5)

Slash-and-burn agriculture, or *tavy*, is the typical feature of the Zahamena socio-ecological system. This land clearing and transformation into crops is slowly eating the national park (Figure 6), home to 13 different species of lemurs [88].

"As we do not own irrigated rice paddies, we maintain our livelihood with *tavy*." (Focus group participant FG2, Ambongabe)

"(We continue to practice *tavy* because) arable land, especially the *horaka* (irrigable rice fields) are insufficient, so one has to do it in the face of demographic growth." (W1P3)

"The population is experiencing strong demographic growth, livelihoods are becoming more and more difficult, land is no longer enough, so this means that people are progressively using the edges of the forests." (W4P5)

Appendix A.4. Perceptions of Farmers on the Best-Suited Forest Management System (Statements of Game Workshop Debriefings)

The consensus amongst the game workshop participants is that local management would still be best suited to serve the local population, who has livelihood security as the highest priority.

"Management by the VOI (out of the three governance systems is best suited) because the forest has already been given to the community; if it is exhausted, then we have to move to other types of forest." (W4P1)

"The VOI system because there we are less afraid. (W4P2)

"It is better to let the local population manage the forests; this way, we have enough arable land and there are no more constraints because of administrative procedures; we no longer need to move far to the *chef de cantonnement*, it is enough to set up a *dina* to handle the cutting of large-diameter wood. Prioritizing land availability over the long term is more important than protecting the forest." (W4P5)

"The actions must be prioritized to improve the standard of living and the livelihoods of local populations if we want to protect the forest." (W6P5)

Still, other participants would prefer a protected area system and/or state control over the forests to ensure their protection.

"The protected area because it is more difficult to discuss and confront the managers. If a person is designated (among the villagers), decisions are generally not respected." (W1P1)

"In addition (to VOI), the central management should take over the work of the *chef de cantonnement*; respect and correctly apply the regulations without distinction or exception." (W1P4)

"The protected area because the managers show no mercy and they do not hesitate to call on gendarme." (W2P2)

Appendix A.5. Authority Perceptions of Forest Governance Issues in the Alaotra Mangoro Region

Statements or information stemming from local forest governance authorities are listed first, followed by statements from regional, more distant offices.

"VOIs are not able to play their role in forest management for various reasons: *dina* (local laws and rules) are not respected. Forest patrols are irregular for lack of compensation; declarations of violations to forest regulations are not reported for fear of being threatened and rejected by the community whose majority are not members of VOI. Members who have registered are either not active and/or do not pay their annual dues. Slash-and-burn agriculture and logging are common practices in VOI-managed forests as controls in the Zahamena Protected Area are more strict." (VOI Presidents, Antanandava)

"(The problem is the) lack of staff and means of the agents to ensure the forest controls especially in the few remaining state forests. In cases where the forest agents notice an infraction in the forest areas, they can appeal to the authorities like the gendarmerie, but that requires financial means." (Staff member of the *cantonnement forestier*, Vavatenina)

"Much of Zahamena National Park is part of the Ambatondrazaka district, while MNP staff in this sector is understaffed. (...) The practice of *tavy* is a traditional mode of cultivation where peasants move to clear one plot after another; forests managed by VOI juxtaposed with the protected area are the most affected by this practice, despite the measures taken, such as the use of firewalls and forest patrols, fires arrive right into the protected area." (MNP, Antanandava)

"The advantages of VOI are that those managing the resources live in proximity to them. Members are volunteers, and therefore, protective and motivated (sic). It is a way to give rights to communities and allow them to manage the resources on which they depend. The royalties from this management (ecotourism, value added from natural resources) allow the communities to finance this management. (. . .) The issues with VOI are that the members are farmers and fishers, so often poor and uneducated. Local rules are not always compatible with regional legislations. (. . .) A big issue is the monitoring of each VOI, as there are 200 of them in the Alaotra region, and the Ministry does not have the funds and staff to do this. There are certain NGOs that help with this task. Many VOIs need to be revitalized, but here again there are financial issues. Since the crisis (January 2009), no new VOI contracts have been signed." (Ministry of the Environment, Ecology, Seas, and Forests, Ambatondrazaka)

"The Ministers are constantly changing, and there is no time of stability, so no long-term vision can take place. Thinking one or two years in the future is not done. Work is always in the short-term." (Ministry of Fisheries, Ambatondrazaka)

"Before, 2000, the Forest Administration distributed *tavy* to communities bordering the forests, but with the new policy no more permits are issued for this activity. The Forest Administration lacks personnel, materials, and equipment; the areas of intervention are far, so we cannot monitor the *savoka*, that is why we decided to no longer issue clearing permits." (DREEF Alaotra-Mangoro, regional director, Ambatondrazaka)

According to the DREEF (Directeur Régional de l'Environnement, de l'Écologie, et des Forêts), problems with the transfer of forest management to VOIs are as follows: (i) the possibility of local conciliation cannot be ruled out; for example, if the written request is inferior to the actual demand, the quantity is distorted and the surplus of the right belongs to the management committee. (ii) There are still gaps where the sanctions are not well defined; there is no big difference between forests transferred or not. (iii) During the management transfer application process, the Forest Administration is only involved by the project promoters at the signature request stage so that many anomalies in the documents can be identified—resulting in the contract not being approved and having no continuation. In some cases, these requests are mainly for operational purposes, which is why the Administration must remain cautious with regard to requests for transfer of management. (iv) There is no close cooperation between the Forest Administration and the MNP; the Forest Administration is usually consulted only in case of force majeure.

"The problem with the management transfer around the Park (Zahamena) is that the consultants who worked on it did not involve the Forest Administration. Therefore, the status and especially the role of the VOI is only in name and the transfer of management is unsuccessful, and the contract is not signed or implemented. (. . .) It is for consideration of the subsistence of the population living in areas without irrigated rice fields that the state has given authorizations for the clearing of dense forests. This authorization has validity—it is as if the community borrowed the land from the state for a period and, after the harvest, these people have no rights over the land they used. However, people conceive that the land becomes their property once they have obtained an authorization. (. . .) The forest is a property belonging to the whole country, everyone must feel responsible for its protection." (DREEF Alaotra-Mangoro, *chef de cantonnement*, Ambatondrazaka)

References

1. Curtis, P.G.; Slay, C.M.; Harris, N.L.; Tyukavina, A.; Hansen, M.C. Classifying drivers of global forest loss. *Science* **2018**, *361*, 1108–1110. [CrossRef] [PubMed]
2. Foley, J.A.; DeFries, R.; Asner, G.P.; Barford, C.; Bonan, G.; Carpenter, S.R.; Chapin, F.S.; Coe, M.T.; Daily, G.C.; Gibbs, H.K.; et al. Global consequences of land use. *Science* **2005**, *309*, 570–574. [CrossRef] [PubMed]
3. Song, X.-P.; Hansen, M.C.; Stehman, S.V.; Potapov, P.V.; Tyukavina, A.; Vermote, E.F.; Townshend, J.R. Global land change from 1982 to 2016. *Nature* **2018**, *560*, 639–643. [CrossRef] [PubMed]

4. Geist, H.J.; Lambin, E.F. Proximate causes and underlying driving forces of tropical deforestation: Tropical forests are disappearing as the result of many pressures, both local and regional, acting in various combinations in different geographical locations. *BioScience* **2002**, *52*, 143–150. [CrossRef]
5. Moat, J.; Smith, P.P. *Atlas of the Vegetation of Madagascar*; Royal Botanic Gardens: London, UK, 2007.
6. Buerki, S.; Devey, D.S.; Callmander, M.W.; Phillipson, P.B.; Forest, F. Spatio-temporal history of the endemic genera of Madagascar. *Bot. J. Linn. Soc.* **2013**, *171*, 304–329. [CrossRef]
7. Ganzhorn, J.U.; Wilmé, L.; Mercier, J.L. Explaining Madagascar's biodiversity. In *Conservation and Environmental Management in Madagascar*; Scales, I.R., Ed.; Routledge: London, UK, 2014; pp. 41–67.
8. Waeber, P.O.; Wilmé, L.; Ramamonjisoa, B.; Garcia, C.; Rakotomalala, D.; Rabemananjara, Z.H.; Kull, C.A.; Ganzhorn, J.U.; Sorg, J.-P. Dry forests in Madagascar: Neglected and under pressure. *Int. For. Rev.* **2015**, *17*, 127–148. [CrossRef]
9. Global Forest Watch. Tree Cover Loss and Gain Area. Available online: https://www.globalforestwatch.org/dashboards/country/MDG (accessed on 20 July 2018).
10. Waeber, P.O.; Schuurman, D.; Wilmé, L. Madagascar's rosewood (*Dalbergia* spp.) stocks as a political challenge. *PeerJ Prepr.* **2018**, e27062v1. [CrossRef]
11. Vieilledent, G.; Grinand, C.; Rakotomalala, F.A.; Ranaivosoa, R.; Rakotoarijaona, J.R.; Allnutt, T.F.; Achard, F. Combining global tree cover loss data with historical national forest cover maps to look at six decades of deforestation and forest fragmentation in Madagascar. *Biol. Conserv.* **2018**, *222*, 189–197. [CrossRef]
12. Horning, N.R. Strong support for weak performance: Donor competition in Madagascar. *Afr. Affairs* **2008**, *107*, 405–431. [CrossRef]
13. Waeber, P.O.; Wilmé, L.; Mercier, J.R.; Camara, C.; Lowry, P.P., II. How effective have thirty years of internationally driven conservation and development efforts been in Madagascar? *PLoS ONE* **2016**, *11*, e0161115. [CrossRef] [PubMed]
14. Scales, I.R. Farming at the forest frontier: Land use and landscape change in western Madagascar, 1896–2005. *Environ. Hist.* **2011**, *17*, 499–524. [CrossRef]
15. Scales, I.R. The drivers of deforestation and the complexity of land use in Madagascar. In *Conservation and Environmental Management in Madagascar*, 1st ed.; Scales, I.R., Ed.; Routledge: London, UK, 2014; pp. 129–150.
16. Casse, T.; Milhøj, A.; Ranaivoson, S.; Randriamanarivo, J.R. Causes of deforestation in southwestern Madagascar: What do we know? *For. Policy Econ.* **2004**, *6*, 33–48. [CrossRef]
17. Moser, C. An economic analysis of deforestation in Madagascar in the 1990s. *Environ. Sci.* **2008**, *5*, 91–108. [CrossRef]
18. Humbert, H. *La Destruction D'une Flore Insulaire Par Le Feu: Principaux Aspects De La Végétation À Madagascar*; Mémoires De l'Académie Malgache 5; Imprimerie Moderne de l'Emyrne, G. Pitot et Cie.: Tananarive, Madagascar, 1927; pp. 1–80.
19. Favre, J.C. Traditional utilization of the forest. In *Ecology and Economy of a Tropical Dry Forest in Madagascar*; Ganzhorn, J.U., Sorg, J.-P., Eds.; Primate Report 46-1: Göttingen, Germany, 1996; pp. 33–40.
20. Laurent, J. The Programme Menabe in Madagascar: A project of the Cooperation Suisse. In *Ecology and Economy of a Tropical Dry Forest in Madagascar*; Ganzhorn, J.U., Sorg, J.-P., Eds.; Primate Report 46-1: Göttingen, Germany, 1996; pp. 5–12.
21. Styger, E.; Rakotondramasy, H.M.; Pfeffer, M.J.; Fernandes, E.C.; Bates, D.M. Influence of slash-and-burn farming practices on fallow succession and land degradation in the rainforest region of Madagascar. *Agric. Ecosyst. Environ.* **2007**, *119*, 257–269. [CrossRef]
22. Poudyal, M.; Jones, J.P.G.; Rakotonarivo, O.S.; Hockley, N.; Gibbons, J.M.; Mandimbiniaina, R.; Rasoamanana, A.; Andrianantenaina, N.S.; Ramamonjisoa, B.S. Who bears the cost of forest conservation? *PeerJ* **2018**, *6*, E5106. [CrossRef] [PubMed]
23. Pollini, J. Agroforestry and the search for alternatives to slash-and-burn cultivation: From technological optimism to a political economy of deforestation. *Agric. Ecosyst. Environ.* **2009**, *133*, 48–60. [CrossRef]
24. Urech, Z.L.; Felber, H.R.; Sorg, J.P. Who wants to conserve remaining forest fragments in the Manompana Corridor? *Madag. Conserv. Dev.* **2012**, *7*, 135–143. [CrossRef]
25. Urech, Z.L.; Zaehringer, J.G.; Rickenbach, O.; Sorg, J.-P.; Felber, H.R. Understanding deforestation and forest fragmentation from a livelihood perspective. *Madag. Conserv. Dev.* **2015**, *10*, 67–76. [CrossRef]

26. Dressler, W.; Büscher, B.; Schoon, M.; Brockington, D.; Hayes, T.; Kull, C.; McCarthy, J.; Shrestha, K. From hope to crisis and back again? A critical history of the global CBNRM narrative. *Environ. Conserv.* **2010**, *37*, 5–15. [CrossRef]

27. Measham, T.G.; Lumbasi, J.A. Success factors for community-based natural resource management (CBNRM): Lessons from Kenya and Australia. *Environ. Manag.* **2013**, *52*, 649–659. [CrossRef] [PubMed]

28. Pollini, J.; Lassoie, J.P. Trapping farmer communities within global environmental regimes: The case of the GELOSE legislation in Madagascar. *Soc. Nat. Resour.* **2011**, *24*, 814–830. [CrossRef]

29. Mercier, J.R. The preparation of the National Environmental Action Plan (NEAP): Was it a false start? *Madag. Conserv. Dev.* **2006**, *1*, 50–54. [CrossRef]

30. Raik, D. Forest management in Madagascar: An historical overview. *Madag. Conserv. Dev.* **2007**, *2*, 5–10. [CrossRef]

31. Borrini-Feyerabend, G.; Dudley, N. Élan Durban: Nouvelles perspectives pour les Aires Protégées de Madagascar 2005. Available online: http://www.equilibriumresearch.com/upload/document/elandurban.pdf (accessed on 20 July 2018).

32. Kull, C.A. Empowering pyromaniacs in Madagascar: Ideology and legitimacy in community-based natural resource management. *Dev. Change* **2002**, *33*, 57–78. [CrossRef]

33. Antona, M.; Motte Biénabe, E.; Salles, J.-M.; Péchard, G.; Aubert, S.; Ratsimbarison, R. Rights transfers in Madagascar biodiversity policies: Achievements and significance. *Environ. Dev. Econ.* **2004**, *9*, 825–847. [CrossRef]

34. Raik, D.B.; Decker, D.J. A multisector framework for assessing community-based forest management: Lessons from Madagascar. *Ecol. Soc.* **2007**, *12*, 1. Available online: http://www.ecologyandsociety.org/vol12/iss1/art14/ (accessed on 20 July 2018). [CrossRef]

35. Fritz-Vietta, N.V.; Röttger, C.; Stoll-Kleemann, S. Community-based management in two biosphere reserves in Madagascar–distinctions and similarities: What can be learned from different approaches? *Madag. Conserv. Dev.* **2009**, *4*, 86–97. [CrossRef]

36. Horning, N.R. Across the great divide: Collaborative forest management. In *The Politics of Deforestation in Africa: Madagascar, Tanzania, and Uganda*, 1st ed.; Horning, N.R., Ed.; Palgrave Mcmillan: Basingstoke, UK, 2018; pp. 135–163. ISBN 978-3-319-76827-4.

37. Neudert, R.; Ganzhorn, J.U.; Wätzold, F. Global benefits and local costs—The dilemma of tropical forest conservation: A review of the situation in Madagascar. *Environ. Conserv.* **2017**, *44*, 82–96. [CrossRef]

38. Gardner, C.J.; Nicoll, M.E.; Birkinshaw, C.; Harris, A.; Lewis, R.E.; Rakotomalala, D.; Ratsifandrihamanana, A.N. The rapid expansion of Madagascar's protected area system. *Biol. Conserv.* **2018**, *220*, 29–36. [CrossRef]

39. Rasolofoson, R.A.; Ferraro, P.J.; Jenkins, C.N.; Jones, J.P. Effectiveness of community forest management at reducing deforestation in Madagascar. *Biol. Conserv.* **2015**, *184*, 271–277. [CrossRef]

40. Rasolofoson, R.A.; Ferraro, P.J.; Ruta, G.; Rasamoelina, M.S.; Randriankolona, P.L.; Larsen, H.O.; Jones, J.P. Impacts of community forest management on human economic well-being across Madagascar. *Conserv. Lett.* **2017**, *10*, 346–353. [CrossRef]

41. Visser, L.E. Reflections on transdisciplinarity, integrated coastal development, and governance. In *Challenging Coasts: Transdisciplinary Excursions into Integrated Coastal Zone Development*; Visser, L.E., Ed.; Amsterdam University Press: Amsterdam, The Netherlands, 2004; pp. 23–47.

42. Béné, C.; Wood, R.G.; Newsham, A.; Davies, M. Resilience: New utopia or new tyranny? Reflection about the potentials and limits of the concept of resilience in relation to vulnerability reduction programmes. *IDS Working Pap.* **2012**, *405*, 1–61. Available online: https://www.ids.ac.uk/publication/resilience-new-utopia-or-new-tyranny (accessed on 20 July 2018). [CrossRef]

43. Sanderson, I. Evaluation, policy learning and evidence-based policy making. *Public Admin.* **2002**, *80*, 1–22. [CrossRef]

44. Lavis, J.N.; Robertson, D.; Woodside, J.M.; McLeod, C.B.; Abelson, J. How can research organizations more effectively transfer research knowledge to decision makers? *Milbank Q.* **2003**, *81*, 221–248. [CrossRef] [PubMed]

45. Gautier, D.; Garcia, C.; Negi, S.; Wardell, D.A. The limits and failures of existing forest governance standards in semi-arid contexts. *Int. For. Rev.* **2015**, *17*, 114–126. [CrossRef]

46. Gardner, C.J. Social learning and the researcher–practitioner divide. *Oryx* **2012**, *46*, 313–314. [CrossRef]

47. Sayer, J.; Bull, G.; Elliott, C. Mediating forest transitions: 'Grand design 'or' muddling through'. *Conserv. Soc.* **2008**, *6*, 320–327. [CrossRef]

48. Rives, F.; Carrière, S.M.; Montagne, P.; Aubert, S.; Sibelet, N. Forest management devolution: Gap between technicians' design and villagers' practices in Madagascar. *Environ. Manag.* **2013**, *52*, 877–893. [CrossRef] [PubMed]

49. Rittel, H.W.; Webber, M.M. Dilemmas in a general theory of planning. *Policy Sci.* **1973**, *4*, 155–169. [CrossRef]

50. Etienne, M. *Companion Modelling, a Participatory Approach to Support Sustainable Development*, 1st ed.; Springer: Dordrecht, The Netherlands, 2014; ISBN 978-94-017-8556-3.

51. Waeber, P.O.; De Grave, A.; Wilmé, L.; Garcia, C. Play, learn, explore: Grasping complexity through gaming and photography. *Madag. Conserv. Dev.* **2017**, *12*. [CrossRef]

52. Rakotoarisoa, T.F.; Waeber, P.O.; Richter, T.; Mantilla-Contreras, J. Water hyacinth (*Eichhornia crassipes*), any opportunities for the Alaotra wetlands and livelihoods? *Madag. Conserv. Dev.* **2015**, *10*, 128–136. [CrossRef]

53. Bousquet, F.; Trébuil, G.; Hardy, B. *Companion Modeling and Multi-Agent Systems for Integrated Natural Resource Management in Asia*; International Rice Research Institute: Los Baños, Philippines, 2005.

54. Dumrongrojwatthana, P.; Barnaud, C.; Gajaseni, N.; Trébuil, G. Companion modeling to facilitate adaptive forest management in Nam Haen sub-watershed, Nan Province, Northern Thailand. In Proceedings of the Towards Sustainable Livelihood and Environment: 2nd International Conference Asian Simulation and Modelling (ASIMMOD 2007), Chiang Mai, Thailand, 9–11 January 2007; pp. 327–332. Available online: http://agritrop.cirad.fr/543099/ (accessed on 20 July 2018).

55. Dumrongrojwatthana, P.; Trébuil, G. *Northern Thailand Case: Gaming and Simulation for Co-Learning and Collective Action; Companion Modelling for Collaborative Landscape Management between Herders and Foresters*; Wageningen Academic Publishers: Wageningen, The Netherlands, 2011; pp. 191–219.

56. Garcia, C.; Speelman, E.N. Landscape approaches, wicked problems and role playing games. Tropenbos International ComMod Workshop. *ForDev Working Pap.* **2015**, *1*, 1–20. Available online: http://www.researchgate.net/publication/320170969 (accessed on 20 July 2018).

57. Speelman, E.N.; van Noordwijk, M.; Garcia, C. Gaming to better manage complex natural resource landscapes. In *Co-Investment in Ecosystem Services: Global Lessons from Payment and Incentive Schemes*; Namirembe, S., Leimona, B., van Noordwijk, M., Minang, P.A., Eds.; World Agroforestry Centre (ICRAF): Nairobi, Kenya, 2017.

58. Etienne, M.; Du Toit, D.R.; Pollard, S. ARDI: A co-construction method for participatory modelling in natural resources management. *Ecol. Soc.* **2011**, *16*, 44. Available online: http://www.ecologyandsociety.org/vol16/iss1/art44/ (accessed on 20 July 2018). [CrossRef]

59. USGS. Available online: https://landsat.usgs.gov/landsat-surface-reflectance-data-products (accessed on 25 July 2018).

60. Banskota, A.; Kayastha, N.; Falkowski, M.J.; Wulder, M.A.; Froese, R.E.; White, J.C. Forest monitoring using Landsat time series data: A review. *Can. J. Remote Sens.* **2014**, *40*, 362–384. [CrossRef]

61. World Bank. *Critical Ecosystem Partnership Fund*; Global program review 2, 1; Washington, DC, USA, 2007; Available online: http://documents.worldbank.org/curated/en/865411468153290094/Critical-Ecosystem-Partnership-Fund-CEPF (accessed on 20 July 2018).

62. Erbek, F.S.; Özkan, C.; Taberner, M. Comparison of maximum likelihood classification method with supervised artificial neural network algorithms for land use activities. *Int. J. Remote Sens.* **2004**, *25*, 1733–1748. [CrossRef]

63. Jensen, L.S.; Mueller, T.; Tate, K.R.; Ross, D.J.; Magid, J.; Nielsen, N.E. Soil surface CO_2 flux as an index of soil respiration in situ: A comparison of two chamber methods. *Soil Biol. Biochem.* **1996**, *28*, 1297–1306. [CrossRef]

64. R Core Team. *R: A Language and Environment for Statistical Computing*; R Foundation for Statistical Computing: Vienna, Austria, 2013; Available online: http://www.R-project.org/ (accessed on 20 July 2018).

65. Turner, I.M. Species loss in fragments of tropical rain forest: A review of the evidence. *J. Appl. Ecol.* **1996**, *33*, 200–209. [CrossRef]

66. Rappaport, D.I.; Morton, D.C.; Longo, M.; Keller, M.; Dubayah, R.; dos-Santos, M.N. Quantifying long-term changes in carbon stocks and forest structure from Amazon forest degradation. *Environ. Res. Lett.* **2018**, *13*, 065013. [CrossRef]

67. Reibelt, L.M.; Woolaver, L.; Moser, G.; Randriamalala, I.H.; Raveloarimalala, L.M.; Ralainasolo, F.B.; Ratsimbazafy, J.; Waeber, P.O. Contact matters: Local people's perceptions of *Hapalemur alaotrensis* and implications for conservation. *Int. J. Primatol.* **2017**, *38*, 588–608. [CrossRef]

68. Timko, J.A.; Waeber, P.O.; Kozak, R.A. The socio-economic contribution of non-timber forest products to rural livelihoods in Sub-Saharan Africa: Knowledge gaps and new directions. *Int. For. Rev.* **2010**, *12*, 284–294. [CrossRef]

69. Ameyaw, J.; Arts, B.; Wals, A. Challenges to responsible forest governance in Ghana and its implications for professional education. *For. Policy Econ.* **2014**, *62*, 78–87. [CrossRef]

70. Sundström, A. Understanding illegality and corruption in forest governance. *J. Environ. Manag.* **2016**, *181*, 779–790. [CrossRef] [PubMed]

71. Pascual, U.; Phelps, J.; Garmendia, E.; Brown, K.; Corbera, E.; Martin, A.; Gomez-Baggethun, E.; Muradian, R. Social equity matters in payments for ecosystem services. *BioScience* **2014**, *64*, 1027–1036. [CrossRef]

72. Waeber, P.O.; Reibelt, L.M.; Randriamalala, I.H.; Moser, G.; Raveloarimalala, L.M.; Ralainasolo, F.B.; Ratsimbazafy, J.; Woolaver, L. Local awareness and perceptions: Consequences for conservation of marsh habitat at Lake Alaotra for one of the world's rarest lemurs. *Oryx* **2017**, 1–10. [CrossRef]

73. Kairu, A.; Upton, C.; Huxham, M.; Kotut, K.; Mbeche, R.; Kairo, J. From shiny shoes to muddy reality: Understanding how meso-state actors negotiate the implementation gap in participatory forest management. *Soc. Nat. Resour.* **2017**, *31*, 74–88. [CrossRef]

74. Berkes, F. Rethinking community-based conservation. *Conserv. Biol.* **2004**, *18*, 621–630. [CrossRef]

75. Rendigs, A.; Reibelt, L.M.; Ralainasolo, F.B.; Ratsimbazafy, J.H.; Waeber, P.O. Ten years into the marshes–*Hapalemur alaotrensis* conservation, one step forward and two steps back? *Madag. Conserv. Dev.* **2015**, *10*, 13–20. [CrossRef]

76. Levins, R. The strategy of model building in population biology. *Am. Sci.* **1966**, *54*, 421–431.

77. Garcia, C.; Dray, A.; Waeber, P. Learning begins when the game is over: Using games to embrace complexity in natural resources management. *GAIA-Ecol. Pers. Sci. Soc.* **2016**, *25*, 289–291. [CrossRef]

78. Reibelt, L.M.; Moser, G.; Dray, A.; Randriamalala, I.H.; Chamagne, J.; Ramamonjisoa, B.; Garcia Barrios, L.; Garcia, C.; Waeber, P.O. Tool development to understand rural resource users' land use and impacts on land type changes in Madagascar. *Madag. Conserv. Dev.* **2017**. [CrossRef]

79. Stoudmann, N.; Waeber, P.O.; Randriamalala, I.H.; Garcia, C. Perception of change: Narratives and strategies of farmers in Madagascar. *J. Rural Stud.* **2017**, *56*, 76–86. [CrossRef]

80. Maret, F. Madagascar. In *Distortions to Agricultural Incentives in Africa*; Anderson, K., Masters, W.A., Eds.; World Bank: Washington, DC, USA, 2009; pp. 101–125.

81. Reibelt, L.M.; Nowack, J. Community-based conservation in Madagascar, the 'cure-all' solution? *Madag. Conserv. Dev.* **2015**, *10*, 3–5. [CrossRef]

82. Fowler, R.; Rockstrom, J. Conservation tillage for sustainable agriculture: An agrarian revolution gathers momentum in Africa. *Soil Till. Res.* **2001**, *61*, 93–108. [CrossRef]

83. Denich, M.; Vlek, P.L.; de Abreu Sá, T.D.; Vielhauer, K.; Lücke, W. A concept for the development of fire-free fallow management in the Eastern Amazon, Brazil. *Agric. Ecosyst. Environ.* **2005**, *110*, 43–58. [CrossRef]

84. Tittonell, P.; Giller, K.E. When yield gaps are poverty traps: The paradigm of ecological intensification in African smallholder agriculture. *Field Crops Res.* **2013**, *143*, 76–90. [CrossRef]

85. Wilmé, L.; Waeber, P.O.; Moutou, F.; Gardner, C.J.; Razafindratsima, O.; Sparks, J.; Kull, C.A.; Ferguson, B.; Lourenco, W.R.; Jenkins, P.D. A proposal for ethical research conduct in Madagascar. *Madag. Conserv. Dev.* **2016**, *11*. [CrossRef]

86. Ravaka, A.; Ramamonjisoa, B.S.; Ratsimba, H.R.; Ratovoson, A.N.A. Circuit court du marché des produits agricoles: Pour une gestion efficace du paysage ouvert, cas du bassin-versant de Maningory, Madagascar. *Madag. Conserv. Dev.* **2017**, *12*. [CrossRef]

87. Stoudmann, N.; Garcia, C.; Randriamalala, I.H.; Rakotomalala, V.A.G.; Ramamonjisoa, B. Two sides to every coin: Farmers' perceptions of mining in the Maningory watershed, Madagascar. *Madag. Conserv. Dev.* **2016**, *11*, 91–95. [CrossRef]

88. Gardner, C.J.; Waeber, P.O.; Razafindratsima, O.H.; Wilmé, L. Decision complacency and conservation planning. *Conserv. Biol.* **2018**. [CrossRef] [PubMed]

![forests logo] *forests*

MDPI

Article

The Impact of Resident Participation on Urban Woodland Quality—A Case Study of Sletten, Denmark

Hanna Fors [1,*], Märit Jansson [1] and Anders Busse Nielsen [2]

[1] Department of Landscape Architecture, Planning and Management, Swedish University of Agricultural Sciences, SE-230 53 Alnarp, Sweden; marit.jansson@slu.se
[2] Department of Geosciences and Natural Resource Management, University of Copenhagen, DK-1958 Frederiksberg C, Denmark; abn@ign.ku.dk
* Correspondence: hanna.fors@slu.se; Tel.: +46-40-415165

Received: 22 September 2018; Accepted: 23 October 2018; Published: 25 October 2018

Abstract: Despite the potential of urban woodlands for recreational use and participatory management, citizens' perception of urban woodland quality, as well as the impact of citizens' co-management on urban woodland quality, have not been thoroughly studied to date. The present study investigated how residents in Holstebro, Denmark define urban woodland quality in their neighborhood named Sletten and how they perceive the quality impact of their participation in the management and maintenance of a transition from private gardens to public urban woodland—the so-called co-management zone. Field survey of participation for all housing units with a co-management zone ($n = 201$) informed strategic selection of residents for individual interviews ($n = 16$). It was found that social, experiential, functional, and ecological dimensions are all part of residents' perception of urban woodland quality, whereby maintenance, accessibility, and nature are dominating aspects of these dimensions. While these aspects are already integrated in quality assessment schemes for other types of urban green space, our study revealed the importance of structural and species diversity between and within woodland stands as central for the perceived woodland quality—a quality aspect that distinguishes woodland from other types of urban green space. Participation in the management and maintenance positively influenced the perceived woodland quality. Residents found that their participation in the co-management zone created functional and ecological, physical qualities in the woodland. Moreover, the active participation provided the residents with a range of social and experiential benefits, many of which they themselves argue that they would have missed out on if they were only allowed to use the woodland "passively". These findings suggest a large—but also largely untapped—potential of participatory urban woodland management to contribute physical qualities to urban woodlands and benefits to its users.

Keywords: green space quality assessment; user participation; urban woodland management

---

1. Introduction

Historically, forests are a natural and indispensable part of most people's lives. People used the forest and left their imprints in the form of winding tracks, traces of work and fire, coppicing of firewood, and so forth. People were dependent on the forests, all year round, and at all times [1]. Only when modern methods of forest management were introduced in the 19th century did forests become a field of action for specialists, i.e., foresters [1]. Nowadays, "ordinary people" are again claiming forests, not only for recreation, but increasingly also to participate as volunteers in their management.

The increased focus on engaging local communities in their "neighbor-wood" is part of a general governance trend toward increased user involvement in the management of local green space, especially in urban settings [2]. This governance approach is due, in part, to widespread agreement on the many benefits citizens gain from using local green spaces such as urban woodlands [3,4], and the importance of involving users in decisions regarding their everyday landscapes [5].

There are several societal trends currently affecting user involvement in the management of urban woodlands (i.e., wooded areas more than 0.5 ha in size, located within an urban context) and other types of urban green space (such as parks, street trees, and neighborhood green spaces). One of these involves the cuts in maintenance budgets forcing authorities to find alternative solutions to maintaining public green space quality, including public–private partnerships and user participation [6]. Woodlands constitute a vital component of urban green infrastructure in terms of both areal cover and ecosystem service provision [7–9]. It is, therefore, particularly relevant and potentially beneficial to involve users in the co-management of urban woodlands. Despite this, there is a lack of knowledge on how participation is affecting urban woodlands and their quality, something which is more studied in other types of urban green spaces. Yet, in most studies, the benefits of participation, including the physical outputs to urban green spaces are assumed rather than empirically evaluated [10].

When local authorities involve users in public green space management, the different actors bring different forms of knowledge to the process and, as a result, their participation may create green spaces that differ from the results of management by local authorities and with other quality norms [11]. The few empirical studies conducted to date show that users participating in management can benefit through an increased sense of satisfaction with their neighborhood [12], greater recreational and social use of green space [13,14], and an increased sense of attachment to the green space [5]. In empirical studies where users' physical participation is part of nature conservation, outcomes benefiting the users involved, e.g., environmental awareness or social cohesion, are labeled "co-benefits", which are either the first step toward more direct benefits to nature conservation or, sometimes, in conflict with these [15]. The majority of empirical studies of physical participation in community woodlands in the United Kingdom (UK) focused on outputs, e.g., number of trees planted, while only 21% studied outcomes, e.g., enhanced neighborhood, or well-being [16]. Green space quality for users becomes a secondary priority when nature conservation, rather than physical qualities in general, is the overall aim. Fors et al. [10], therefore, argue that more empirical studies are needed to develop the knowledge base on how user participation may impact green space quality for both users and the physical environment. In the present study, a "co-management zone" (see definition in Section 2) in a publicly accessible neighborhood woodland in the residential area of Sletten, Holstebro, Denmark was used as a case study in an exploration of how urban woodland quality is affected by residents' participation in management.

"Quality" of urban woodland and other types of green space is a contested concept. Practically all definitions and models of quality for a specific green space are debatable, and include some values and world views of particular actors and interests while excluding others [17]. No global or universal definition of quality exists; rather, the definition that is most appropriate varies depending on the specific situation and context [18]. Claims that quality lies in the eyes of the beholder [11] and that it is context-dependent [18] indicate that the concept is very subjective and practically impossible to measure. That said, there are some aspects of public green space quality, in focus in this study, that reflect general preferences, such as being well-maintained, safe places with vegetation [19]. These aspects are also reflected in existing assessment schemes and tools designed to measure green space quality. The majority of these schemes and tools deal specifically with how green space quality is associated with physical activity, an aspect also noted in earlier reviews [20,21]. However, quality assessments of green spaces on a regional or national scale become synoptic by nature, focusing on, e.g., tree canopy cover, and they are unable to capture what benefits local green space users. On searching the literature dealing with quality assessments schemes for local green spaces in relation to use, some overall aspects emerge repeatedly, namely "maintenance", "accessibility", "nature", and "facilities" [20–24].

While woodland is generally distinguished from other types of urban green space (e.g., References [25,26]), quality assessment schemes focusing specifically on "urban woodland" do not—to our knowledge—exist. Several studies claim that user-perceived quality influences park use more than objectively expert-measured green space quality (see, e.g., Ries et al. [27]), which could potentially also be valid in relation to user participation. Aspects of urban woodland quality as perceived by local communities or the general public have nonetheless been studied. A study in the UK of attitudes toward urban woodland vegetation showed that people of all ages associated meanings such as "relaxation", "peacefulness", "seasonal change", "scenery", and "education" with urban woodlands, whereas people aged over 65 particularly valued the woods for their links with the past and opportunities for deeply engaging in nature, and were more concerned about their personal security in the woodland [28]. A study of local woodland use in Scotland identified freedom from rubbish as being the most important physical quality to people, and directional signs, good information boards, variety of trees, and tidiness of appearance as being the most decisive physical qualities for woodland visits/use [29]. Most respondents in that study reported that they feel at peace in woodlands. Ode and Fry [30] developed a model for quantitative assessment of visitor pressure on urban woodlands on a regional scale in Sweden. Distance and access to woodland were found to be the main factors; however, the woodland qualities "size" (large enough to provide a forest feeling), "forest structure" (broad-leaved forests preferred for their diversity) "path density", and "protection status" (since protected areas have recognized botanical, cultural, or recreational qualities) also affected visitation rates [30]. Based on both preference studies and expert assessments, it was concluded that visual aspects that are important for urban woodland management can be reduced to scale, structural and species diversity, naturalness/continuity, stewardship, visual accessibility, and coherence, with all dimensions except coherence being well represented in management guidelines from the UK and Sweden included in that review [31]. Nielsen and Jensen [32] developed this further, concluding, from an expert perspective, that different planting designs for urban woodlands had different visual qualities, focusing on scale, diversity, naturalness, and visual accessibility, with mature woodland generally having higher levels of visual qualities than young woodland. In Finland, survey respondents commonly associated their favorite green spaces (mainly urban woodlands) with peacefulness, the feeling of forest, naturalness, and functionality [33]. A study in the UK showed that residents saw the following qualities in local urban woodlands: important for nature and wildlife conservation and human co-existence with nature; making residents aware of natural cycles and seasonal change; connecting them with nature giving them existential experiences; relaxation, contentment, and stress relief, and the feeling of being in a rural idyll [34]. At the same time, woodlands with valued qualities were also perceived as unsafe, due to a perceived lack of woodland management among some people who, therefore, probably derive less restorative benefits from urban woodland use [34].

The findings regarding urban woodland quality identified through the perception and preference studies referred to above reflect and confirm the four dimensions of urban woodland design as defined by Bell et al. [35], "the social", "the experiential", "the functional", and "the ecological", and that they all need to be considered when assessing urban woodland quality. Acknowledging the importance of employing a situation and context-specific quality definition [18], we, therefore, converted design dimensions into an assessment scheme for "urban woodland quality" (Table 1). The scheme resembles existing quality schemes for urban green spaces [20–24], as it also includes the aspects of maintenance, accessibility, nature, and facilities.

In the assessment schemes referred to above, quality is described as consisting of different aspects that need to be present for a high-quality local green space (e.g., maintenance, accessibility, and so forth) as opposed to quality assessments of green spaces on a regional or national scale. In this study of the impact of participation on urban woodland quality, such aspects were included, but with a focus on the result of *outcomes* of participation [15,16] as *perceived* by residents. Participation outcomes affecting users could be seen as *benefits* to users, while outcomes affecting the physical woodland could be seen as *physical qualities* being created.

Table 1. Assessment scheme for "urban woodland quality", based on dimensions presented by Bell et al. [35].

Dimension	Aspects of Dimension	Indicators for Urban Woodland Quality
Social	Escape	Is the possibility to escape the urban scene provided? Impression of naturalness/wildness? Are cultural references incorporated to help people identify with their community?
	Social activities	Are there possibilities for social activities (e.g., walking, sitting, socializing with friends, children's play)? Is there a mix of larger and smaller spaces for different activities?
	Safety and security	Is there greater visibility along paths and beneath trees? Are there more obvious signs of management presence? Is there clear signposting?
Experiential	Aesthetics	Are multi-sensory experiences available? Is seasonal change perceivable?
	Design style	What degree of control or active presence of people is shown in the design of paths, planting patterns, and open spaces? Do they affect the user experience?
	The role of the urban forest in urban life	Does the woodland provide a non-urban experience? Is there a sense of timelessness and continuity? Does the urban woodland act as a stepping stone between built city and nature?
Functional	Accessibility	Is the woodland accessible to all societal groups?
	Carrying capacity	Is the woodland designed to satisfy both physical and visual carrying capacity? Are there winding paths among trees or straight paths in the open?
	Climate	Do woodland trees provide the site-specific desired climate-regulative functions (e.g., shade, shelter from the wind, and moderation of extreme temperatures)? Is year-round use possible?
Ecological	Urban ecology	Does the urban woodland help improve or revitalize the natural capital of an urban area (e.g., increase of ground water infiltration, soil amelioration, or erosion control)? Are new habitats developed?
	Landscape ecology principles	Were landscape ecology principles employed as a key part of the design process (e.g., linking corridors to connect scattered habitat fragments, and allowing wildlife species to move in between)? Do woodland design and management promote habitat diversity (not necessarily only natural habitats)? Is it possible for people to get close to nature in their everyday lives?

Using the scheme for the assessment of "urban woodland quality" (Table 1) as an analytical framework, the present study explores how urban woodland quality, in general and as affected by residents' participation in management and maintenance, is described by residents. The research was guided by the following research questions:

- How do residents perceive (residential) "urban woodland quality"?
- How do residents describe "urban woodland quality" as affected by participation?

2. Materials and Methods

2.1. The Case Area: Sletten

The study of urban woodland quality, in general and as affected by local residents' participation in management and maintenance, was conducted in north-western Denmark in the 160-ha large urban neighborhood Sletten (The Plain), Holstebro city (Figure 1). Sletten was developed in 1999–2004, including commercial areas (20 ha) and 400 housing units arranged in eight forest villages, six fortress villages, and a retirement home (21 ha). The housing is set in a matrix of new woodland plantings (32 ha) and pastures (30 ha), intersected by the road infrastructure (27 ha), existing shelterbelts, wetlands, and natural brooks (30 ha) that flow into the lake. The present study focused on the so-called "forest villages", i.e., the eight housing areas within Sletten that are surrounded by woodland (*n* = 201 housing units). The residents in the forest villages are a rather homogeneous societal group of middle-class people.

The woodlands in Sletten were established as a publicly accessible "landscape laboratory" in three phases, in parallel to residential development, in the period of 1999–2004. Landscape laboratories are experimental woodland areas in a local landscape context where innovative design and management concepts for urban forests are tested in full scale [36]. The woodland design comprised 52 stand types and 85 tree and shrub species, resulting in differing appearance (e.g., tree height, planting distance, vegetation structure, and species composition) between different parts of Sletten [32].

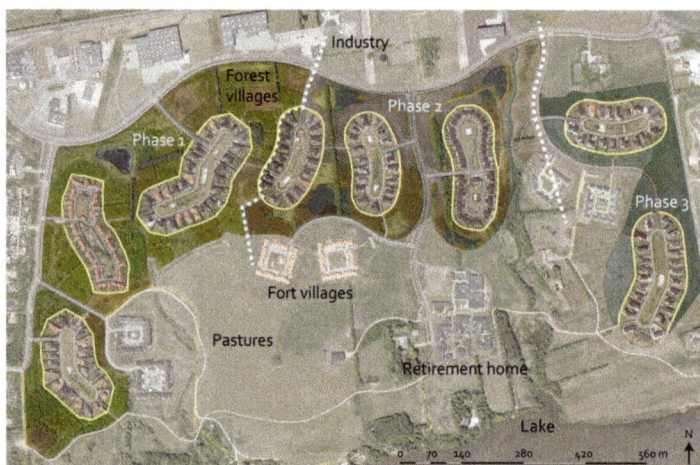

Figure 1. Plan of Sletten. Varied woodland surround the forest villages. The colored fields in the woodland correspond to the 52 different stand types. The yellow border around each forest village shows the stipulated width of the co-management zone, i.e., the first 4 m of the public woodland. Based on an aerial photo, ®GST.

Early on, some individual residents in the forest villages on their own initiative started weeding around the planted seedlings or growing flowers and vegetables at the woodland edge. As the tree canopy started closing, residents engaged in pruning and thinning amongst the trees, planting their own plants, providing nesting and feeding boxes for birds, setting up hammocks, placing garden furniture, making paths or huts as part of children's play, and so forth. These activities were tolerated and even encouraged by the local authorities as they created a gradual transition from the plant communities, maintenance levels, and activities in private gardens to those of the public woodland. The local authority green space manager regarded this transition and resident engagement as positive for the long-term integration of residential housing and woodland and for the residents' attitudes to having a neighboring woodland, in particular as the trees grow taller and shade the gardens. In 2010, these resident activities became formalized into collaboration in a so-called "co-management zone' (Figure 1) with guidelines set by the local authority green space managers:

- The co-management zone extends 4 m into the woodland (three planting rows) and must be accessible to the public.
- Each household may choose whether and to what extent to participate in the section of woodland edge that borders its property (i.e., the width of its garden).
- A minimum of 30% of the originally planted trees and shrubs (planted with a spacing of 1.5 m × 1.5 m) must be retained.
- Up to 40% of the trees may be replaced with other trees or shrubs.
- Up to 30% of the trees may be replaced with herbaceous plants, etc.
- Weeding, pruning of trees and shrubs, removal of field layer vegetation, and other management and maintenance activities should respect and maintain a forest character.
- Establishment of permanent structures such as sheds and greenhouses is not permitted, nor is keeping storage space for firewood, tools, garden compost, etc.

The guidelines were distributed to all residents and meetings were arranged to give inspiration and clarify questions and uncertainties. The guideline document also provided inspiration in the form of a list of suitable woody plants, summer flowers, vegetables, and woodland herbs. Procedures for guideline enforcement were not described; rather, residents were encouraged to contact the local authority when in doubt about whether a specific management action was permitted. Since then, manager presence in the neighborhood and enforcement of the co-management zone guidelines were limited, and information about the co-management zone was not distributed to newcomers.

Participation in Sletten

Field surveys of physical signs of resident participation in the woodland management in Sletten conducted in 2010 and 2015 showed an increase in participation, from 41% in 2010 to 65% in 2015, out of the 201 households with gardens bordering on the woodland. From the field surveys, four main types of resident participation were distinguished: plant maintenance, plant establishment, function establishment, and misuse, i.e., all actions prohibited in the guidelines [37].

2.2. Individual Interviews with Residents

In green space where participation occurs, participation affects both participants' and non-participants' recreational experiences. Acknowledging this, we aimed to include both non-participating and participating residents as interviewees. Information on participants and non-participants was retrieved from the 2015 field survey of participation in the co-management zone. The local authorities of Holstebro assisted in booking interviews with the residents in their homes during four consecutive days in October 2017. In total, 16 residents were interviewed, of which only two were non-participants, as non-participants were less eager to participate in the study. Residents were approached in each of the eight forest villages, in order to obtain an even spatial distribution in the neighborhood and to capture potential local variations caused by, e.g., differences in woodland attributes. Eventually, residents from one to three households from each forest village were interviewed. It varied between interviews whether one or two family members were at home. Because of this, some of the interviews were conducted with two family members in the household, resulting in 21 interviewees in total (Figure 2). Interviewees were between 31 and 79 years old, with a mean age of 55.1 years. This can be compared with the mean age of all Sletten residents, which was 46.8 years (standard error (SE) = 1.491 years) in 2015.

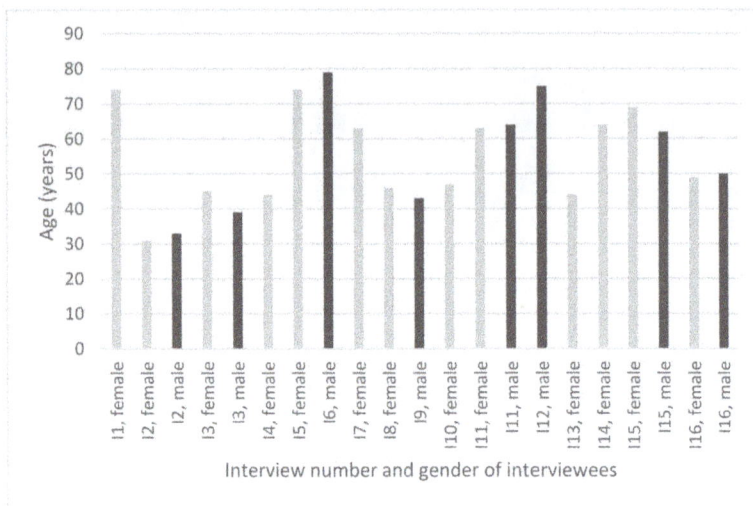

Figure 2. Age and gender of the 21 residents interviewed in the 16 interviews. I1 = Interview 1, etc.

Interviews were semi-structured and focused on residents' definitions of urban woodland quality, descriptions of their own participation, and views on the impact of resident participation on urban woodland quality. Each interview lasted 27–77 min and was audio-recorded and transcribed verbatim. Interviews were qualitatively analyzed by coding, followed by categorization of codes [38]. Finally, findings were structured according to the quality assessment scheme for urban woodland quality (Table 1).

3. Results

3.1. Resident Definition of "Urban Woodland Quality"

3.1.1. Nature Experience

The residents interviewed were asked to describe what urban woodland quality in Sletten meant to them. Quality aspects mentioned motivated a further development of the experiential dimension of the urban woodland quality scheme by adding the aspect "structural and species diversity" and "management and maintenance style" to the existing "design style", the latter since both design and management and maintenance influence user experience (Table 2). Many interviewees (hereinafter referred to as "I" for interview followed by interview number, e.g., I12) mentioned the possibility of experiencing nature and wild animals such as birds, squirrels, and roe deer, i.e., part of the social and experiential quality dimensions, or natural woodland for its own sake, i.e., part of the ecological dimension (I1, I2, I4, I6, I7, I11, I13, I15, I16). An interviewee enjoyed following the animals throughout the year: "In the wintertime, you can see all the animals inside the forest, and, in the summertime, they simply come out. It is fantastic to sit and look at the squirrel flying around in all the trees" (I1). Interviewees found it important that the woodland was not too "plantation-like", but rather, looked like "wild nature" (I4), as well as functioned like "natural nature", since it led to increased biodiversity: "the trees that die, they die, and then some insects can live in the half-dead trees" (I11).

Table 2. Urban woodland quality as perceived by residents in relation to the urban woodland quality assessment scheme based on Bell et al. [35]. "Management and maintenance style" and "structural and species diversity" were added to make the scheme better reflect user perceptions.

Dimension	Aspects of Dimension	Resident-Perceived Urban Woodland Quality
Social	Escape Social activities Safety and security	Nature experience
Experiential	Aesthetics Design, management and maintenance style The role of the urban forest in urban life Structural and species diversity	Management and maintenance Nature experience Structural and species diversity
Functional	Accessibility Carrying capacity Climate	Accessibility Facilities (paths)
Ecological	Urban ecology Landscape ecology principles	Management and maintenance Nature experience

3.1.2. Structural and Species Diversity

Some residents mentioned structural and species diversity of the woodland stands as important for quality, including diversity and density of the woodland and species characteristics (I2, I3, I5, I9, I10, I13, I14, I15). One interviewee thought that woodland density was a quality aspect that affected use during walks: "that the forest is dense, that it is nice to walk in, and that there is a path to walk on ... that it is not too open" (I3). Some appreciated species diversity and said that it meant that wild animals such as birds, squirrels, and roe deer kept coming close to the gardens (I13 and I15), while others described diversity more in terms of variety in experiences along a walk than in species diversity per se, e.g.,

I think it [the woodland] is very diverse ... Different forests in one way or another. Different trees. You walk out there, and all of a sudden you are out in something open, you turn right and then you are inside something, so different forests ... You get different experiences. (I10)

Interviewees even linked the experience of woodland diversity to human well-being: "I believe it is good for the soul ... because it is a great sense-experience to get the impressions from the different woodland stands" (I14). An aspect of species characteristics is the age of trees. An interviewee looked forward to the succession of the woodland, since older trees have a higher play value for children than the black thorn they had in their co-management zone: "With small children, it is no fun to make a den in black thorn. It is better with some old oak trees" (I2). Possibility to pick edible berries and fruits was also perceived as a quality associated with woodland diversity and species characteristics (I2 and I9), something that could make it fun for children to come along to the woodland "because there is something to come for" (I9).

3.1.3. Accessibility

Other aspects of urban woodland quality that were repeatedly mentioned by the interviewees were that they were allowed to use the woodland and that it was accessible at their doorstep, to look at from inside, as well as for use, where residents especially emphasized the importance of functional paths (I1, I3, I4, I6, I7, I8, I10, I12), and that "it would be a disaster to fence in the forest" (I6). This means that, under the functional dimension, they mentioned accessibility and the single facility *paths*. An interviewee viewed the accessible co-management zone as part of urban woodland quality and woodland access: "I just think it is nice that it is there and that you are allowed to use it. That there is no barrier against each garden, but the co-management zone instead of a large hedge in the border.

It is just the freedom to use it as it is" (I8). Another interviewee described how the paths and the proximity made nature accessible to her:

> The trampled paths let you get into the nature rooms with trees and water and where you are completely in nature, where you completely shut the rest of the world out, I think that is incredibly nice, so that I see as high quality ... The best thing about taking the trampled path is that if you walk from the right end, you get to gaze over the lake several times. If you go the opposite direction you have to turn around to be able to see it, because of where the trees stand and how the path turns. That I find unique, to have so close by. That when you walk down there, you have the trees in the background and then you have [the view] over the lake. Sometimes it is motionless and the sun is about to set over it or about to rise or mirrors in the lake, and other times it is nasty weather and rainy and windy and a restless water surface. But both are just as life-affirming. It is when it becomes life-affirming that I think it is high quality. (I4)

3.1.4. Management and Maintenance

Two of the interviewees thought that, to achieve a high-quality woodland in Sletten, better management or maintenance would be needed, i.e., thinning of the dense woodland to allow better development of remaining trees (I11), i.e., part of the ecological dimension, and weeding and more frequent mowing of the high grass between the trees in some parts of the neighborhood for a better appearance, i.e., part of the design, management and maintenance style of the experiential dimension (I16).

3.2. Impact of Resident Participation on "Urban Woodland Quality"

Table 3 shows the impact of resident participation on urban woodland quality, as described by the residents interviewed, charted through the assessment scheme for urban woodland quality based on Bell et al. [35] in the version where the experiential dimension was further developed as described in Section 3.1.1. Some households extended their participation in the woodland beyond the stipulated 4-m-wide co-management zone. Outcomes of participation primarily benefiting users are rarely visible in the physical landscape, but could nonetheless be very valuable for the individual. For this type of benefit, there was a predominance of outcomes benefiting participating residents (14 positive outcomes) over outcomes benefiting all residents (10 positive outcomes). The majority of the physical qualities created as an outcome of participation benefited all residents (45), while 16 outcomes benefited individuals. Outcomes impacting users are henceforth termed *benefits*, while outcomes impacting the urban woodland are termed *physical qualities*. Figure 3 shows four examples of resident-created environments and lists the outcomes participation had for the interviewee participating in that particular part of the co-management zone.

Table 3. Outcomes of participation in Sletten compared against the assessment scheme for "urban woodland quality", based on Bell et al. [35]. "Management and maintenance style" and "structural and species diversity" were added to make the scheme better reflect user perceptions. Resident-described outcomes of their own participation, as well as outcomes seen (or believed to result) from participation by other residents. Examples of codes: 2 (I) means that two residents mentioned this participation outcome and that it benefits the individual; 4 (A) means that four residents mentioned this participation outcome and that it benefits all/many residents in the neighborhood.

Dimension	Aspects of Dimension	Participation Outcome Affecting the Physical Environment	Participation Outcome Affecting Residents	Participation Outcome Affecting Participating Residents
Social	Escape		Nature experience 4 (I)	Sense of community 4 (I)
	Social activities	Better usability 4 (I), 1 (A)	Social interaction 2 (I), 1 (A) Increased use of urban woodland 6 (I)	Empowerment 4 (I) Increased participation from inspiration 2 (I)
	Safety and security		Improved safety 1 (I)	Bird boxes as pest control
Experiential	Aesthetics	Experiencing domesticated animals 1 (I), 1 (A)		Happiness and pleasure 4 (I) Relaxation 4 (I)
	Design, management and maintenance style	Better appearance 5 (I), 4 (A)	Enjoyable experiences during walks 3 (A)	Participation as personal hobby 1 (I) Enhancement of private garden 7 (I)
	The role of the urban forest in urban life		Nature experience	Memories stored in resident-planted trees 2 (I) Enhanced view of woodland from inside 3 (I) Recreational experiences for participants 1 (I)
	Structural and species diversity	Better appearance 2 (I), 1 (A)	Enjoyable experiences during walks 4 (A)	
Functional	Accessibility		Better accessibility to woodland 1 (I), 1 (A)	Food 7 (I), 2 (A)
	Carrying capacity	Paths 2 (I), 10 (A)		Firewood 1 (I)
	Climate	Wind-sheltered environment 1 (I) Better usability 1 (I)		Storing firewood in woodland 2 (I)
Ecological	Urban ecology	Fertilizing the woodland/creating mold 7 (A) Increased biodiversity 15 (A) Better tree development 5 (A) Bird boxes as pest control 1 (A)	Clean air 1 (A)	Environmental awareness 2 (I)
	Landscape ecology principles	Increased biodiversity	Nature experience	
Sum of participation outcomes and whom it benefited		Individual participants: 16 All residents: 45	Individual participants: 14 All residents: 10	Individual participants: 44 All residents: 2

205

Figure 3. Exemplifying photos of the co-management zone. For each photo, the outcomes that the resident at the address described of their own participation are listed: (**a**) enhancement of private garden, empowerment, food, increased use of urban woodland, relaxation, increased biodiversity, environmental awareness, nature experience, and better accessibility to woodland; (**b**) wind-sheltered environment, food, increased use of urban woodland, relaxation, memories stored in resident-planted trees, increased biodiversity, better appearance, and environmental awareness; (**c**) happiness and pleasure, food, fertilizing the woodland/creating mold, and better usability; (**d**) happiness and pleasure, enhancement of private garden, increased biodiversity, and better appearance.

3.2.1. Participation Outcomes—The Social Dimension

Several residents, through participation, adapted the woodland to suit their recreational needs and create possibilities for social activities, leading to the outcome *better usability*. Residents, e.g., put out benches in the woodland, making it possible to sit there. Others pruned trees to make room for a hammock, made room for social activity by maintaining a space in the zone keeping it open to provide space for play activities with grand children, or made a glade in the woodland for barbecue (BBQ) parties with neighbors. An example of the same participation outcome, but benefiting more residents than only participants, was when a couple, together with their two neighbors on one side, made a glade in the zone with a table, where the four of them ate lunch together every now and then during the summer; however, they allowed anyone who wanted to use the table (I15).

The benefit *nature experience*, e.g., being able to watch roe deer and squirrels right outside the garden and finding it nice that wild birds use residents' bird houses, corresponds to aspects of the social, experiential, and ecological dimensions of urban woodland quality. Although the co-management zone guidelines allow for individual participation without necessarily coordinating or collaborating with neighbors, participation led to *increased social interaction and sense of community* between some participants. Even though they did not work together in the zone, they appreciated knowing that others participated as well, enjoying the fellowship between participants and the possibility to share ideas and to show each other what they did. Participation also led to more socializing with non-participating neighbors, e.g., a participant sometimes met neighbors when she was thinning among the trees in the zone. Another participant regularly talked with his neighbor in the zone, instead of over the too high hedge between their gardens. A third interviewee organized a trail run around the forest village for the neighboring children—two laps on a path some boys created in the zone. Social interaction between neighbors was also achieved when a resident who was good at growing tree seedlings helped others plant trees in their zones. Participation also led to *increased woodland use*, among both adults and children. One interviewee said that he would not have used it as much if he was not allowed to also influence the woodland (I2). One interviewee described how participation in the zone was important for children's play:

[Residents living on the other side of the forest village] used a lawn mower to make some mowed grass paths, allowing for not only a single entrance and exit, but possibility to get in and out in several different places. I believe that gives them a greater sense of community . . . I have the feeling that the children living over there . . . they had a great deal of pleasure out of being able to run [through the woodland] to each other and meet in the co-management zone and play there. (I4)

However, for some children, the possibility to participate did not lead to *increased woodland use*, due to woodland characteristics. Early on, the trees were too small to climb and children living next to black thorn could not build dens there. Outputs of participation in this category included *improved safety* through well-balanced pruning so vegetation would not grow too large, while remaining rather closed to screen the public path, since a burglar once went into this resident's house when having more open vegetation (I14). There were also a few participation outcomes affecting the physical participation process itself for individual residents (six positive outcomes), part of the social quality dimension: Firstly, the *empowerment* of residents was mentioned by some residents, appreciating the possibility to influence the woodland as they wanted, instead of watching the woodland outside the garden grow too dense and dark. Secondly, a few residents *increased their own participation after being inspired* by seeing other residents' actions in the zone. An interviewee described how she got really inspired from once seeing some residents living on the other side of her forest village pruning their trees, which she thought looked really nice (I8). However, an interviewee who saw lots of things other residents made in the zone, had not yet thought: "Wow! That's something I will do as well!", i.e., their participation did not have high enough quality for her to become *inspired to participate* more herself.

3.2.2. Participation Outcomes—The Experiential Dimension

Participation provided the possibility to *experience domesticated animals*, e.g., a resident enjoying when a neighbor, for a period of time, had pheasants which he let out in the woodland. Several residents said that participation led to *better woodland appearance*. The small original woodland trees were planted in rows; thus, a resident described how he removed some of them to create a more natural path flow, and planted some new, more interesting tree species in between the woodland trees, making a path system of his own (I9). In some cases, this outcome benefited all residents, e.g., where a resident planted hundreds of trees in Sletten from seedlings collected in the woodland and other places, as well as sowed lupins and planted lily of the valley in areas far beyond his own zone (I12). These are two examples of *better woodland appearance* leading to structural and species diversity (Table 3). Other residents "beautified" the woodland by pruning trees, planting winter aconite and flower bulbs, tidying up, removing dead nurse trees, or mowing the grass in the woodland, i.e., *better woodland appearance* as a part of design, management and maintenance style. Low accessibility from paths to the co-management zone or residents feeling uncomfortable walking too close to private gardens meant that some of them did not see other residents' management and maintenance actions, and therefore, they did not benefit from it; for them, participation did not lead to *better woodland appearance*. Another example of this is a resident who thought that, instead of fluent transitions, many residents made ornamental gardens with bark chips in the entire zone and pots with annual flowers, which she found too unnatural and not beautifying (I1). What looked like non-participation for a passerby was sometimes, in fact, conscious resident participation in the management, creating invisible qualities. A woman and her husband created *better woodland appearance* by removing the fruit trees and some other plants that the former owner planted in the zone, considering these to be too gardenlike, and therefore, not suitable for a wild, natural woodland when striving to create a nice transition from well-maintained garden to the wild woodland with no hedge in between (I7). Apart from removing, invisible participation also took the form of refraining from ornamental gardening in the zone. While a man said that they did not have a gardening interest, his wife said they did, and further explained that they did not find it suitable to have an ornamental garden that close to "nature" (i.e., the woodland). They liked to have an ordered, but at the same time naturalistic garden, and aimed for a fluent transition from garden to woodland. Therefore, they made conscious choices of natural materials, e.g., used wooden posts to hang their hammock and made a fireplace with natural stumps to sit on, the idea being that wood goes well with the woodland. They also trimmed the pine trees that they planted in their garden together with some pillar fruit trees, thinking that the woodland trees should be higher than their garden trees (I16).

Residents reported having interesting, exciting, surprising, fun, and diverse *enjoyable experiences during walks* from seeing other residents' actions in their zones. One interviewee enjoyed the zone characteristic diversity, widely varying between neighbors, e.g., with regards to open vs. closed woodland appearance and the number of trees being replaced by other species, i.e., part of structural and species diversity (I13). Another interviewee described surprises during her walks along woodland paths: "I often think that, all of a sudden: 'Oh! It seems like some trees have just popped up here, someone has been working, and here is a new path as well!'" (I10). A man thought that paths running through parts with uniform, thicket-like parts of the woodland were hardly used, since it was too boring to walk there, while finding it exciting to pass by places where residents influenced the woodland through replacing trees and planting new plants (I9). Experiential outcomes also included getting *a personal hobby from participation* and feelings of *happiness*, *pleasure*, and *relaxation*. A man described the relaxation he gains from participating and simply spending time in the physical environment he created through participation:

> I have noticed that some time passes by when I am out there [in the co-management zone] and just enjoy the quietness ... out there. It's more clinical in here [in the garden] while there is more peace out there, with birds and insects. So I really like just walking around out there. I spend a lot of time there! (I2)

One of the most common outcomes to users directly linked to the physical environment was participation leading to *extension or enhancement of the private garden* by pruning or thinning among the trees for more evening sun, or weeding in the zone to limit weeds from spreading into the garden and attempting to "get the forest into the garden" (I10). Another household also managed to do this, saying that participation gave them a totally different garden: "When we sit on our terrace and look [toward the woodland], we almost think that we have Amalienborg royal park! Nothing less! We just have a plain boring garden like most people, but the [woodland] trees in our backyard, they take it all to the next level" (I16). Other outcomes related to experiential and recreational values were when *participants gained recreational experiences*, and *the view of the woodland from inside residents' houses was enhanced*. Another example of the invisible participation described further above, is interviewees wanting to improve their view from the garden, resulting in them tidying up the woodland after the former owner left trees and branches after thinning, as well as several old Christmas trees on the ground, and also resulting in them removing an old deserted children's den (I3). For the two households that had *memories stored in the resident-planted trees*, participation had a symbolic value. A man felt connected to the trees he planted through participation: "I know every single tree that I have planted. They are my grandchildren—I have many grandchildren!" (I12). Another man enjoyed following the growth of the trees he planted, especially the little spruce his child sowed when still in kindergarten: "It grows in the forest today and that's nice to see. We cherish it because it has a symbolic value to us that it stands there" (I9).

3.2.3. Participation Outcomes—The Functional Dimension

Many residents either maintained *paths* for more private use close to home or longer paths further into the woodland, keeping a hand pruner in the pocket cutting twigs along paths during walks, or regularly mowed a longer grass path for everyone to use, or regularly used paths they knew other residents created. One interviewee mentioned that two teenage boys pruned trees and trimmed grass to make a 1.2-km-long mountain bike path around an entire forest village, a path that later was mostly used by adults going for a run (I9). A resident created a more *wind-sheltered environment* by putting straw around exposed trees. The physical quality *better usability* was created by an interviewee who adapted the woodland for it to provide the climate-regulative functions she desired (e.g., shade and shelter from the wind), making a nice place for herself: "I have actually made something cave-like over here, where I have felled or pruned some trees" (I1). Some reported that participation led to *better accessibility* to and within the woodland, where, e.g., a resident meant that, when people participated, woodland vegetation became less dense, facilitating ease of movement during walks. However, for one interviewee, participation did not increase her *accessibility* to woodland, since her thicket-like part of the woodland was practically impossible to participate in (I4). Functional outcomes also included concrete outputs such as getting *firewood* from trees residents felled in the zone, *storing private firewood in the woodland*, and *food* for participants (fruits, berries, hazelnuts, blackcurrants, potatoes, rhubarb, and beetroot) and for all residents (possibility to pick apples from trees other residents planted in their zone and harvest ramson others planted). However, a resident missed picking apples after the neighbors cut down the apple tree in their part of the zone, thereby missing out on the output *food*.

3.2.4. Participation Outcomes—The Ecological Dimension

Several residents neatly spread out garden waste in the woodland, thereby *fertilizing the woodland and creating mold*. Residents also contributed to *increased plant and animal biodiversity* by planting with the intent of attracting animals, also creating diverse vegetation and feeding wild birds, squirrels, and other animals. Participation led to *better tree development* when residents thinned in the dense woodland aiming to give remaining trees a better chance to develop properly. However, some residents did not consider their neighbors' participation as proper forestry work, missing out on this quality aspect. As an example, a resident did not like when neighbors pruned trees to get a better view from their garden, instead of thinning among the trees, which was needed to give trees room to develop properly. Residents in one of the forest villages jointly put up *bird boxes for starlings as pest control*, since many of the households got their garden lawns destroyed by garden chafers, which starlings like eating. The joint activity makes this outcome belong to the social dimension of urban woodland quality as well. A concrete output under the ecological dimension was *clean air*, which a resident thought the trees he planted contributed to achieving. A man reported that, when his children pick berries and fruits in the zone, they learn where food comes from and see that butterflies and insects are supposed to be there, which can be interpreted as *environmental awareness*. Another example of this was the value a man saw in conversations between him and his children, initiated thanks to apple picking in the zone:

> When we pick fruits, harvest something, you talk about it with the children. Saying: "Here [in the co-management zone] you can eat the fruit straight from the tree. You do not have to wash it." . . . This dialogue makes them more conscious about the difference between going to the forest and picking something, and buying something from the store and how it has been treated and why you have to wash it. (I9)

4. Discussion

4.1. Residents' Definition of Urban Woodland Quality

The residents interviewed generally shared the image of high-quality woodland in Sletten. To them, high quality was (1) to have a natural woodland making nature experiences possible; (2) related to the structural and species diversity of woodland stands, including diversity and density, and species characteristics; (3) accessibility to the woodland, both physical (available at the doorstep and from the paths) and mental (that residents were allowed to use it); and (4) for a few, something for which better management or maintenance would be needed. The majority of the mentioned quality aspects related to residents' passive and active use of the woodland, e.g., experiential values during walks. Residents' definition is in line with the positive qualities Finnish and UK residents associate with local urban woodlands according to other studies, i.e., peacefulness, the feeling of forest, naturalness [33], woodlands being important for nature conservation, providing residents nature experiences making them aware of natural cycles and seasonal change and giving them existential experiences; relaxation, contentment, and stress relief [34], and relaxation, peacefulness, seasonal change, scenery, and nature experience [28].

In all of the general green space quality schemes described in the literature, four overall quality aspects occur repeatedly: maintenance, accessibility, nature, and facilities [20–24]. Except for facilities, these were all also central to Sletten residents' urban woodland quality definition. The facilities aspect was less important, possibly because facilities such as toilets, sports fields, and play equipment are more associated with urban parks and not naturally available in an urban woodland. Furthermore, the urban woodland being situated so close to interviewees' homes made facilities somewhat redundant for their woodland use, with toilets at home and playgrounds in the middle of forest villages. People have also been found to dislike constructed facilities in forests even when placed there in order to support recreational forest use [39]. The only facility mentioned was *paths*, which were also identified by Ode and Fry [30] as important for urban woodland quality.

Structural and species diversity of woodland stands are aspects unique to urban woodland quality studies, both in the literature and in Sletten, as opposed to studies of green space quality in general. Within this category, residents mentioned woodland density, diversity in species, variety in experiences, and the species characteristic age of trees, and plants with edible berries, fruits, and nuts. Resident-perceived urban woodland quality is, in this respect, in line with qualities identified by users or experts in earlier studies, i.e., variety of trees [29], structural and species diversity [31,32], variation between stands [40], size (large enough to provide a forest feeling), and forest structure (broad-leaved forests preferred for their diversity) [30].

At the time of interviews, the Sletten woodland was 17–21 years old, varying between parts. While the interior of young woodlands is generally perceived as visually unattractive and not appreciated for recreational use [41], and mature woodland generally has higher levels of visual qualities than young [32], the age limit for experiential and recreational qualities seems to have been crossed at that point in the Sletten case.

4.2. Impact on Urban Woodland Quality of Physical Participation

The urban woodland quality obtained from public management partly differs from that also obtained from resident participation. Outcomes of participation in Sletten affected both the users, i.e., the residents, and the physical environment (Table 3). The majority of the effects on the physical environment benefited a larger group of the residents in the neighborhood, not only participants. The effects on the physical landscape corresponded much to the functional and ecological quality dimensions of the urban woodland quality scheme based on Bell et al. [35]. The effects on users corresponded predominantly with quality aspects within the social and experiential dimensions for a publically managed urban woodland. The urban woodland quality assessment scheme was further developed to better reflect resident-perceived quality. Together, the effects of participation in Sletten on users and the physical environment covered all dimension aspects of the adapted urban woodland quality assessment scheme, suggesting that these dimensions and aspects could work well for the purpose of future urban woodland quality assessments.

While the previously mentioned outcomes of participation could also potentially be a result of woodland management by the local authority, benefiting woodland users, it is not given that they do. It depends on the public management and maintenance intensity and whether this is performed with the intent to meet user needs or rather aimed at nature conservation. For the resident-reported outcomes from participation, their passive and active use is practically always in focus, as well as when it comes to physical qualities serving nature conservation. As an example, residents feeding wild animals or planting different trees and shrubs with the aim of increased biodiversity also benefit users, rendering enjoyable experiences during walks and better woodland appearance. User participation in urban green space management has been found to affect urban biodiversity values positively [42]. When it comes to urban woodlands and Sletten, it remains to be studied whether residents create more positive, ecological qualities in the limited co-management zone than would be possible to create through active, systematic, public management of the entire woodland.

Participation in urban woodland management had additional benefits for participants, showing a difference between the urban woodland quality for participants vs. all residents, as well as between participation and woodland use alone (Table 3). The only exception was *food*, which, to some extent, benefited the resident group as a whole. This type of benefit included, above all, experiential qualities, e.g., residents obtaining happiness and pleasure from the act of participating, sense of community between participants, and an enhanced view of the woodland from inside. Due to only two non-participating residents agreeing to be interviewed, the study could not add much knowledge on how their recreational experience was affected by other residents' participation, apart from the finding that participation had fewer benefits for them than for participants. How non-participants are affected by participation, therefore, remains an interesting topic for future studies. In Sletten, participation led to social qualities for both participants and residents in general, despite the guidelines for the co-management zone not demanding participating neighbors to collaborate with each other. In other words, individual participation can also bring social values.

Sometimes, both participating and non-participating residents missed out on physical qualities and benefits that would have been possible outcomes of participation if it was not for participation of other residents or hindering physical environment characteristics, e.g., low accessibility from paths to the woodland or residents living next to a dense, thicket-like part of the woodland. This means that participation does not only affect urban woodland quality positively. Furthermore, this implies that urban woodland quality that can be obtained from participation is affected by the original urban woodland quality as affected by design (e.g., species selection), and public management (e.g., long-term local authority strategies). The latter has the possibility to respond to new user needs that arise, such as improvement of path systems. Moreover, urban woodland quality in Sletten is affected by the qualities of the neighborhood at large. Features of a high-quality built environment at a neighborhood scale identified by Dempsey [23], such as connectedness and permeability or legibility, are intrinsic qualities in the Sletten landscape plan benefiting all residents, qualities not affected by resident participation.

Aalbers and Sehested stated that, when users are involved in green space management, they create green spaces of a different kind, with other qualities [11]. The fact that there are corresponding quality aspects in the scheme based on Bell et al. [35] for all physical qualities created from participation suggests that little difference exists between physical qualities created through participation and those created through public woodland management. However, a close look at the outcomes shows a number of differences. While managers and residents can both create paths, it is more likely that residents place their path exactly where they are needed to support woodland use. Managers could adapt the woodland for social activities, e.g., create a glade for children's play or prune trees making them ready for someone to install a hammock; however, they are unlikely to identify such needs without asking residents. To put straw around exposed trees to create a more wind-sheltered environment is a small-scale way of influencing the woodland. Public management does not have the resources to perform such frequent, small-scale management and maintenance contributions. Participation mainly occurs within the limited area of the co-management zone, and, for the creation of some physical qualities, this is not an advantage. Rational, large-scale thinning among the trees, performed by public managers, leads to better tree development for the entire remaining woodland, while a few residents' small-scale thinning only supports a small number of trees.

In sum, Sletten residents created physical qualities better adapted to local user needs, both in regards to the actual needs and the placement of the physical quality. This leads to better urban woodland quality for woodland use, especially for participating residents, while some ecological qualities, such as better tree development, are likely more efficiently created when performed by public managers.

Residents sometimes refrained from ornamental gardening in favor of a more natural woodland character, or removed too garden-like plants, thereby creating "invisible" physical qualities. These could only be identified through participant interviews, since such resident-created qualities cannot be measured in the physical environment. They can be regarded as other types of qualities [11] with regards to the detail and rationale behind residents' "invisible" participation compared to public managers'. While the risk of privatization of public land increases due to the proximity between the garden and the area where the residents participate, the likelihood of residents caring for and protecting their environment also increases, simply because they participate in their local landscape. This was reflected in the finding that the interviewed residents generally seemed to have a sound and conscious nature view and opinion about plants that are suitable in a woodland and how to maintain them, thereby preserving ecological urban woodland quality. Guidelines and municipal control are still needed to prevent misuse, especially since some residents participated in a larger area of the woodland than the 4-m-wide zone stated in the guidelines. Residents sometimes disliking other residents' actions in the zone supports the idea of limiting participation and keeping it within a co-management zone.

With regards to the type of environments and physical qualities created through participation, many residents transferred garden characteristics to the woodland, while few did the opposite and transferred woodland characteristics into their own garden. However, the participation benefits *enhanced view of woodland from inside* and *enhancement of private garden* are examples of residents *visually* bringing the woodland into their gardens. Defining the quality impact of participation on an urban woodland in terms of enhanced nature conservation only, and labeling benefits to users as co-benefits [15], is not reasonable for a woodland integrated with a neighborhood, highly used by residents. Participation in Sletten mainly benefited the people, in particular participating residents; however, those benefits partly build on physical qualities being created, suggesting that benefits to users and physical qualities cannot always be easily separated.

5. Conclusions

The present study explored how residents perceive "urban woodland quality". It was demonstrated that social, experiential, functional, and ecological dimensions are all part of residents' perception of urban woodland quality. Maintenance, accessibility, nature, and, to a small extent, facilities, are quality aspects of these dimensions which are well integrated in existing expert assessment schemes for other types of green space quality [20–24]. Our results provide support to such assessment schemes by showing that these aspects are also important for local residents' quality perception of the specific green space type "urban woodlands". Additionally, results may add to the development of quality assessment schemes focusing on urban woodlands in so far that it points to the structural and species diversity of and between woodland stands and between forested and open habitats as being central for the perceived quality. The relative importance of this quality aspect distinguishes woodland from other types of urban green space. In a wide perspective, this limits the usability of existing quality assessment schemes, since these were mainly developed for other types of urban green space, such as parks. The study also sought to explore how urban woodland quality is affected by resident participation in woodland management and maintenance, according to residents. The main contribution of the present study is that it demonstrates that residents' participation in their "neighbor-wood", mainly but not only, contributes positively to perceived woodland quality. In plain words, participation in the co-management zone physically affected the woodland's functional and ecological qualities as perceived by the residents. Moreover, the active participation zone provided a range of social and experiential benefits to participating residents, as well as to the residents as a community group, part of which they themselves argue that they would have missed out on if they were only allowed to use the woodland "passively". Participation had additional benefits for participants, showing a difference between the urban woodland quality for participants vs. all residents, as well as between participation and woodland use alone. These findings suggest a large—but still largely untapped—potential of participatory urban woodland management to contribute physical qualities to urban woodlands and benefits to its users.

That said, the present study has some noteworthy limitations. It was explorative in its nature and was confined to a site-specific context and to a limited number of participants (16 respondents, including only two non-participants); further research is needed in order to produce stronger evidence as the basis for recommendations and future development of urban woodland quality assessment methods and their application in practice.

Author Contributions: Conceptualization, H.F. Formal analysis, H.F. Funding acquisition, A.B.N. Investigation, H.F. Methodology, H.F. and M.J. Project administration, H.F. Supervision, M.J. and A.B.N. Validation, H.F., M.J., and A.B.N. Visualization, H.F. Writing—Original Draft, H.F. Writing—Review and Editing, H.F., M.J. and A.B.N.

Funding: This research was funded by the European Commission, 7th Framework Program Grant GREEN SURGE collaborative project, FP7-ENV.2013.6.2-5-60356 and by the Movium Partnerskap project 91 12.

Conflicts of Interest: The authors declare no conflicts of interest.

References

1. Fritzbøger, B.; Søndergaard, P. A short history of forest uses. In *Multiple-Use Forestry in the Nordic Countries*; Hytönen, M., Ed.; METLA, The Finnish Forest Research Institute, Helsinki Research Centre: Vantaa, Finland, 1995; pp. 11–41.
2. Mattijssen, T.; Buijs, A.; Elands, B.; Arts, B. The 'green' and 'self' in green self-governance—A study of 264 green space initiatives by citizens. *J. Environ. Policy Plan.* **2017**, *20*, 1–18. [CrossRef]
3. Bastian, O.; Haase, D.; Grunewald, K. Ecosystem properties, potentials and services—The EPPS conceptual framework and an urban application example. *Ecol. Indic.* **2012**, *21*, 7–16. [CrossRef]
4. Jansson, M. Green space in compact cities: The benefits and values of urban ecosystem services in planning. *Nord. J. Archit. Res.* **2014**, *26*, 139–160.
5. Van Herzele, A.; Collins, K.; Tyrväinen, L. Involving people in urban forestry—A discussion of participatory practices throughout Europe. In *Urban Forests and Trees*; Konijnendijk, C.C., Nilsson, K., Randrup, T.B., Schipperijn, J., Eds.; Springer: Berlin, Germany, 2005; pp. 207–228.
6. Van der Jagt, A.P.N.; Elands, B.H.M.; Ambrose-Oji, B.; Gerőházi, E.; Steen Møller, M. Participatory governance of urban green space: Trends and practices in the EU. *Nord. J. Archit. Res.* **2016**, *28*, 11–40.
7. Haase, D.; Kabisch, N.; Strohbach, M.; Klemen, E.; Železnikar, Š.; Cvejić, R.; Pintar, M. *Inventory of Quantitative and Qualitative Functional Linkages between UGI Components, BCD and Impact*; Humboldt-Universität zu Berlin: Berlin, Germany, 2016.
8. Gulsrud, N.; Nielsen, A.B.; Bastrup-Brik, A.; Olafsson, A.S.; Lier, M.; Fischer, C.; Zalkauskas, R.; Hedblom, M.; Sievanen, T.; Nordh, H.; et al. *Urban Forests in a European Perspective: What can National Forest Inventory tell us?* Department of Geosciences and Natural Resource Management, University of Copenhagen: Copenhagen, Denmark, 2018.
9. Nielsen, A.B.; Hedblom, M.; Olafsson, A.S.; Wiström, B. Spatial configurations of urban forest in different landscape and socio-political contexts: Identifying patterns for green infrastructure planning. *Urban Ecosyst.* **2017**, *20*, 379–392. [CrossRef]

10. Fors, H.; Molin, J.F.; Murphy, M.A.; van den Bosch, C.K. User participation in urban green spaces—For the people or the parks? *Urban For. Urban Green.* **2015**, *14*, 722–734. [CrossRef]

11. Aalbers, C.B.E.M.; Sehested, K. Critical upscaling. How citizens' initiatives can contribute to a transition in governance and quality of urban greenspace. *Urban For. Urban Green.* **2018**, *29*, 261–275. [CrossRef]

12. Nannini, D.K.; Sommer, R.; Meyers, L.S. Resident involvement in inspecting trees for Dutch elm disease. *J. Arboric.* **1998**, *24*, 42–46.

13. Glover, T.D.; Shinew, K.J.; Parry, D.C. Association, sociability, and civic culture: The democratic effect of community gardening. *Leis. Sci.* **2005**, *27*, 75–92. [CrossRef]

14. Jones, R. Enticement: The role of community involvement in the management of urban parks. *Manag. Leis.* **2002**, *7*, 18–32. [CrossRef]

15. Mattijssen, T.; Buijs, A.; Elands, B. The benefits of self-governance for nature conservation: A study on active citizenship in the Netherlands. *J. Nat. Conserv.* **2018**, *43*, 19–26. [CrossRef]

16. Lawrence, A.; Ambrose-Oji, B. Beauty, friends, power, money: Navigating the impacts of community woodlands. *Geogr. J.* **2015**, *181*, 268–279. [CrossRef]

17. Lindholst, A.C.; Sullivan, S.G.; van den Bosch, C.C.K.; Fors, H. The inherent politics of managing the quality of urban green spaces. *Plan. Pract. Res.* **2015**, *30*, 376–392. [CrossRef]

18. Reeves, C.A.; Bednar, D.A. Defining quality: Alternatives and implications. *Acad. Manag. Rev.* **1994**, *19*, 419–445. [CrossRef]

19. Dempsey, N.; Bramley, G.; Brown, C.; Watkins, D. *Understanding the Links between the Quality of Public Space and the Quality of Life: A Scoping Study*; CABE Space: London, UK, 2008.

20. Gidlow, C.J.; Ellis, N.J.; Bostock, S. Development of the neighbourhood green space tool (NGST). *Landsc. Urban Plan.* **2012**, *106*, 347–358. [CrossRef]

21. Rigolon, A.; Németh, J. A QUality INdex of Parks for Youth (QUINPY): Evaluating urban parks through geographic information systems. *Environ. Plan. B Urban Anal. City Sci.* **2018**, *45*, 275–294. [CrossRef]

22. Lindholst, A.C.; van den Bosch, C.C.K.; Kjøller, C.P.; Sullivan, S.; Kristoffersson, A.; Fors, H.; Nilsson, K. Urban green space qualities reframed toward a public value management paradigm: The case of the Nordic Green Space Award. *Urban For. Urban Green.* **2016**, *17*, 166–176. [CrossRef]

23. Dempsey, N. Quality of the built environment in urban neighbourhoods. *Plan. Prac. Res.* **2008**, *23*, 249–264. [CrossRef]

24. Van Herzele, A.; Wiedemann, T. A monitoring tool for the provision of accessible and attractive urban green spaces. *Landsc. Urban Plan.* **2003**, *63*, 109–126. [CrossRef]

25. Randrup, T.B.; Konijnendijk, C.; Dobbertin, M.K.; Prüller, R. The concept of urban forestry in Europe. In *Urban Forests and Trees*; Konijnendijk, C.C., Nilsson, K., Randrup, T.B., Schipperijn, J., Eds.; Springer: Berlin, Germany, 2005; pp. 9–21.

26. Urban Atlas. Available online: https://www.eea.europa.eu/data-and-maps/data/copernicus-land-monitoring-service-urban-atlas (accessed on 19 August 2018).

27. Ries, A.V.; Voorhees, C.C.; Roche, K.M.; Gittelsohn, J.; Yan, A.F.; Astone, N.M. A quantitative examination of park characteristics related to park use and physical activity among urban youth. *J. Adolesc. Health* **2009**, *45*, S64–S70. [CrossRef] [PubMed]

28. Jorgensen, A.; Anthopoulou, A. Enjoyment and fear in urban woodlands—Does age make a difference? *Urban For. Urban Green.* **2007**, *6*, 267–278. [CrossRef]

29. Thompson, C.W.; Aspinall, P.; Bell, S.; Findlay, C. "It gets you away from everyday life": Local woodlands and community use—What makes a difference? *Landsc. Res.* **2005**, *30*, 109–146. [CrossRef]

30. Ode, Å.; Fry, G. A model for quantifying and predicting urban pressure on woodland. *Landsc. Urban Plan.* **2006**, *77*, 17–27. [CrossRef]

31. Ode, Å.K.; Fry, G.L.A. Visual aspects in urban woodland management. *Urban For. Urban Green.* **2002**, *1*, 15–24. [CrossRef]

32. Nielsen, A.B.; Jensen, R.B. Some visual aspects of planting design and silviculture across contemporary forest management paradigms—Perspectives for urban afforestation. *Urban For. Urban Green.* **2007**, *6*, 143–158. [CrossRef]

33. Tyrväinen, L.; Mäkinen, K.; Schipperijn, J. Tools for mapping social values of urban woodlands and other green areas. *Landsc. Urban Plan.* **2007**, *79*, 5–19. [CrossRef]

34. Jorgensen, A.; Hitchmough, J.; Dunnett, N. Woodland as a setting for housing-appreciation and fear and the contribution to residential satisfaction and place identity in Warrington New Town, UK. *Landsc. Urban Plan.* **2007**, *79*, 273–287. [CrossRef]

35. Bell, S.; Blom, D.; Rautamäki, M.; Castel-Branco, C.; Simson, A.; Olsen, I.A. Design of urban forests. In *Urban Forests and Trees*; Konijnendijk, C.C., Nilsson, K., Randrup, T.B., Schipperijn, J., Eds.; Springer: Berlin, Germany, 2005; pp. 149–186.

36. Tyrvainen, L.; Gustavsson, R.; Konijnendijk, C.; Ode, A. Visualization and landscape laboratories in planning, design and management of urban woodlands. *For. Policy Econ.* **2006**, *8*, 811–823. [CrossRef]

37. Fors, H.; Nielsen, A.B.; van den Bosch, C.C.K.; Jansson, M. From borders to ecotones—Private-public co-management of urban woodland edges bordering private housing. *Urban For. Urban Green.* **2018**, *30*, 46–55. [CrossRef]

38. Creswell, J.W. Qualitative inquiry and research design: Choosing among five approaches. *Health Promot. Prac.* **2015**, *16*, 473–475.

39. Nielsen, A.B.; Heyman, E.; Richnau, G. Liked, disliked and unseen forest attributes: Relation to modes of viewing and cognitive constructs. *J. Environ. Manag.* **2012**, *113*, 456–466. [CrossRef] [PubMed]

40. Filyushkina, A.; Agimass, F.; Lundhede, T.; Strange, N.; Jacobsen, J.B. Preferences for variation in forest characteristics: Does diversity between stands matter? *Ecol. Econ.* **2017**, *140*, 22–29. [CrossRef]

41. Ryan, J.; Simson, A. 'Neighbourwoods': Identifying good practice in the design of urban woodlands. *Arboric. J.* **2002**, *26*, 309–331. [CrossRef]

42. Dennis, M.; James, P. User participation in urban green commons: Exploring the links between access, voluntarism, biodiversity and well being. *Urban For. Urban Green.* **2016**, *15*, 22–31. [CrossRef]

forests

MDPI

Article

Cocoa and Climate Change: Insights from Smallholder Cocoa Producers in Ghana Regarding Challenges in Implementing Climate Change Mitigation Strategies

Lord K. Ameyaw [1,*], Gregory J. Ettl [1], Kristy Leissle [2] and Gilbert J. Anim-Kwapong [3]

[1]   School of Environmental and Forest Sciences, University of Washington, P.O. Box 352100, Seattle,
       WA 98195-2100, USA; ettl@uw.edu
[2]   African Studies, University of Washington, Bothell, WA 98011-8246, USA; kleissle@uw.edu
[3]   Cocoa Research Institute of Ghana, P. O. Box 8, New Tafo-Akim, Eastern Region, Ghana;
       gjanimkwapong@yahoo.com
*   Correspondence: lkameyaw@uw.edu; Tel.: +1-334-498-1372

Received: 30 October 2018; Accepted: 27 November 2018; Published: 28 November 2018

Abstract: This study investigates the knowledge and perception of smallholder cocoa farmers on the potential impacts of climate change on cocoa production in Ghana. It addresses opinions on the inclusion of climate change mitigation strategies (such as Reducing Emissions from Deforestation and Forest Degradation—REDD+) into cocoa production, and potential obstacles and roles of stakeholders in ensuring community acceptance of such strategies in a unique multiple land use area—the Krokosua Hills Forest Reserve. Data from the Ghana Meteorological Agency and through survey of 205 cocoa farmers were assessed with Mann-Kendall, Kruskal Wallis and Mann-Whitney tests. Farmers' perceptions of changes in climate were notably diverse and did not always match historic weather data, but accurately described increases in temperature and drought which are linked to cocoa productivity. Farmers appreciate the importance of tree maintenance for ecosystem services but were skeptical of financially rewarding climate change strategies which favor tree protection. Cultural practices associated with cocoa production encourage carbon release and may pose a threat to the objectives of REDD+. Farmers' experience on the land, interactions with other farmers, government extension agents and cocoa buyers all influence cocoa agroforestry practices in the area, and communication through existing entities (particularly extension agents) presents a pathway to community acceptance of climate change mitigation strategies. The study recommends reforms in REDD+ strategies to adopt flexible and participatory frameworks to facilitate adoption and acceptability due to pronounced heterogeneity in community perceptions and knowledge of climate change and related issues.

Keywords: cocoa; Ghana; smallholder; perceptions; climate change; REDD+; stakeholders; participatory

1. Introduction

Cocoa (*Theobroma cacao* L.) cultivated under the shade of forest trees, in combination with annual food crops (i.e., cocoa agroforestry) on the same piece of land, is common for smallholder farmers across the cocoa-forest mosaic of tropical Ghana. Currently, Ghana is the world's second largest cocoa producing nation (behind Côte d'Ivoire). The cocoa industry employs about 3.2 million people along its commodity chain and accounts for 25% of foreign exchange earnings [1]. It is estimated that 800,000 smallholder cocoa farmers in Ghana derive between 70%–100% of their yearly income solely from cocoa production [2]. Benefits from cocoa agroforestry are multifaceted and include greater biodiversity than monocultures; societal and economic benefits of continuous food supply (food crops/staples); annual income from cocoa; and long-term financial reserves in timber.

Cocoa generally requires high temperatures, precipitation and humidity to achieve optimum productivity, and cultivation is restricted to the "cocoa belt" (20° N and 20° S of the Equator). Specifically, cocoa trees need temperatures between 21–23 °C and rainfall between 1000–2500 mm annually to achieve optimum yield. Cocoa production is sensitive to precipitation and is reduced by drought which may increase in Ghana under climatic changes. A temperature increase of about 2 °C and a 1% decrease in precipitation (1467 mm to 1455 mm) is projected by the year 2050 in Ghana with potential decreases in cocoa cultivation [3], particularly in areas bordering the cocoa growing suitability area to the north and south respectively [3,4]. Long-term trends in precipitation are lacking due to high variability along both inter-annual and inter-decadal timescales [5]. The impacts of the severe El Niño years of the early 1980's on cocoa yield in the entire West African sub region [6], provides a reference point for potential future impacts of increased drought under climate change projections.

A "climate-smart" [7] approach (i.e., agricultural strategies that foster sustainable production, resilience, mitigation, food security and development) is needed to counter the potential impacts of climate change on global cocoa production. Non-governmental organizations (NGOs) have made significant efforts in developing sustainable practices related to cocoa production and climate change across the West African sub region and other developing nations in the cocoa belt. However, the development of cocoa varieties with tolerance for higher temperature and low precipitation is needed [8], particularly in Ghana, where strategic climate change ameliorating strategies are essential to sustaining cocoa production [9]. Current and emerging climatic trends could render smallholder cocoa farmers vulnerable and pose a significant threat to livelihoods centered on cocoa production [10].

In 1995 Ghana ratified the United Nations Framework Convention on Climate Change—UNFCCC global alliance to reduce carbon emissions [11] and in 2008 adopted the Reducing Emissions from Deforestation and Forest Degradation (REDD+) program to foster carbon goals [12]. REDD+ aims to create financial value and incentive for activities which lead to sustainable natural resource management in developing nations [13]. REDD+ reinforces conservation, sustainable management of forests and enhancement of forest carbon stocks. Potential benefits envisaged include conservation of biodiversity, water and soil regulation, and direct human benefits including enhancing opportunities for participatory natural resource management. Ghana has made significant strides toward a national scale implementation of REDD+ and has submitted its Readiness Preparation Proposal (RPP) to the World Bank's Forest Carbon Partnership Facility (FCPF) in 2010 [14,15]. The National Forest and Wildlife Policy (2012) and National Climate Policy (2013) were passed by Ghana to offer a favorable policy pathway for climate change strategies, including REDD+. The integration of cocoa agroforestry within REDD+ (Cocoa Forest REDD+ program, [16]) aims at improving net carbon gains through the integration of trees on crop lands and subsequently providing an opportunity toward climate change mitigation. REDD+ funding differs from mainstream project funding where funds are provided before the initiation of a project. REDD+ is rather performance based with a built-in component of demonstrating the impact(s) of the project before funds are released [12]. Expectedly, REDD+ has a strict set of criteria which are essential to its implementation. Although Ghana's REDD+ pathway has received accolades, globally applicable issues pertaining to tree and land tenure, benefit sharing mechanisms, technical capacity and governance [12,17–19] are yet to be fully resolved.

Numerous studies have examined the perceptions of farmers on such topics as the impact of climate change on cocoa yields [8,9,20,21], smallholder choice of cocoa production systems [22–24], the potential benefits of cocoa agroforestry [25,26] and advantages of REDD+ in cocoa production [13,27]. There is however, limited information on how farmers perceive the inclusion of climate change mitigation strategies into their land/farm management objectives. Since agroforestry emphasizes "people" as its key element [28], understanding cocoa farmers' perceptions of issues such as tree planting, and local/indigenous knowledge on the role of climate on sustainable forest management and environmental conservation is important in answering questions on land use, land-use change (deforestation) and cocoa production. This suggests that, agroforestry is not just about the cocoa and associated shade trees [29], as there is a strong linkage between farmers' perception and management decisions on tree retention on cocoa farms in Ghana; positive perceptions of shade trees increase the probability that a farmer will retain trees on cocoa farms [30]. The importance of stakeholder perception on the success of conservation projects has been previously demonstrated; for example, in Kenya, stakeholder perceptions influenced adoption of new and improved strategies [31].

Smallholders perceptions also take into account interaction between their farming activities and changes in microclimate, and their perceptions may determine whether mitigation/adaptation strategies are implemented [32]. In fact, the social acceptability of the agroforestry system at the individual farmer level, is influenced by: Community heterogeneity, perceptions towards trees, land and tree tenure arrangements, gender and other socio-cultural factors like age, labor and cultural habits [33]. Remarkably, a strong correlation between climate change, the level of concern for associated implications, and ultimately, farmers' decision to subscribe to climate change mitigation policies and projects exists, irrespective of the accuracy of farmers' experience with regards to individual perceptions of climate change and actual historical climatic trends [34]. For instance, people who believed that climatic changes were occurring and that changes were a result of human activities, were more likely to perceive temperature increases despite inconsistencies with available climate records. In the end, perceptions about climate patterns effectively determine actions of farmers irrespective of patterns determined through analysis of empirical climate data [35].

Based on previous studies, there is empirical reason to suggest that the aforementioned demographic profiles have an impact on perceptibility of climate change, and consequently, actions to be taken. In Ghana, for instance, smallholder farmers (both men and women) in different communities hold specific views of climate change which ultimately influences coping strategies [36]. Elsewhere, farmer age is a significant determinant of overall farming and climate experience [37,38]. Differences in access to information on climate change also correlates with climate change perception among male- and female-headed households, with the former more likely to be educated on climate-related issues [39]. Women, on the other hand are considered more susceptible to the impacts of climate change because they are generally less informed [40]. This connotes that education in general influences how farmers perceive climate change [22,41]. Additionally, the accumulation of knowledge and experience with both farming and climate makes farmer age an important factor in climatic change perception inquiry [42]. In Ghana, marital status among smallholder cocoa farmers influences access to information on climate change and ultimately, how individuals perceive climate change and adaptation strategies [22]. Lastly, comparisons between indigenous and migrant farmers, indicate that the former have a higher tendency to subscribe to long-term climate ameliorating programs and strategies. Lack of property rights is highlighted as a significant cause of this observation [43].

This study reports on findings of a survey conducted in smallholder cocoa communities in a major cocoa-producing area of Ghana. A semi-structured questionnaire was employed to collect demographic profiles, relevant information and opinions of individual farmers. In consonance with the study objectives, it is expected that different communities, and the gender, age, educational status, migrant status, and family/household status within communities will be tied to farmers' perceptions of climatic changes, potential causes, receptiveness to climate change mitigation projects and general opinions about climate change.

This paper investigates the perceptions of smallholder cocoa farmers on the inclusion of climate change mitigation strategies and payment for ecosystem services into land/farm management objectives. Specifically, this study:

1. Examines cocoa farmers' knowledge and perceptions of climate change in contrast with climate data and potential impacts of climate change on cocoa production;
2. Investigates the perceptions of smallholder farmers on the feasibility of including climate change mitigation strategies in cocoa farming;

3. Explores the roles of scientific and non-scientific actors (cocoa farmers and non-cocoa farmers) in promoting the implementation of climate change mitigation strategies in combination with cocoa production; and

4. Examines potential obstacles to incorporating climate change mitigation strategies into cocoa production.

2. Materials and Methods

2.1. Study Area

The study was carried out in the Krokosua Hills Forest Reserve (KHFR), in the Juaboso District of the Western Region of Ghana (Figure 1). Specifically, the study was conducted among smallholder cocoa farmers resident in communities that fringe KHFR, one of the major forest reserves in the Western Region. KHFR covers an area of about 481.61 km^2 (48,160 ha), situated at the east bank of River Bia and bisected by the Sefwi-Wiawso—Côte d'Ivoire border road (6°15′–6°40′ N and 2°40′–3°00′ W) [44]. Cocoa farming (agriculture) is the main source of livelihood for people living around the reserve [45]. Between 2006 and 2009, the population of thirteen large fringe communities was 66,766. For management purposes, KHFR has been designated into two major zones: (1) the production zone (where harvesting of timber and non-timber forest products is officially permitted for prospective Timber Utilization Contracts (TUC, a written contract signed by the Sector Minister and ratified by the Parliament of Ghana granting a timber harvesting right to its holder upon a successful competitive public bidding process) and permit holders respectively—23,639 ha) and (2) the protection zone (includes areas of high biodiversity conservation priority, areas recovering from past disturbances and no timber harvesting areas—24,521 ha) [44].

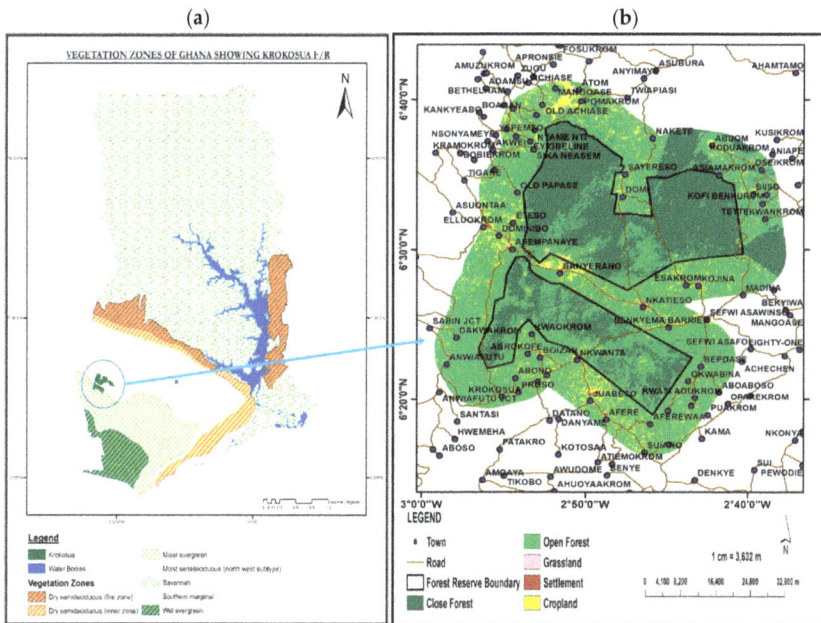

Figure 1. Ecological zones in Ghana highlighting the location of the Krokosua Hills Forest Reserve (KHFR) (**a**). Ecological zones correspond with the legend, with KHFR located in the moist semi-deciduous zone. Map of KHFR showing approximate locations of study communities is also shown on the right side of the figure (**b**).

Prior to the official designation of the KHFR as a Forest Reserve in 1948, fringe communities utilized the land for agriculture and cocoa farming. After the reservation status was conferred, cocoa farms were given legal status to remain in the forest and were termed admitted farms. Per the most recent management plan, there are 38 admitted farms in the reserve. These admitted farms have footpaths as routes connecting farms and huts scattered within the forest. Over time, population growth and land scarcity have forced cocoa farmers to extend their farms further into the forest and outside the area demarcated for cocoa production (admitted farms) [44]. Cocoa farming is the leading driver of deforestation in the region [46].

2.2. Farmer Selection, Data and Analysis

The study targeted cocoa farming communities within a range of 2 and 5 km away from the KHFR. The distance varied to sample cocoa farmers who interacted with or specifically had cocoa farms within the KHFR. A mixed method approach (qualitative and quantitative methodologies) was used. Apart from the inherent trait of complementarity of qualitative and quantitative procedures, using mixed methods provides a platform for cross-checking and validation of collected data [47]. A list of farmers in the target communities was not readily available so the study identified farmers through community heads and leaders. Farmers were then stratified based on gender and randomly selected for interviews. A purposive sampling approach employing snowballing (i.e., respondent referrals) was also used to increase sample size and heterogeneity of respondents. A total of 205 face-to-face interviews were administered in 30 communities surrounding the KHFR between December 2016 and February 2017. Identifying information for responses given were not taken as per the instructions of the Institutional Review Board of the University of Washington. In each community, unequal samples were obtained, with at least 2 interviews in selected communities. Interviews were conducted at home and on farms and lasted between 60 to 90 min. Notes were taken as the interview progressed with corresponding answer choices checked as well. Questions were prepared in English but the local language, Twi, was used during the interview except in situations where respondents could understand English. Survey enumerators were given prior training in translation and the survey instrument.

Survey questions were structured into four different themes: (1) knowledge on climate change, (2) perceptions about climate change mitigation strategies, (3) roles of local and external stakeholders, and (4) potential setbacks to climate change mitigation strategies, and examined for demographic trends (e.g., gender, age, level of education and migrant status). Questions were designed to collect mostly quantitative data (structured) but also included qualitative data collection through semi-structured (open ended) questions to allow farmers to expatiate on opinions and in so doing verify answers given on structured questions.

Answers to survey questions were first summarized in Microsoft Excel and R statistical software was used for statistical analysis. Descriptive analysis such as modes, frequencies and percentages were used to summarize data. As responses from farmers did not follow a normal distribution, a Mann-Whitney and Kruskal-Wallis (KW) test were employed where variables had only two levels (gender, migrant and household status) and three or more levels (community, age and educational level) respectively. Individual questions on knowledge and perception of climatic changes, causes and related impacts (Table A1) were used as dependent variables. For statistically significant results on variables with three or more levels, a post-hoc Dunn Test using the Bonferroni-type adjustment of p-values (to reduce type I error) was used to determine which group(s) accounted for the significance.

Based on the four themes of the survey, Likert scale questions were utilized. Questions on five-point scales were converted to three groups; group one combined responses for agree and strongly agree, disagree and strongly disagree on group two and neutral responses on group three. Questions on 4-point Likert scales were converted into binary variables; e.g., not at all worried and not very worried were recorded as zero whiles somewhat worried and very worried were recorded as one for the binary variable [34]. Open-ended questions were categorized under the survey question themes and further sorted for recurring words and phrases (open coding, [48]).

Monthly means of maximum and minimum temperature and monthly precipitation climate data were collected from the Ghana Meteorological Agency for the period 1970–2017 from the closest weather station (Sefwi Bekwai), which is about 55 miles from the study area. Data were used to describe the physical environment and compare perceptions of climate change to climatic records. A Mann-Kendall (MK) test was used to determine monotonic trends in climatic variables over the period [49,50]. A cutoff of $\alpha = 0.05$ was used to determine significance in trends [51]. The weather data allows for comparisons with the experiences and observations of individual farmers and the overall community experience. To verify answers on perceptions regarding the length of dry and wet seasons (drought), the Standardized Precipitation-Evapotranspiration Index (SPEI) was used. SPEI is based on a combination of Palmer Drought Severity (PDSI) and Standardized Precipitation Indexes (SPI) [52]. SPEI incorporates temperature by finding the difference between precipitation (P) and potential evapotranspiration (PET) (using Thornthwaite's equation [53]) to produce an adjusted log-logistic distribution. Upon choosing an appropriate time scale, standard deviations of average values are calculated [54]. SPEI lends from SPI to classify drought severity in a range between no drought (≥ 0) and extreme drought (≤ -2) [55]. Estimation of SPEI was done using the SPEI package in R [56].

3. Results

3.1. Historical Climate Trends

3.1.1. Temperature

Based on the MK test on data from 1970 to 2017, temperature has significantly increased, a probable manifestation of climate change. Analysis of mean monthly minimum ($\tau = 0.285$, $p < 0.001$) and maximum ($\tau = 0.168$, $p < 0.001$) temperature both indicated statistically significant increased trends (Figure 2a,b). The mean temperature observed for the period 1970–2017 ranges from 22.6 °C to 32 °C for minimum and maximum respectively. Seasonal MK tests (SMK) also revealed an increasing seasonal temperature trend for minimum ($\tau = 0.395$, $p < 0.001$) and maximum ($\tau = 0.460$, $p < 0.001$). For recordings of mean minimum temperature, the lowest record for the period (1970–2017) was 18.2 °C which was in January 1975, while the highest record was taken as 27.7 °C in February 2011. Maximum temperature on the other hand, had its lowest record as 28 °C in August 1982 and highest record for the period as 37.2 °C in February 1995. Generally, low temperatures were mostly between August to January while February to June were the hottest months.

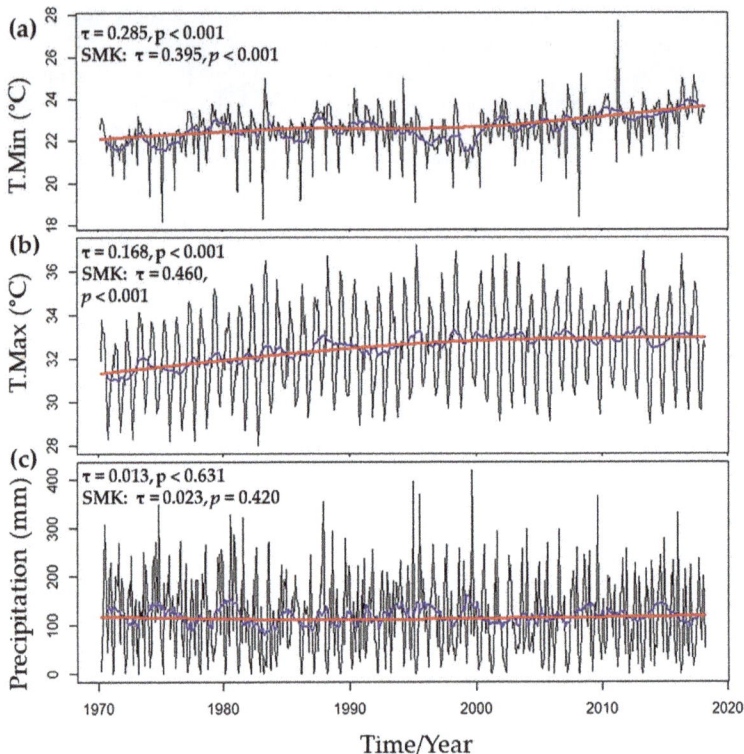

Figure 2. Monthly trend analysis of minimum and maximum temperature (**a,b**) in °C and precipitation in mm (**c**) at the Sefwi Bekwai weather station. Data was obtained from the Ghana Meteorological Agency (Kumasi, Ashanti Region, Ghana). Test for overall trend in data is shown by the red line with statistical test results indicated by Kendal's Tau (τ) and a resultant *p*-value. Tests for seasonality in data is indicated in blue and results shown with the Seasonal Mann-Kendall test (SMK). The "Kendall" package in R statistical software was used for the analysis.

3.1.2. Precipitation

Weather records indicated that since 1970, mean precipitation has consistently been above 1000 mm. Trend analysis however indicates an erratic rainfall pattern which is confirmed by the MK test. Specifically, the MK test detected no specific trend ($\tau = 0.013$, $p = 0.631$). Although Kendall's tau remained positive ($\tau = 0.013$), that is overall rainfall increased, the increase was not significantly different from zero. A seasonal MK test (SMK) further confirmed no seasonality trend in precipitation over the period ($\tau = 0.023$, $p = 0.420$) (Figure 2c). Apart from 1977, 1981, 1982, 1983, 1986 and 2016, all other years recorded rainfall greater than 1250 mm, the minimum value required for optimum cocoa production. The lowest and highest precipitation were recorded in 1983 (1071 mm) and 1980 (1826.5 mm) respectively. The most significant drought event in Ghana occurred in 1983, reinforcing the significantly low precipitation level for that year.

3.1.3. SPEI (Drought Severity)

Figure 3 shows SPEI values (using monthly data) for the study area, showing a decrease in extended wet periods. Periods of dryness on the other hand have increased, particularly from 2000–2017. There is indication that the area witnessed its worst drought (SPEI < −2) between 2014 and 2016.

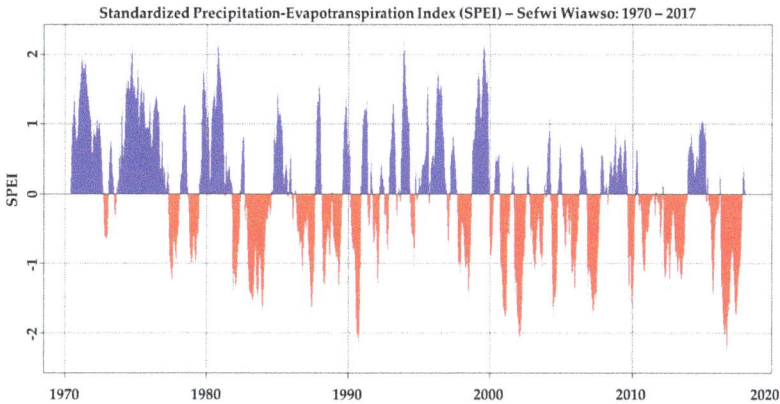

Figure 3. Time series of Standardized Precipitation-Evapotranspiration Index—SPEI values for Sefwi Wiawso (55 miles from study site): 1970–2017. Areas in blue indicate periods of no drought and red depicts periods of drought and corresponding SPEI values.

3.2. Characteristics of Respondents

Survey respondents were all adults (>18 years) mostly between 30–59 years; 67% were male and 33% female, a proportion epitomizing cocoa farming as a predominantly male-dominated activity. Approximately 65% of respondents were over 45 years-old. Educational levels ranged from basic (46%), to secondary (16%), to tertiary (7%) and no formal education (31%). Male farmers were generally more educated, 74% (compared to 61% of females) having received either basic, secondary or tertiary education. While females had similar basic education (51%) to males (49%), more males (19%) than females (6%) had received secondary education. Among male respondents, 87% were household heads. Only 23% of females were heads of their households. Natives of communities were generally more educated (71%) than non-natives (66%). Most respondents were married (86%), Christian (80%) and natives of their respective communities (71%) (Table A2).

3.3. Farmer Knowledge and Perceptions of Climate Change and Impact on Cocoa Production

Most farmers had perceived changes in climatic patterns over the last 20 years. Notably, farmers perceive rising temperature (88% of famers) and reduction in the amount of rainfall (89% of farmers) in recent times. Within the same period, the length of the wet season had reduced (81%), with a resultant increase in dry spells (89% of farmers) (Table 1). A small proportion (<1%) of farmers indicated they had witnessed spikes in cases of wildfires within the same period. Farmers overwhelmingly (95%) agreed that observed and experienced climatic changes over the past decade have had a negative impact on cocoa yields. The remaining respondents were evenly split on yields: some said they remained the same (2%) or improved (2%), respectively over the last decade.

Table 1. Respondent perceptions of climatic changes (*n* = 205).

Climatic Variables	Increased (%)	Decreased (%)
Temperature	88	5
Rainfall	2	89
Length of Wet Season	9	81
Length of Dry Season	89	6

Descriptive examination of survey responses indicated that farmers believe that climatic changes are mostly as a result of human activities; precipitation (181: 88%) and temperature (175: 86%). This finding is consistent with farmers' belief that climatic changes are not just isolated climatic anomalies. Farmers mostly disagreed that both precipitation (71%) and temperature (75%) were climatic anomalies. Superstitious (curses/spells) association with climatic changes were mostly dispelled by respondents with respect to changes in precipitation (85%) and temperature (81%). On the possible human attributable causes of climatic changes, illegal logging (95%) was the most highlighted. Slash and burn agricultural practice in the area (84%) as well as pollution from vehicles (83%) were similarly pointed out as detrimental to the environment. According to 80% of respondents, widespread woodfuel harvesting could also contribute to climate change. Despite previous research on the potentially harmful environmental impacts of implementing full-sun cocoa systems (cocoa monocultures), the majority of farmers (65%) believed such systems do not contribute to climatic changes, as opposed to 30% who perceive a change to full-sun cocoa systems, a plausible climate change driver (Figure 4).

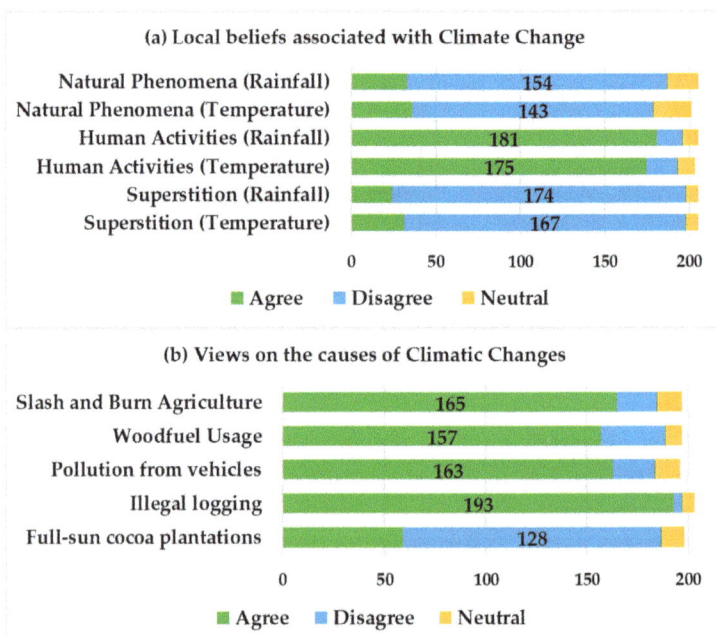

(a) Local beliefs associated with Climate Change

Natural Phenomena (Rainfall)	154
Natural Phenomena (Temperature)	143
Human Activities (Rainfall)	181
Human Activities (Temperature)	175
Superstition (Rainfall)	174
Superstition (Temperature)	167

■ Agree ■ Disagree ■ Neutral

(b) Views on the causes of Climatic Changes

Slash and Burn Agriculture	165
Woodfuel Usage	157
Pollution from vehicles	163
Illegal logging	193
Full-sun cocoa plantations	128

■ Agree ■ Disagree ■ Neutral

Figure 4. Number of survey responses to two questions on beliefs associated with climate change (**a**) and causes of climatic changes (**b**). A total of 205 responses were collected with few respondents choosing not to respond. Groups were created from a 5-point Likert scale in 3 categories: Agree, Disagree and Neutral. Each stacked bar depicts responses into 3 groups with the corresponding question listed in the left strip. Labelled bars indicate only the majority response.

Farmers' experience of climatic changes in recent times has heightened their fears about the future outlook of their main income earner, cocoa farming. Almost all farmers, 202 out of 205, expressed worry about changing climate. Concerns were mostly about reduction in cocoa yield (195), food crop loss due to droughts (181), increased rate of crop disease and pest infestation (135), increased wildfire incidents (127), and increased cocoa tree mortality associated with flooding (76). As a climate ameliorating mechanism, respondents held similar opinions on the role of trees in regulating temperature (94%) and precipitation (93%). According to respondents, specific climate change initiatives will mostly require a participatory approach (94%) as well as governmental interventions (78%). When farmers were asked to elaborate on what participatory approach they were specifically referring to, comments like "it is not any government, but we must get involved in the fight against climate change", "government and citizenry must make concerned effort in the fight against climate change", "government alone cannot fight the climate change menace so we must get involved", "government and stakeholder should all rise against climate change", and "all should participate in effort to combat climate change" were made. This suggests that farmers believe their active participation in developing/implementing climate change mitigation efforts is paramount to its eventual success. Farmers believe climate change is a reality and also agree, unanimously, that, interventions are vital to avert potentially devastating impacts on future generations (92%).

3.4. Information Related to Climate Change Mitigation Strategies and Perceptions of Farmers

Hybrid cocoa is the predominantly cultivated variety in the communities surveyed. More than 80 percent of farmers claim to plant this variety. Farm observations revealed that, such farms had a combination of trees, cocoa and food crops, with significant open patches in tree canopy to allow optimum level of sunlight. Some farmers (10%) had some of their farms with closed overhead tree canopies being described as traditional shaded cocoa stands. Other farmers (7%) also noted that they had some farm lands dedicated to cocoa monocultures. Cocoa lands were mostly acquired from relatives who were still alive or through inheritance (46% of farmers). Some farmers procured their own lands (42%), while others leased (15%) or had lands in multiple ownership categories. Farm sizes depicted the overarching Ghanaian smallholding cocoa farm/cultivation technique, revealing specifically that farmers mostly had 2 tracts of lands between sizes 0.4–2 ha, with cocoa trees ranging between one year to 19 years.

According to farmers, the choice of cocoa variety was mostly as a result of time to maturity (90%), resistance to pests and diseases (67%), expert advice from extension agents (62%), and availability (49%). Maintaining trees on cocoa lands seems plausible to many farmers (92%) in the study area by virtue of the inherent benefits to the environment. Interestingly, the idea of direct monetary benefits for tree maintenance on cocoa lands met a lesser response than preserving trees for their inherent value. Although the majority of farmers agreed to maintain trees on their farms for direct monetary benefits, a reduction of 17% was observed (a total of 75%) for farmers who answered yes. Farmers were almost even regarding their views on current tree tenure, which allocated ownership/benefits from trees differently based on location of tree within the forest reserve and outside it. It was specifically observed that 51% of 205 farmers were satisfied with the current tree tenure, in sharp contrast to the remaining 44%, who held a dissatisfied opinion of tree tenure.

3.5. Investigation into the Role of External Stakeholders in Cocoa Farming and Pathways to Increased Acceptability of Climate Change Mitigation Strategies and Potential Setbacks

Results revealed that communities in the study area had witnessed substantial presence of extension agents from governmental organizations and initiatives on cocoa as well as that of non-governmental organizations. Overall, farmers were satisfied (84%) with the influence such stakeholders had had on cocoa farming. Farmers in general indicated their inclusion in any direct income earning climate change mitigation strategy will hinge on the details of such a strategy. A total of 81% shared this opinion. Farmers (89%) suggested that the presence of extension agents will be vital as a bond of trust for any such project. Ultimately, the provision of farming incentives (68%) and assurances on the sustainability (63%) of climate-related strategies are equally important to enhance farmers' interest.

The study shows that cultural practices of cocoa farmers tend to release carbon due to vegetation removal. Although this was not evident on the farms that were visited, 76% of farmers interviewed revealed they mostly cut down cocoa trees when they see significant reduction in yield. Cutting down illegally cultivated cocoa trees in the area has mostly been done by staff of the Forestry Commission (the government body in Ghana tasked with management and regulation of forest and wildlife resources) to combat further encroachment of farm lands into the forest reserve, making this finding surprising. Farmers (27%) also indicated that they removed some timber trees to open the canopy and subsequently allow more incident sunlight to cocoa trees, when cocoa trees begin to decline in yield. The remaining farmers either leave the cocoa farm/trees or abandon the land completely when cocoa yield declines beyond commercially acceptable levels.

During preparation of lands for cultivating cocoa and other agricultural crops, slash and burn is the preferred strategy (88% and 32% respectively). Illegal logging is quite prevalent in the study area and cocoa farms with economically attractive timber species are the prime targets. Farmers indicated that illegal logging is reported to the Forestry Commission (57%). Despite this observation, farmers indicated that response and actions to such offenses are not always effectively dealt with. Farmers are sensitive about this topic, since illegal loggers do not conform to any logging standards and almost always leave significant damage to cocoa trees in the process. As a safeguard strategy, farmers resort to inducing mortality (through girdling, burning and pouring hot water on roots) of economic tree species before they reach maturity or merchantable structure/form (23%).

3.6. Statistical Variations among Responses Based on Demographic Attributes

We found differences in responses among farmers based on socio-demographic attributes including: community, age, gender, educational status, migrant status, and family/household status. Variations among communities were the most prevalent. All thirty communities generally shared similar climate experiences, with the majority response of increasing temperature/length of dry season and decreasing rainfall/length of wet season contributing 87% of responses, in comparison to 13% of other responses ($p < 0.05$). Gender also showed statistical differences among responses to observed temperature ($p = 0.01$), and rainfall ($p = 0.03$) variations; male respondents were more likely to have experienced increasing temperature (93%) and rainfall (93%) than females (78% and 81% respectively). On the other hand, age, migrant, and household status appears not to be a significant contributor to responses on climate experience ($p > 0.10$ for all responses).

Despite near unanimous agreement (95% of community responses) on the potentially negative impacts of climatic changes, concern regarding climate change impacts on cocoa yields varied significantly ($p < 0.001$). Most communities (77%) believe that collection/harvesting of woodfuel contributes to climate change. Other respondents (23%) had an opposing view which was statistically different from the majority response ($p < 0.001$). Responses to questions on slash and burn as an agricultural practice that contributes to climate change (80% of responses) also varied significantly among communities ($p = 0.03$). The age of a respondent may also contribute to farmers' views on the contribution of illegal logging to climate change ($p = 0.06$); with respondents 18–29 years old more likely to believe this (100%) than those in other age groups (94%). Cocoa monocultures have been highlighted as a potential environmental degrading agent, however native farmers (59%) found this to be less of a problem than non-native farmers (72%; $p = 0.02$).

The association of traditional beliefs and myths (such as curses and natural causes) to climate change appears to be community specific ($p < 0.01$). However, farmers' thoughts on human activities effects on rainfall patterns were statistically uniform ($p = 0.22$). Gender ascription of superstition (e.g., curses) to changes in rainfall patterns as a result of climate change was found to be statistically significant ($p = 0.01$). Females were found to be more superstitious than men. To a lesser extent, gender also influenced perceptions on rainfall just being a natural weather anomaly ($p = 0.09$). Overall, there was a general consensus (98% of responses) on the immediate concern about climate change impacts on their livelihoods and social well-being. However, farmers' view of future implications of climate change varied significantly among the study communities ($p < 0.001$); 92% of community responses pointed to a high likelihood of negative impacts.

Farmers' responses showed no differences in the perceived ability of trees to regulate temperature ($p = 0.88$) but there was a difference in response to trees' role in regulating rainfall ($p = 0.04$). There was a general acceptability of cocoa production which includes tree maintenance for ecosystem services exclusively, and one that remunerates farmers for maintaining trees on their farms, however, answers varied significantly among communities ($p < 0.001$). Tree tenure has historically been a contentious issue particularly at the community level. Communities have different opinions on current tree tenure arrangements ($p < 0.001$); (51%) believed that current tenure patterns are satisfactory in contrast to those (44%) who hold a dissatisfied opinion about tree tenure. The educational and migrant status of farmers plays a significant role ($p = 0.02$ and $p = 0.04$ respectively) in respondents' opinions on existing tree tenure regulations. Post-hoc tests indicate statistical differences among respondents with basic education (58% satisfied) and those without any formal education (38% satisfied; $p = 0.01$). Native farmers (56%) were generally more satisfied with current tree tenure than migrant farmers (40%). Age ($p = 0.07$) and gender ($p = 0.09$) also contributed to respondent thoughts on tree tenure. Younger age groups were more satisfied with existing tree tenure arrangements than older generations. Lastly, although farmers believe the influence of external help (through extension services) has mostly resulted in positive on-farm cocoa production (84%), a section of responses disagreed (13%; $p = 0.03$). Education also influenced farmer impressions on the input of extension and related services ($p = 0.04$). Respondents with basic education (42%) were more satisfied whereas those with tertiary education (6%) were the least satisfied with the influence of external help.

4. Discussion

4.1. Variations in Climate Change Knowledge/Perceptions Based on Social Indicators and Potential Impacts

4.1.1. Accuracy in Climate Change Knowledge/Perceptions

Changes in climatic pattern were widely recognized by resident cocoa farmers in communities surrounding the Krokosua Hills Forest Reserve. Perceptions about increasing temperature patterns are consistent with trend analysis (Figure 2). Trend analysis of rainfall, however, did not conform with respondents' views. The trend analysis indicated no significant change in the amount of rainfall over the same period (1970–2017), contrary to popular respondent belief that rainfall amounts have reduced. These observations have been previously stated by [4] with regards to temperature. The problems associated with rainfall observations have also been highlighted in the literature [5].

The respondents' perceptions appear responsive to drought, which was captured by SPEI values. Prolonged dry spells, in particular are a major concern to respondents due to their close association to cocoa yield and productivity. Respondents suggest that plummeting cocoa yield in the area is a manifestation of climatic changes. These concerns have been raised previously by [3,8–10]. The correct observation of increasing temperature trends and length of dry season in this study also brings into perspective that farmers accurately perceive weather patterns in relation to crop production and tend to amend their farming practices accordingly [34].

Since the link between climate change perceptions and people's likelihood of subscribing to environmental protection strategies (in general) has already been established [34,35], we evaluated the level of accuracy with which local indigenous small holder farmer climate change knowledge compares with empirical weather data. Farmers' reliance on their indigenous knowledge and associated perceptions, leading into a defined climate experience, leads them into taking core decisions regarding their farming/cultural practices [30]. Ultimately, the accuracy of climate change perceptions, or specifically the potential for negative impacts of climate variability, are essential for maintaining cocoa agriculture [20]. Our findings suggest farmers do not always perceive climatic changes accurately, leaving room for further efforts to relay climate information to them.

4.1.2. Interplay between Social Indicators, Climate Experience and Potential Outcomes

A major observation of this study was that although communities were situated in the same geographic location, opinions on climate change and related occurrences vary considerably. Thus, community heterogeneity is of paramount concern in the enactment of any climate-based initiative since farmers' opinions and experiences differ within the least temporal and spatial differentiation, irrespective of geographical location [36]. Contrary to other studies [28,37,42], age did not have a strong influence on perception or knowledge on climate change. Age may however influence opinions on the causes of climate change, biological climate control methods, and tree tenure. Gender, on the other hand, significantly influenced overall farmer climate experience, demonstrating that gender influences knowledge and perceptions on climate change [39,40]. Similar to findings that determined male farmers generally have more access to climate information than female counterparts, male respondents were found to be more educated at higher levels than females. Although this study did not find a strong linkage between education and perception/knowledge on climate change among male and female farmers, education played a significant role in determining opinions and differences regarding current tree tenure mechanisms and ratings of external/extension help with cocoa farming. It was noted that these differences were mostly between educated and non-educated farmers.

The sensitivity of tree and land tenure was also observed in this study. Migrant farmers (non-natives) offered harsher criticism of existing tree tenure arrangement than natives. This is however not a surprising finding. A lack of property rights among migrant farmers may influence interest in long-term investments [43], like tree planting and management in this case. In addition, migrant farmers are more likely than native farmers to engage in cocoa monoculture. Since migrant farmers lack property rights, it heightens their propensity to engage in activities or cocoa farming practices (in this case), that may be detrimental to the environment.

4.2. Cocoa Farming for Livelihood and Climate Change Mitigation: Views of Smallholders

This study investigated how farmers felt about incorporating climate change mitigation mechanisms, like REDD+ into their farming activities. As seen in [13,27], the main attribute of climate change mitigation mechanisms is to improve livelihoods and concurrently, enhancement of environmental protection goals. Farmers in general, acknowledged the importance of environmental protection and correctly noted aspects of their farming activities that are detrimental to the environment. Farmers also appreciate the climate ameliorative ability of trees and tree maintenance on their farms. The addition of monetary incentives, however, was marked with skepticism among farmers. Fewer farmers appeared to understand how a system of tree maintenance on their cocoa farms was going to provide them direct income as against one that prescribes maintenance for environmental protection. The only direct mechanism known to farmers is one in which the tree is eventually harvested, and some proceeds are extended to a farm/tree owner (per prescribed benefit sharing arrangements; see [57]). Such a system clearly navigates away from the goals of REDD+ but seems to be the only plausible explanation, apart from maintaining trees for ecosystem benefits exclusively. Elsewhere in one of the very first REDD+ project sites in the Brazilian Amazon region, farmers' perceptions and eventual participation in the project was significantly improved with a decentralized approach. This approach fostered active farmer participation in the planning phase of REDD+, a move that promoted equity at both the community and individual farmer level regarding information on REDD+. This significantly influenced acceptability and success of the project in the area [58]. In essence, this study corroborates the findings of [33] which asserted that socio-cultural attributes significantly influence social acceptability of agroforestry systems. For purposes of this study, community heterogeneity and migrant status appear to be the significant factors for the adoption of climate change mitigation mechanisms in cocoa farming. Community heterogeneity in particular has been discovered to have a strong connection with social capital, which is the driving force for improved performance of mainstream developmental initiatives (like REDD+). Specifically, pronounced community heterogeneity may influence social capital [59].

4.3. Pathways to Integration of Climate Change Mitigation Mechanisms: The Role of Stakeholders (Extension, Forestry Commission, Farmers' Cooperatives/Community Based Organizations, Civil Society Organizations, Cocoa Buying Companies)

The essential role of agricultural extension has been reported in implementing a Climate-Smart Cocoa (CSC) approach [7]. The approach recommends broadening the scope of cocoa-related extension efforts to increase and improve the capacity of cocoa farmers [12] to adopt environmentally friendly mechanisms which also fulfill socio-economic goals. Within the CSC framework, this study noted that farmers generally held a satisfactory opinion regarding the influence of other stakeholders when it comes to cocoa farming.

Stakeholder influence has mostly been towards helping farmers make sound decisions from cocoa cultivation to final cocoa bean sale. The already established cordiality between cocoa farmers and other stakeholders presents an opportunity to disseminate information on climate change mitigation and could also act as a social safeguard for farmers willing to invest time and other resources towards mitigation strategies. An added benefit of such safeguards will see a reduction in costs of implementation and monitoring of climate change mitigation strategies [19]. The active participation of civil society organizations in general has been highlighted as a proponent of REDD+ related activities [19].

4.4. Potential Roadblocks to the Successful Incorporation of Climate Change Mitigation Strategies into Cocoa Production

Climate change related strategies generally prescribe mechanisms aimed at carbon neutrality. REDD+ in particular hinges on specific implementation criteria: (1) simplification of tree tenure and benefit-sharing mechanisms, (2) clear demonstration of the impact of REDD+ in comparison to scenarios without it (additionality), (3) assurance of adherence to REDD+ goals for as long as the project lasts (permanence and risk assessment) and (4) guarantees that REDD+ project sites do not promote carbon release in other areas (leakage) [12–14]. Cocoa farming in general, has had a checkered history when it comes to its association with forests, in fact spearheading massive deforestation since it became a mainstay of the Ghanaian economy [60,61]. This study showed that although cocoa farming communities around the KHFR recognize the importance of environmental protection and its relation to climate, their activities, per se, pose potential challenges for the implementation of a full-scale REDD+ project. Illegal logging on cocoa farms has necessitated an historical imperative for farmers to take preventative actions by inducing mortality in trees of economic importance before they reach maturity. Farmers often have several portions of fragmented cocoa farms. It is unclear how to make sure farmers maintain carbon neutrality on other portions of lands, especially in cases where they put only portions of their lands under REDD+. Cultural practices of farmers tend to favor carbon release in cases where cocoa productivity declines significantly due to age or shade. Farmers noted that tree removal was essential in such cases to reduce the level of shading to optimum levels. Undesirable cocoa trees are also removed during shade tree removal. The importance of tree tenure needs to be emphasized. A REDD+ project will need to develop ways to address all these challenges to facilitate full scale implementation and realization of its goals.

We identified a lack of willing participation in programs which provide payments for ecosystem services (PES). The skepticism of farmers on PES could lead to an eventual removal of intrinsic and altruistic characteristics relating to environmental conservation/protection. The motivation behind general environmental stewardship could potentially be reduced to how much can be earned, and becomes even more complicated when you place people receiving payments for ecosystem services in close proximity to those who are not [62].

5. Conclusions

The novelty of climate change mitigation strategies has been heralded by the international scientific community as a gateway to the implementation of desirable forest governance mechanisms with significant potential to influence the livelihoods of developing economies. This notwithstanding, such mechanisms are prone to several obstacles which could work against implementation. Since the inception of REDD and REDD+ ideologies into scientific platforms in Ghana, several attempts have been made to move beyond pilot projects towards full scale implementation. While these attempts are laudable, this study provides data to support a bottom-up approach to effectively manage the challenges surrounding climate ameliorating strategies in general.

There is reason to recommend a REDD+ strategy that takes into consideration specific community needs. This study revealed that even within the same geographical location, perceptions and knowledge on climate change vary significantly. The cocoa growing mosaic presents an entirely new challenge to the implementation of REDD+. Cocoa has historically driven the economies of communities in these areas and though farmers welcome other livelihood and environment enhancing opportunities, there is ripe skepticism about the potential success and sustainability of such 'new ideas'. As farmers suggested in this study, climate change mitigation efforts need to effectively ensure the participation of farmers in initial project designs. A "think big, but start small" approach has the potential to help formulate community- or location-specific strategies towards implementation of REDD+. With that in mind, the global community needs to deliberate on measures to implementing an adaptable REDD+ program moving away from the strictly national-scaled orientation of strategies. There are several ecological zones in Ghana, each with unique physiography, biological assemblages, and agroforestry capacity. It remains to be determined if a single definition of a forest is equally applicable to all such ecosystems. This study suggests REDD+ and other climate change mitigation strategies may need to adopt a significant degree of flexibility and focus more on the human dimensions aspect, especially in areas where cocoa production is interwoven into general forestry practice. See Table A3 for a summary of major findings of the study.

Author Contributions: Conceptualization, L.K.A. and G.J.E.; Methodology, L.K.A. and G.J.E.; Software, L.K.A.; Validation, L.K.A. and G.J.E.; Formal Analysis, L.K.A.; Investigation, L.K.A., K.L. and G.J.E.; Resources, G.J.E., L.K.A. and G.J.A.-K.; Data Curation, L.K.A. and G.J.E.; Writing—Original Draft Preparation, L.K.A., G.J.E. and K.L.; Writing—Review & Editing, L.K.A., G.J.E., K.L. and G.J.A.-K.; Visualization, L.K.A. and G.J.E.; Supervision, G.J.E. and K.L.; Project Administration, L.K.A.; Funding Acquisition, G.J.E. and L.K.A.

Funding: This research was funded by the Corkery Family Fund and the Center for Sustainable Forestry at Pack Forest (http://www.packforest.org/), both from University of Washington (http://www.washington.edu/).

Acknowledgments: This research acknowledges the immense support of Stanley Asah (University of Washington), Araba Sey (United Nations University), Ivan Eastin (University of Michigan), Martha Groom (University of Washington-Bothell), Rebecca Ashley Asare (Nature Conservation Research Centre—Ghana), Kwakye Ameyaw (Forestry Commission—Ghana), Henry Kwakye Ameyaw (Ghana COCOBOD), Henry Kudiabor (Forest Services Division—Ghana), Yakubu Mohammed (Resource Management Support Center—Ghana), Baafi Frimpong (Forest Services Division—Ghana) and the Technical Officers of Forest Services Division, Juaboso District, Ghana. Your support in diverse ways contributed to the success of this research.

Conflicts of Interest: The authors declare no conflict of interest.

Appendix A

Table A1. Kruskal Wallis and Mann Whitney test results indicating statistical differences ($p < 0.05$) between farmer groups/responses based on variables: community, age, gender, education, migrant status and household status.

Description of Variable	Groups	Community	Age	Gender	Education	Migrant Status	Household Status
Section 1							
Observed climatic changes							
Temperature		0.004 **	0.588	0.009 **	0.464	0.703	0.178
Rainfall	Increased	<0.001 ***	0.852	0.034 *	0.425	0.184	0.124
Length of Wet Season	Decreased Same	0.007 **	0.897	0.801	0.343	0.089 ·	0.187
Length of Dry Season	Don't know	<0.001 ***	0.972	0.936	0.935	0.840	0.260
Climatic impacts on cocoa yield							
	Positive Negative No impact Not sure	<0.001 ***	0.920	0.088 ·	0.677	0.228	0.125
Views on causes of climatic changes							
Cocoa monocultures		<0.001 ***	0.838	0.248	0.354	0.024 *	0.518
Illegal logging	Agree Disagree	0.015 *	0.056 ·	0.811	0.988	0.764	0.277
Vehicular pollution	Don't know	0.081 ·	0.885	0.952	0.391	0.141	0.198
Woodfuel usage		<0.001 ***	0.356	0.653	0.658	0.140	0.544
Slash and burn agriculture		0.032 *	0.150	0.837	0.835	0.984	0.798

Table A1. *Cont.*

Description of Variable	Groups	Community	Age	Gender	Education	Migrant Status	Household Status
Climate change beliefs							
Curses (Temperature)		0.009 **	0.568	0.223	0.357	0.545	0.880
Curses (Rainfall)		0.002 ***	0.288	0.009 **	0.806	0.913	0.348
Human activities (Temperature)	Agree Disagree Don't know	0.002 ***	0.899	0.833	0.780	0.479	0.481
Human activities (Rainfall)		0.223	0.486	0.416	0.419	0.637	0.884
Natural occurrence (Temperature)		<0.001 ***	0.657	0.261	0.520	0.566	0.237
Natural occurrence (Rainfall)		<0.001 ***	0.488	0.088 ·	0.510	0.529	0.218
Tree regulation of climate change							
Temperature regulation	Agree Disagree Don't know	0.877	0.091 ·	0.501	0.596	0.210	1.000
Rainfall regulation		0.042 *	0.098 ·	0.289	0.325	0.416	0.216
Concern about climate change	Yes No	0.691	0.364	0.193	0.785	0.859	0.136
Future implications of climate change	Yes No	0.001 ***	0.256	0.130	0.109	0.489	0.987
Section 2 *Tree maintenance for ecosystem services*	Yes No Maybe	0.002 ***	0.580	0.916	0.253	0.106	0.734
Tree maintenance for payment	Yes No Maybe	<0.001 ***	0.190	0.697	0.403	0.030 **	0.563
Rating of tree tenure	Good Bad Neither	<0.001 ***	0.071 ·	0.099 ·	0.013 **	0.044 *	0.103
Section 3 *Rating of external help*	Good Bad Neither	0.029 *	0.468	0.925	0.044 *	0.282	0.262

Significant difference: *** = $p < 0.001$; ** = $p < 0.01$; * = $p < 0.05$; · = $p < 0.1$

Table A2. Demographic characteristics of cocoa farmers in communities surrounding the KHFR ($n = 205$).

Attribute	Category	Percentage of Total Respondents
Gender	Male	67
	Female	33
Age	Less than 18	0
	18–29	3
	30–44	32
	45–59	49
	>60	16
Highest Level of Education	Basic	46
	Secondary	16
	Tertiary	7
	No Formal Education	31
Marital Status	Married	86
	Single	6
	Divorced	4
	Widowed	4
Migrant Status	Native	71
	Non-native	29
Religion	Christian	80
	Muslim	10
	Traditionalist	5
	Other	5

Table A3. Summary of major findings.

Findings	Implications/Recommendations for Climate Change Mitigation
Farmers accurately perceive changes in climate (particularly temperature and drought).	Perceptions guide farmers in choosing farming practices.
Farmers' perceptions on precipitation and length may not always be consistent with empirical weather data.	It is prudent to accurately inform farmers about climate since this may have implications for environmental protection in general.
Population/community demographics play a major role in climate perceptibility and subsequent actions to take regarding cocoa farming.	Mitigation strategies need to zero in on specific community/population attributes to foster effective implementation.
The concept of payment for ecosystem services, which has been adopted by most climate change mitigation strategies, has not been fully explained.	There is a need to adopt strategies that engage farmers in designing climate change mitigation strategies or better still, improve their capacity to understand the concept.
Cocoa farmers share a cordial relationship with extension services and other stakeholders associated with cocoa farming.	This presents a practical opportunity to relay information on climate change mitigation strategies to cocoa farmers.
The current situation of illegal logging on cocoa farms may exacerbate carbon release.	Pertinent measures are needed to curb illegal logging on cocoa farms.
Cultural practices favor removal of overhead shade to facilitate productivity/yield of cocoa.	Strategies need to emphasize practices that favor tree retention on cocoa farms.

References

1. Essegbey, G.O.; Ofori-Gyamfi, E. Ghana Cocoa industry—An analysis from the innovation system perspective. *Technol. Invest.* **2012**, *3*, 276–286. [CrossRef]
2. Anim-Kwapong, G.J.; Frimpong, E.B. Climate Change on Cocoa Production. In *Ghana Climate Change Impacts, Vulnerability and Adaptation Assessments under the Netherlands Climate Assistance Programme (NCAP)*; Agyeman-Bonsu, W., Ed.; Environmental Protection Agency: New Tafo Akim, Ghana, 2008; pp. 263–298.
3. Schroth, G.; Läderach, P.; Martinez-Valle, A.; Bunn, C.; Jassogne, L. Vulnerability to climate change of cocoa in West Africa: Patterns, opportunities and limits to adaptation. *Sci. Total Environ.* **2016**, *556*, 231–241. [CrossRef] [PubMed]
4. Läderach, P.; Martínez-Valle, A.; Schroth, G.; Castro, N. Predicting the future climate suitability for cocoa farming of the world's leading producer countries, Ghana and Côte d'Ivoire. *Clim. Chang.* **2013**, *119*, 841–854. [CrossRef]
5. McSweeney, C.; New, M.; Lizcano, G. UNDP Climate Change Country Profiles: Ghana. 2010. Available online: https://digital.library.unt.edu/ark:/67531/metadc226664/m2/1/high_res_d/Ghana.hires.report.pdf (accessed on 28 May 2018).
6. Ruf, F.; Schroth, G.; Doffangui, K. Climate change, cocoa migrations and deforestation in West Africa—What does the past tell us about the future? *Sustain. Sci.* **2015**, *10*, 101–111. [CrossRef]
7. Asare, R.A. Understanding and Defining Climate-Smart Cocoa: Extension, Inputs, Yields, and Farming Practices. Available online: https://www.forest-trends.org/wp-content/uploads/imported/climate-smart-cocoa_extension-yields-inputs-practices_v12_02-18-14-pdf.pdf (accessed on 28 May 2018).
8. Wiah, E.N.; Twumasi-Ankrah, S. Impact of climate change on cocoa yield in Ghana using Vector Autoregressive Model. *Ghana J. Technol.* **2017**, *1*, 32–39.
9. Hutchins, A.; Tamargo, A.; Bailey, C.; Kim, Y. Assessment of Climate Change Impacts on Cocoa Production and Approaches to Adaptation and Mitigation: A Contextual View of Ghana and Costa Rica. Available online: https://elliott.gwu.edu/sites/g/files/zaxdzs2141/f/World%20Cocoa%20Foundation.pdf (accessed on 19 February 2018).
10. Rainforest Alliance. Preparing Cocoa Farmers for Climate Change. Available online: https://www.rainforest-alliance.org/article/preparing-cocoa-farmers-for-climate-change (accessed on 18 March 2018).
11. Awetori, Y.J. Stakeholder analysis of climate change in Ghana. In *Stakeholders on Climate Change: North & South Perspectives*; Report nr., 3; Midttun, A., Ed.; CERES21-Creative Responses to Sustainability: Oslo, Norway, 2009; pp. 49–84.

12. Asare, A.R.; Kwakye, Y. A Guide to Implementing REDD+ in Ghana: Criteria and Modalities for Developing a REDD+ Project. Available online: http://www.itto.int/files/itto_project_db_input/3046/Technical/RED-PD093-12Rev.3(F)_Progress-Report%2031-Jan-2014%20Annex1-Guide-Implementing-REDD-Ghana.pdf (accessed on 1 October 2018).

13. Agyei, K.; Agyeman, V.K.; Asante, W.A.; Benefoh, D.T.; Blaser, J.; Damnyag, L.; Deppeler, A.; Feurer, M.; Foli, E.G.; Heeb, L.; et al. REDD+ in Agricultural Landscapes: Evidence from Ghana's REDD+ Process. Available online: https://www.bfh.ch/fileadmin/data/publikationen/2014/1_Blaser_REDD-_in_Agricultural_Landscapes_in_Ghana.pdf (accessed on 10 February 2018).

14. Ministry of Lands and Natural Resources. Readiness Preparation Proposal Ghana. Available online: https://theredddesk.org/sites/default/files/Ghana%20R-PP.pdf (accessed on 1 October 2018).

15. Ghana Forestry Commission. Ghana REDD+ Strategy. Available online: https://www.fcghana.org/userfiles/files/REDD+/Ghana%20REDD+%20Strategy.pdf (accessed on 2 October 2018).

16. Ghana Forestry Commission. Ghana Cocoa REDD+ Program. Available online: http://fcghana.org/userfiles/files/redd/GCFRP_draft_Implementation_Plan_2016.pdf (accessed on 1 October 2018).

17. Thompson, M.C.; Baruah, M.; Carr, E.R. Seeing REDD+ as a project of environmental governance. *Environ. Sci. Policy* **2011**, *14*, 100–110. [CrossRef]

18. Vatn, A.; Vedeld, P. National governance structures for REDD+. *Glob. Environ. Chang.* **2013**, *23*, 422–432. [CrossRef]

19. Chhatre, A.; Lakhanpal, S.; Larson, A.M.; Nelson, F.; Ojha, H.; Rao, J. Social safeguards and co-benefits in REDD+: A review of the adjacent possible. *Curr. Opin. Environ. Sustain.* **2012**, *4*, 654–660. [CrossRef]

20. Ehiakpor, D.S.; Danso-Abbeam, G.; Baah, J.E. Cocoa farmer's perception on climate variability and its effects on adaptation strategies in the Suaman District of Western Region, Ghana. *Cogent Food Agric.* **2016**, *2*, 1210557.

21. Ofori-Boateng, K.; Baba, I. The impact of climate change on cocoa production in West Africa. *Int. J. Clim. Chang. Strateg. Manag.* **2014**, *6*, pp–296. [CrossRef]

22. Denkyirah, E.K.; Okoffo, E.D.; Adu, D.T.; Bosompem, O.A.; Yildiz, F. What are the drivers of cocoa farmers' choice of climate change adaptation strategies in Ghana? *Cogent Food Agric.* **2017**, *3*, 1334296. [CrossRef]

23. Gyau, A.; Smoot, K.; Kouame, C.; Diby, L.; Kahia, J.; Ofori, D. Farmer attitudes and intentions towards trees in cocoa (*Theobroma cacao* L.) farms in Côte d'Ivoire. *Agrofor. Syst.* **2014**, *88*, 1035–1045. [CrossRef]

24. Codjoe, F.N.Y.; Ocansey, C.K.; Boateng, D.O.; Ofori, J. Climate change awareness and coping strategies of cocoa farmers in rural Ghana. *J. Biol. Agric. Healthc.* **2013**, *3*, 19–30.

25. Asare, R.; Afari-Sefa, V.; Osei-Owusu, Y.; Opoku, P. Cocoa agroforestry for increasing forest connectivity in a fragmented landscape in Ghana. *Agrofor. Syst.* **2014**, *88*, 1143–1156. [CrossRef]

26. Cerda, R.; Deheuvels, O.; Calvache, D.; Niehaus, L.; Saenz, Y.; Kent, J.; Vilchez, S.; Villota, A.; Martinez, C.; Somarriba, E. Contribution of cocoa agroforestry systems to family income and domestic consumption: Looking toward intensification. *Agrofor. Syst.* **2014**, *88*, 957–981. [CrossRef]

27. Baruah, M. Cocoa and Carbon: Remedying Forest Governance through Community Participation in a REDD+ Pilot in Ghana. Available online: http://scholarcommons.sc.edu/etd/4262 (accessed on 19 March 2018).

28. Huxley, P. *Tropical Agroforestry*; Blackwell Science: Oxford, UK, 1999.

29. Jerneck, A.; Olsson, L. More Than Trees! Understanding the agroforestry adoption gap in subsistence agriculture: Insights from narrative walks in Kenya. *J. Rural Stud.* **2013**, *32*, 114–125. [CrossRef]

30. Atkins, J.E.; Eastin, I. Seeing the trees: Farmer perceptions of indigenous forest trees within the cultivated cocoa landscape. *For. Chron.* **2012**, *88*, 535–541. [CrossRef]

31. Wiesmann, U. *Sustainable Regional Development in Rural Africa: Conceptual Framework and Case Studies from Kenya*; Geographica Bernensia: Berne, Switzerland, 1998.

32. Ogalley, A.S.; Vogl, C.R.; Eitzinger, J.; Hauser, M. Local perceptions and responses to climate change and variability: The case of Laikipia District, Kenya. *Sustainability* **2012**, *4*, 3302–3325. [CrossRef]

33. Atangana, A.; Khasa, D.; Chang, S.; Degrande, A. 2014, Socio-cultural aspects of agroforestry and adoption. In *Tropical Agroforestry*; Springer: Dordrecht, The Netherlands, 2014; pp. 323–332.

34. Niles, M.T.; Mueller, N.D. Farmer perceptions of climate change: Associations with observed temperature and precipitation trends, irrigation, and climate beliefs. *Glob. Environ. Chang.* **2016**, *39*, 133–142. [CrossRef]

35. Meze-Hausken, E. Contrasting climate variability and meteorological drought with perceived drought and climate change in northern Ethiopia. *Clim. Res.* **2004**, *27*, 19–31. [CrossRef]

36. Derkyi, M.; Adiku, S.G.K.; Nelson, V.; Dovie, B.D.; Codjoe, S.; Awuah, E. Smallholder farmers' perception of climatic and socio-economic factors influencing livelihoods in the transition zone of Ghana. *AAS Open Res.* **2018**, *1*, 7. [CrossRef]

37. Debela, N.; Mohammed, C.; Bridle, K.; Corkrey, R.; McNeil, D. Perception of climate change and its impact by smallholders in pastoral/agropastoral systems of Borana, South Ethiopia. *SpringerPlus* **2015**, *4*. [CrossRef] [PubMed]

38. Deressa, T.; Hassan, M.; Ringler, C. Perception of and adaptation to climate change by farmers in the Nile basin of Ethiopia. *J. Agric. Sci.* **2011**, *149*, 23–31. [CrossRef]

39. Asfaw, A.; Admassie, A. The role of education on the adoption of chemical fertilizer under different socioeconomic environments in Ethiopia. *Agric. Econ.* **2004**, *30*, 215–228. [CrossRef]

40. Food and Agriculture Organization FAO. The State of Food and Agriculture 2010–2011: Women in Agriculture, Closing the Gender Gap for Development. Available online: http://www.fao.org/docrep/013/i2050e/i2050e.pdf (accessed on 11 February 2018).

41. Mustapha, S.B.; Sanda, A.H.; Shehu, H. Farmers' perception of climate change in central agricultural zone of Borno state, Nigeria. *J. Environ. Earth Sci.* **2012**, *2*, 21–28.

42. Juana, J.; Kahaka, Z.; Okurut, F. Farmers' perceptions and adaptations to climate change in sub-Sahara Africa: A synthesis of empirical studies and implications for public policy in African agriculture. *J. Agric. Sci.* **2013**, *5*, 121–135. [CrossRef]

43. Antwi-Agyei, P.; Dougill, A.J.; Stringer, L.C. Impacts of land tenure arrangements on the adaptive capacity of marginalized groups: The case of Ghana's Ejura Sekyedumase and Bongo districts. *Land Use Policy* **2015**, *49*, 203–212. [CrossRef]

44. Forestry Commission. *Krokosua Hills Forest Reserve Management Plan*; Forestry Commission: Accra, Ghana, 2010.

45. Ghana Statistical Service. Population and Housing Census. Available online: http://www.statsghana.gov.gh/docfiles/2010_District_Report/Western/Juaboso.pdf (accessed on 2 March 2018).

46. Noponen, M.R.; Mensah, C.; Schroth, G.; Hayward, J. A Landscape Approach to Climate-Smart Agriculture in Ghana. Available online: https://www.rainforest-alliance.org/sites/default/files/2016-08/A-landscape-approach-to-climate-smart-agriculture-in-Ghana.pdf (accessed on 20 February 2018).

47. Bernard, R. *Research Methods in Anthropology: Qualitative and Quantitative Approaches*, 4th ed.; AltaMira Press: Oxford, UK, 2006; p. 522.

48. Strauss, A.; Corbin, J. *Basics of Qualitative Research: Techniques and Procedures for Developing Grounded Theory*, 2nd ed.; Sage: Thousand Oaks, CA, USA, 1998.

49. Mann, H.B. Non-parametric tests against trend. *Econometrica* **1945**, *13*, 163–171. [CrossRef]

50. Kendall, M.G. *Rank Correlation Methods*, 4th ed.; Charles Griffin: London, UK, 1975.

51. Helsel, D.; Hirsch, R. Statistical Methods in Water Resources. In *Techniques of Water-Resources Investigations Book 4*; Chapter A3; US Geological Survey: Reston, VA, USA, 2002; p. 522.

52. Vicente-Serrano, S.M.; Begueria, S.; Lopez-Moreno, J.I. A multiscalar drought index sensitive to global warming: The standardized precipitation evapotranspiration index. *J. Clim.* **2010**, *23*, 1696–1718. [CrossRef]

53. Thornthwaite, C.W. An approach toward a rational classification of climate. *Geogr. Rev.* **2010**, *38*, 55–94. [CrossRef]

54. Vicente-Serrano, S.M.; Beguería, S.; Lorenzo-Lacruz, J.; Camarero, J.J.; López-Moreno, J.I.; Azorin-Molina, C.; Revuelto, J.; Morán-Tejeda, E.; Lorenzo, A.S. Performance of drought indices for ecological, agricultural, and hydrological applications. *Earth Interact.* **2012**, *16*, 1–27. [CrossRef]

55. Charusombat, U.; Niyogi, D. A hydroclimatological assessment of regional drought vulnerability: A case study of Indiana droughts. *Earth Interact.* **2011**, *15*, 1–65. [CrossRef]

56. Beguería, S.; Vicente-Serrano, S.M. Calculation of the Standardised Precipitation—Evapotranspiration Index. Available online: https://cran.r-project.org/web/packages/SPEI/SPEI.pdf (accessed on 12 July 2018).

57. Ministry of Lands and Natural Resources, MLNR. Tree Tenure and Benefit Sharing Framework in Ghana. Available online: https://www.fcghana.org/userfiles/files/MLNR/Tree%20Tenure%20final%20(2).pdf (accessed on 4 February 2018).

58. West, T.A.P. Indigenous community benefits from a de-centralized approach to REDD+ in Brazil. *Clim. Policy* **2016**, *16*, 924–939. [CrossRef]

59. Coffé, H. Social capital and community heterogeneity. *Soc. Indic. Res.* **2009**, *91*, 155–170. [CrossRef]

60. Gockowski, J. *Cocoa Production Strategies and the Conservation of Globally Significant Rainforest Remnants in Ghana*; ODI Presentation, Sustainable tree crops program, International Institute of Tropical Agriculture: Accra, Ghana, 2007.

61. Kolavalli, S.; Vigneri, M. Cocoa in Ghana: Shaping the success of an economy. In *Yes, Africa Can: Success Stories from a Dynamic Continent*; World Bank Publication: Washington, DC, USA, 2011; pp. 201–217. Available online: http://siteresources.worldbank.org/AFRICAEXT/Resources/258643-1271798012256/Ghana-cocoa.pdf (accessed on 3 October 2018).

62. Chan, K.M.A.; Anderson, E.; Chapman, M.; Jespersen, K.; Olmsted, P. Payments for ecosystem services: Rife with problems and potential—For transformation towards sustainability. *Ecol. Econ.* **2017**, *140*, 100–122. [CrossRef]

MDPI

St. Alban-Anlage 66

4052 Basel

Switzerland

Tel. +41 61 683 77 34

Fax +41 61 302 89 18

www.mdpi.com

Forests Editorial Office

E-mail: forests@mdpi.com

www.mdpi.com/journal/forests

www.ingramcontent.com/pod-product-compliance
Lightning Source LLC
Chambersburg PA
CBHW051728210326
41597CB00032B/5648